MICROBIOLOGICAL QUALITY ASSURANCE

A Guide Towards Relevance *and* Reproducibility *of* Inocula

MICROBIOLOGICAL QUALITY ASSURANCE

A Guide Towards Relevance *and* Reproducibility *of* Inocula

Edited by
Michael R. W. Brown
Professor of Pharmaceutical Microbiology
Aston University
Birmingham, England

Peter Gilbert
Senior Lecturer in Pharmacy
University of Manchester
Manchester, England

CRC Press
Taylor & Francis Group
Boca Raton London New York

CRC Press is an imprint of the
Taylor & Francis Group, an informa business

First published 1995 by CRC Press
Taylor & Francis Group
6000 Broken Sound Parkway NW, Suite 300
Boca Raton, FL 33487-2742

Reissued 2018 by CRC Press

Library of Congress Cataloging-in-Publication Data

Microbiological quality assurance : a guide towards relevance and
reproducibility of inocula / edited by Michael R.W. Brown and Peter
Gilbert.
p. cm.
Includes bibliographical references and index.
ISBN 0-8493-4752-1
1. Microbial biotechnology. 2. Microbial inoculants--Quality
control. I. Brown, Michael R. W. (Michael Robert Withington) 1931–
. II. Gilbert, Peter.
TP248.27.M53M54 1995
660'.62—dc20 94-43936

A Library of Congress record exists under LC control number: 94043936

Publisher's Note
The publisher has gone to great lengths to ensure the quality of this reprint but points out that some imperfections in the original copies may be apparent.

Disclaimer
The publisher has made every effort to trace copyright holders and welcomes correspondence from those they have been unable to contact.

ISBN 13: 978-1-315-89545-1 (hbk)
ISBN 13: 978-1-351-07455-1 (ebk)

Visit the Taylor & Francis Web site at http://www.taylorandfrancis.com and the
CRC Press Web site at http://www.crcpress.com

Preface

The activities of microorganisms in the natural environment, in animals, plants, and in food, cosmetics, and pharmaceuticals have enormous ecological and commercial significance.

For obvious reasons of ease, economy, and ethics, the great majority of studies involve test tube investigations and use relevant isolates. Predictions about the behavior of these microbes are extrapolated from behavior in the test tube. Evidence has accumulated (but is as yet hardly acted upon) that growth in standard laboratory media using batch cultures often produces microbes greatly different from those *in vivo,* in animals, and *in situ,* in the natural environment. Information about such properties is limited and what is known is often overlooked. The purpose of this book is to draw attention to this area and also to the manner in which such information could be used to improve inocula preparation (relevance and reproducibility) for laboratory tests intended to mimic the *in vivo/in situ* behavior of bacteria.

It is as much the ability of microorganisms to adapt their physiology to match the growth environment that makes them such an outstanding success as it is the failure of many *in vitro* experiments to mimic correctly these phenotypes that render them, on occasions, so spectacularly irrelevant. There are several widely overlooked but crucial causes contributing to such irrelevances, all related to *in vivo/in situ* phenotypic adaptation.

- First, the surface adherent, biofilm mode of growth (sessile) predominates in natural ecosystems (as opposed to the test tube), both within infected hosts and in the environments which serve as reservoirs for infection (hospital sinks, cooling towers, etc). Biofilms are often found associated with infected implanted devices and also in chronic infections of the lung, gut, and urogenital tract. Biofilms are also important in non-medical situations such as oil pipelines, fuel lines, river beds, and soil. There is increasing evidence that biofilm growth per se influences microbial physiology. The population dependent production of inter-cell signals may characteristically occur in biofilm populations.

- Second, individual vegetative bacteria are able to exhibit remarkable plasticity with respect to the cell envelope, through which there is a constant interaction between cell and environment. Its composition, structure, and hence numerous surface properties adaptively reflect the nature of the growth environment. Any static description of the microbial surface should be interpreted as only a "snapshot", resulting from particular growth circumstances. The influence on microbial physiology of temperature, hydrostatic pressure, pH, Eh, and osmolarity has long been recognized in the design of cultures intended to mimic *in vivo/in situ* conditions. Still widely overlooked are the very profound influences, especially on the envelope, of specific nutrient deprivations and growth rate of the cells per se. The presence of dormant vegetative microbes can also contribute to lack of reproducibility.

- Third, during the cell cycle numerous developments take place resulting in many changes in biological properties. Differing growth conditions can result in cultures with different proportions of cells at the various stages of the division cycle and thus with varying properties.

- Fourth, handling and standardization procedures for inocula can greatly alter cell properties. Thus, cell harvesting involving centrifugation can cause cell damage with consequent changes in properties and/or death. Changes in menstrua, for example, affecting pH, temperature, or osmolarity, can also damage or kill.

- Last, all environments are selective. Consequently, several subcultures (and thirty plus generations) in a medium different to the original environment may select a variant more suited to the new circumstances.

In addition to providing inappropriate cells grown inappropriately, standard laboratory procedures for inoculum standardization may well be a test of the organisms ability to survive. These standardization procedures, perceived as leading to increased test reproducibility, may in truth give rise to irrelevant and artifactual results. The results themselves may well be reproducible and wrong, precisely and consistently.

Michael R.W. Brown
Peter Gilbert

The Editors

Michael R.W. Brown, M.Sc., Ph.D., D.Sc., F.R.Pharm.S., F.I.Biol., is Professor of Pharmaceutical Microbiology and head of the Pharmaceutical Microbiology Research Group at the Department of Pharmaceutical and Biological Sciences, Aston University in Birmingham, England. He has held this position since 1970 and has also held the positions of Dean of the Faculty of Life and Health Sciences and of Pro-Vice-Chancellor for Research. Prior to this he was Senior Lecturer in Pharmaceutical Microbiology at the University of Bath, England, Lecturer in Pharmacy at the University of London, and Research Associate at the University of Florida, Gainesville. He has held visiting research positions at Sandoz Research Institute in Vienna, Austria, Ciba-Geigy in Basel, Switzerland, and at the Universities of Calgary in Canada, Oxford in England, and Lille in France.

He qualified as a pharmacist at the University of Manchester in 1954 and subsequently obtained Master and Doctor of Science degrees from the University of Manchester and his Doctor of Philosophy from the University of London. He was elected Fellow of the Royal Pharmaceutical Society of Great Britain and of the Institute of Biology. Professor Brown has published over 180 research papers in the areas of antimicrobial resistance and the physiological status of cells *in vivo, in vitro,* and *in situ*. He acts as adviser and consultant to a number of pharmaceutical companies at an international level.

Michael Brown is a member of the American Society for Microbiology, the Society for General Microbiology, the Society for Applied Bacteriology and the British Society for Antimicrobial Chemotherapy.

Peter Gilbert, B.Sc., Ph.D., is Senior Lecturer in pharmacy at the University of Manchester, England. He obtained his bachelors degree in bacteriology from the University of Newcastle-upon-Tyne in 1972 and his doctorate in 1975 from the School of Pharmacy, Sunderland, England. Following a postdoctoral fellowship with Professor Brown at the University of Aston, he took up an appointment of Lecturer in Pharmacy at the University of Manchester in 1978. He was promoted to Senior Lecturer in that University in 1989.

Dr. Gilbert has a long standing research interest in the mechanisms of action and resistance towards chemical antimicrobial agents. In recent years such research has concentrated on the effects of environment, particularly attachment and biofilm formation, upon antimicrobial susceptibility and pathogenicity of microorganisms.

He has published more than 130 scientific papers and has made more than 80 presentations to scientific meetings. He is frequently invited to address international and national conferences. He is co-organizer of the British Biofilm Club, a confederation of some 450 scientists working in the area of bacterial attachment to surfaces. He is a consultant to several industrial companies.

Peter Gilbert is a member of the American Society for Microbiology, the Society for General Microbiology, the Society for Applied Bacteriology, and the British Society for Antimicrobial Chemotherapy, and editorial advisor to *Pharmaceutical Sciences.*

Contributors

Gerald Adams, M.I. Biol.
Freeze-Drying Research and Development
Centre for Applied Microbiology and Research
Salisbury, England

Ibrahim Al-Adham, Ph.D.
Faculty of Pharmacy
University of Jordan
Amman, Jordan

Julie Andrews, Ph.D.
Department of Pharmacy
University of Manchester
Manchester, England

Rosamund Baird, Ph.D.
Summerlands House
Summerlands
Yeovil, England

Johan S. Bakken, M.D.
The Duluth Clinic, Ltd.
Duluth, Minnesota

Sally F. Bloomfield, Ph.D.
Department of Pharmacy
Kings College London
London, England

Michael R. W. Brown, D.Sc.
Department of Pharmaceutical and Biological Sciences
University of Aston
Birmingham, England

Philip J. Collier, Ph.D.
Department of Molecular and Life Sciences
University of Abertay
Dundee, Scotland

J. William Costerton, Ph.D.
Center of Biofilm Engineering
Montana State University
Bozeman, Montana

Rene Courcol, M.D.
Bacteriology Laboratory
Calmette Hospital
Lille Cedex, France

Peter Gilbert, Ph.D.
Department of Pharmacy
University of Manchester
Manchester, England

Peter Hambleton, Ph.D.
Production Division
Centre for Applied Microbiology and Research
Salisbury, England

Norman Hodges, Ph.D.
Department of Pharmacy
University of Brighton
Brighton, England

Anne E. Hodgson, Ph.D.
Department of Pharmacy
University of Manchester
Manchester, England

Jana Jass, Ph.D.
Department of Biological Sciences
Exeter University
Exeter, England

Roberto Kolter, Ph.D.
Department of Microbiology and Molecular Genetics
Harvard Medical School
Boston, Massachusetts

Hilary Lappin-Scott, Ph.D.
Department of Biological Sciences
Exeter University
Exeter, England

Brian Perry, Ph.D.
Proctor & Gamble Ltd.
(Health & Beauty Care) Ltd.
Egham, England

Andrew Robinson, Ph.D.
Research Division
Centre for Applied Microbiology and Research
Salisbury, England

Christine C. Sanders, Ph.D.
Creighton University School of Medicine
Omaha, Nebraska

Kenneth S. Thomson, Ph.D.
Creighton University School of Medicine
Omaha, Nebraska

Howard S. Tranter, Ph.D.
Production Division
Centre for Applied Microbiology and Research
Salisbury, England

Christopher N. Wiblin, Ph.D.
Quality Control Department
Centre for Applied Microbiology and Research
Salisbury, England

Maria M. Zambrano, Ph.D.
Instituto de Immunologia
Universidad Nacional
Bogotá, Colombia

Table Of Contents

GROWTH CONDITIONS

Chapter 1.1
Influence of the Environment on the Properties of Vegetative Microorganisms: An Overview
Michael R.W. Brown and Peter Gilbert
Introduction 3
Phenotypic Flexibility 5
Growth Rate and Nutient Deprivation 5
Growth as Adherent Biofilms 7
Solid or Liquid Culture for Conventional Susceptibility Testing 7
Conclusions 8
References 9

Chapter 1.2
Definition of Phenotype in Batch Culture
Michael R.W. Brown, Philip J. Collier, Rene J. Courcol, and Peter Gilbert
Introduction 13
Specific Nutrient Depletion in Batch Culture 14
Defined Nutrient-Depleted Media 15
An Empirical Approach 15
A Quantitative Approach 16
USP Preservative Test Organisms 17
References 19

Chapter 1.3
Changes in Bacterial Cell Properties in Going from Exponential Growth to Stationary Phase
Maria M. Zambrano and Roberto Kolter
Introduction 21
Changes that Result from Starvation 22
Regulation of Gene Expression in Stationary Phase 23
Population Changes during Starvation 25
Acknowledgments 26
References 27

Chapter 1.4
Continuous Culture of Microorganisms in Liquid Media: Effects of Growth Rate and Nutrient Limitation
Peter Gilbert and Michael R.W. Brown
Introduction 31
Continuous Culture: Theoretical Considerations 32
Carbon Limitation 35

Magnesium and Iron Limitation ... 36
Oxygen Limitation ... 37
Turbidostats ... 37
Continuous Culture: Practical Considerations ... 37
Culture Homogeneity ... 38
Foaming ... 38
Floccules ... 38
Biofilms ... 38
Continuous Culture Fermenter Design ... 39
Modulation of Antimicrobial Susceptibility in Continuous Culture ... 40
Susceptibility Towards Chemical Biocides ... 41
Susceptibility Towards Antibiotics ... 42
References ... 44

Chapter 1.5
The Influence of Nutrient Environment During Vegetative Growth
on the Properties of Bacterial Endospores
Peter Gilbert, Michael R.W. Brown, Ibrahim S.I. Al-Adham, and Norman Hodges
Introduction ... 49
Sporulation Medium and Definition of Spore Phenotype ... 51
Sporulation Temperature ... 54
Growth and Sporulation at Various pH ... 56
Culture Age Prior to Harvesting ... 57
References ... 58

Chapter 1.6
Influence of the Environment on the Properties of Microorganisms Grown
in Association with Surfaces
Peter Gilbert, Anne E. Hodgson, and Michael R.W. Brown
Introduction ... 61
Surface Colonization and Growth ... 63
Resistance Properties of Bacterial Biofilms ... 64
Influence of Surface Attachment and Growth upon Microbial Activity ... 64
Role of the Glycocalyx ... 65
The Glycocalyx as a Moderator of Drug/Biocide Resistance ... 66
Attachment-Specific Physiologies ... 67
Quorum Sensing Transcriptional Activation ... 68
In Vitro Models of Biofilms ... 71
Closed Growth Models ... 71
Solid:Air Interfaces ... 71
Solid:Liquid Interfaces ... 72
Open Growth Models (Non-Steady-State) ... 73
Chemostat-Based Models ... 73
The Robbins Device ... 74
Annular Reactor ... 75
Open Growth Models (Steady-State) ... 75
Constant Thickness Biofilms ... 76
Perfused Biofilms ... 76
Conclusions ... 77
References ... 77

Chapter 1.7

Sources of Biological Variation and Lack of Inoculum Reproducibility: A Summary

Michael R.W. Brown, Sally F. Bloomfield, and Peter Gilbert

Culture Origins 83
Growth Conditions of the Inoculum 84
Dormancy 84
Post-Growth Handling Procedures 85
Procedure/Test Conditions 85

POST-GROWTH CONDITIONS

Chapter 2.1

The Preservation of Inocula

Gerald D.J. Adams

Introduction 89
Reproducibility 89
Biochemical Properties 90
Genetic Stability 90
Preservation by Cooling 90
Behavior of Solutes During Freezing 91
Rates of Cooling 92
"Slow" Cooling Injury 92
Optimal Cooling Rate 93
Rapid Cooling Injury 93
Rate of Sample Thawing 93
Factors Affecting Survival 93
Organism Species 93
Influence of Stages in Growth Cycle on Cryosensitivity 94
Cell Concentration and Cryotolerance 94
Cryoprotection 94
Toxic Effects of Cryoprotectants 95
Freeze-Preservation in Practice 95
Methods of Cooling 96
Storage Temperatures 96
Domestic Deep Freezers 96
Storage at –60°C to –80°C 97
Liquid/Gas-Phase Nitrogen Storage 97
Cold Storage in the Absence of Ice Formation 97
Freeze-Drying 98
Freeze-Drying Equipment 98
Rate of Freezing Prior to Freeze-Drying 98
Heat and Mass Transfer 99
Shelf Temperatures During Primary Drying 99
Secondary Drying (Desorption Drying) 100
Sample Prefreezing — Collapse 100
Containers and Sample Removal 101
Factors Affecting the Survival of Freeze-Dried Microorganisms 101
Microorganism Species 101

Cell Concentration and Culture Age ... 103
Suspending Medium Composition ... 103
Ablation ... 104
Reconstitution ... 104
Lyophilization Damage and Storage Conditions ... 105
Factors Influencing Stability During and After Drying ... 105
Residual Moisture Content ... 105
Sealing Gas ... 106
Free-Radical Damage ... 106
Temperature of Sample Storage ... 106
Collapse in the Solid State ... 107
Maillard Reactions ... 108
Changes in the Characteristics of Freeze-Dried Inocula ... 108
Mutation Caused by Freeze-Drying or Drying ... 108
Damage by Electromagnetic Radiation ... 109
Visible Light ... 109
Gamma, X-Irradiation, or Background Radiation ... 109
Glow Discharge Testing ... 109
The Influence of Cell Dormancy on Postpreservation Survival ... 109
Alternatives to Vacuum Freeze-Drying ... 111
Atmospheric Freeze-Drying ... 111
Drying in the Absence of Freezing ... 111
Culture Collection Practice ... 111
References ... 112

Chapter 2.2
Influence of Post-Growth Procedures on the Properties of Microorganisms
Peter Gilbert, Philip J. Collier, Julie Andrews, and Michael R.W. Brown
Introduction ... 121
Pretreatment Objectives ... 121
Methods of Harvesting Cultured Bacteria ... 122
Centrifugation ... 122
Filtration ... 126
Washing and Resuspension of Harvested Cells ... 128
Injury Resulting form Cold-Osmotic Shock ... 128
Starvation Storage of Cell Suspensions ... 129
References ... 130

APPLICATIONS

Chapter 3.1
Factors Affecting the Reproducibility and Predictivity of Performance Tests
Peter Gilbert and Michael R.W. Brown
Introduction ... 135
Reproducibility and Predictivity of Inocula Grown in Batch Culture ... 135
Reproducibility of Logarithmic Phase Inocula ... 136
Predictivity of Logarithmic Phase Inocula ... 138
Temperature ... 139
Carbon Substrate ... 140

Nitrogen Source ... 140
Other Nutrients ... 140
Reproducibility of Stationary Phase Inocula ... 141
Predictivity of Stationary Phase Inocula ... 142
Reproducibility and Predictivity of Inocula Grown in Continuous Culture ... 142
Reproducibility of Inocula Generated in Continuous Culture ... 143
Predictivity of Inocula Generated in Continuous Culture ... 145
Solid versus Liquid Culture ... 146
Biofilms or Planktonic Inocula ... 146
References ... 147

Chapter 3.2
Preservative Efficacy Testing in the Pharmaceutical Industries
Rosamund M. Baird
Development of Testing Methods ... 149
Test Method Limitations ... 150
Test Method Modifications ... 152
Factors Causing Test Variability ... 153
Cultivation of Wild Strains ... 153
Influence of Nutritional Status of Challenge Inocula ... 154
Stability of Test Strains ... 154
Nature of the Product ... 155
Addition of Organic Matter ... 156
Neutralization of Antimicrobial Activity ... 157
Recovery Systems and Stressed Cells ... 158
Overview ... 160
References ... 160

Chapter 3.3
Preservation Efficacy Testing in the Cosmetics and Toiletries Industries
Brian F. Perry
Introduction ... 163
European Community Cosmetics Directive ... 163
Preservation ... 165
Preparation of Inocula ... 167
Selection of Preservative Efficacy Test Microorganisms ... 168
Establishment and Maintainance of Master Cultures ... 175
Preservation (Conservation) ... 176
Authentication ... 176
Development ... 176
Preparation of Stock Culture ... 178
Preparation of Challenge Inocula from Stock/Working Cultures ... 178
Inoculation of Product ... 180
Levels of Inoculum ... 181
Number of Challenges ... 182
In-Use Tests ... 182
Special Considerations ... 182
Concluding Remarks ... 183
Acknowledgments ... 184
References ... 184

Chapter 3.4
Reproducibility and Predictivity of Disinfection and Biocide Tests
Sally F. Bloomfield

Introduction 189
Repeatability and Reproducibility of Disinfectant Testing 191
End-Point Tests 191
Quantitative Tests 192
Intralaboratory Variability of Test Results 194
Standardization of Test Suspensions 194
Operator Skill 197
Variations Between Test Organisms and Products 197
Within-Day and Between-Day Variability 197
Inter- and Intralaboratory Variability of Test Results 198
Capacity Tests 202
Relevance to Practice 202
Choice of Test Organism 203
Cultivation of Test Organisms 204
Effect of Nutrient Depletion in Batch Culture 204
Effect of Growth Rate and Nutrient Limitation in Continuous Culture 207
Preparation of Test Suspensions 207
Contact Time and Temperature 208
Presence of Interfering Substances 208
Surface Testing 209
Application of Laboratory Tests for Approval of Disinfectants 214
Appendix 215
References 215

Chapter 3.5
Reproducibility and Performance of Endospores as Biological Indicators
Norman Hodges

Development and Current Status of Sterilization Indicators 221
Selection of Organisms for Use as Sterilization Indicators 222
Factors Affecting Biological Indicator Performance 226
Spore Resistance Resulting from Environmental Influences during Sporulation and Harvesting 226
Levels of Definition 226
Post-Harvesting Modification of Spore Resistance 227
Stability of Biological Indicators 229
Quick-Response Biological Indicators 230
References 231

Chapter 3.6
Development and Production of Vaccines
Andrew Robinson, Howard S. Tranter, Christopher N. Wiblin, and Peter Hambleton

Introduction 235
Variation of Bacterial Composition 236

Phase Variation ... 237
Fimbriae (Pili) ... 237
Coordinated Phase Variation ... 238
Specific Nutrient Uptake Systems ... 238
Antigenic Variation ... 239
Pili (or Fimbriae) ... 239
Major Outer Membrane Proteins (OMP) ... 239
Loss of Toxicity ... 240
Recombinant Strains ... 241
Control of Vaccine Seed Stocks ... 242
The Seed Lot System ... 242
Monitoring of the Seed Source ... 242
Fermentation ... 243
Product Characterization ... 244
References ... 245

Chapter 3.7
Screening for Novel Antimicrobial Activity/Compounds in the Pharmaceutical Industry
Peter Gilbert and Michael R.W. Brown
Introduction ... 247
Screening Strategies ... 249
Primary Screens ... 249
Target-Directed, Cell-Free Screens ... 249
Increasing the Predictivity of Cell-Free Screens ... 250
Dormancy and Slow Growth Rates ... 251
Cell-Density-Dependent Physiology ... 251
Whole Cell Screens (Planktonic) ... 251
Whole Cell Screens (Attached Cells and Biofilms) ... 252
In-Vitro Models for Secondary Screening of Antibiotics ... 255
Exposure of Chemostats to Antibiotics ... 257
Antibiotic Perfusion of the Biofilm Fermenter ... 257
References ... 258

Chapter 3.8
Screening for Novel Compounds/Activity in the Environmental Protection Industries
Hilary Lappin-Scott, Jana Jass, and J. William Costerton
Introduction ... 261
The Industrial Processes ... 261
Screening of Antimicrobials for Industry ... 262
Mixed Cultures ... 263
Starved Bacteria ... 263
Sessile versus Planktonic ... 265
Test Methods ... 266
Biofilm Reactors ... 266
RotoTorque Biofilm Reactor ... 266
Constant Depth Film Fermenter ... 268

Biofilm Sampling Devices 269
The Robbins Device 269
Submerged Test Pieces 271
Concluding Remarks 271
References 272

Chapter 3.9
Antimicrobial Susceptibility Testing within the Clinic
Kenneth S. Thomson, Johan S. Bakken, and Christine C. Sanders
Introduction 275
The Need for Susceptibility Tests 276
Selection of a Susceptibility Test Method 277
The Inoculum 277
Disk Diffusion Tests 279
Dilution Tests 281
Agar Dilution Tests 282
Broth Dilution Tests 282
Recent Developments 284
Antibiotic Selection and Reporting of Results 285
Antibiotics to be Tested 285
Results to be Reported 285
Interpretation of Results 286
Summary 287
References 287

Index 289

Growth Conditions

1.1 Influence of the Environment on the Properties of Vegetative Microorganisms: An Overview

Michael R.W. Brown and Peter Gilbert

Microorganisms grown in vitro *may be incomplete as regards all the determinants of virulence, since the genetic basis for virulence may be expressed completely only under the conditions of the test for virulence, namely during growth* in vivo (Smith, 1977).

The quotation above applies equally to determinants of antimicrobial susceptibility or to microbial properties in nature or indeed in any environment other than the test tube. Scientists interested in quality assurance, screening, vaccine production, or bioassay require reproducible inocula that are relevant for the purpose. Relevance is especially desirable when test tube simulations are extrapolated either to *in vivo* infections or to *in situ* situations in nature or man-made environments.

This chapter will give an overview of the influence of environment on cell properties (especially the surface).

INTRODUCTION

The United States Pharmacopoeia has no monograph for any microorganism, yet the possibility of such monographs is implied by the detailed texts for the regulatory tests of efficacy for antimicrobial systems. The assumptions here and in other forms of antimicrobial testing are that their performance characteristics are unvarying, provided cultures can be maintained in a pure state and have been cultured in media allowing rapid growth and high yields (i.e., they are "healthy"). This likens them perhaps to laboratory reagents of such a high

0-8493-4752-1/95/$0.00+$.50

grade that monographs on their purity and identity are superfluous. This fallacy might well be at the root of many problems of reproducibility and relevance for such biological testing procedures.

In the physical sciences rigid definitions of all aspects of the testing systems create relatively high precision and reproducibility in end results. For biological sciences, however, variation (see Chapter 1.7) is all too readily accepted as an inevitable consequence of using living systems. While physicochemical control of the test system is usually applied, often this does not extend to the environmental history of the test strain. In animal studies, where numbers are limited, the age, health, weight, and diet of the individuals used are regulated. In principle, microbiological systems are similar. However, the use of large sample populations causes neglect of such factors, with the assumption that the mean responses of a population will be constant. Generally populations of microorganisms will respond as collections of individuals from related backgrounds, rather than as heterogeneous mixtures of all possible biotypes. It is imperative that, if reproducibility and relevance are to be attained, homogeneous populations of an appropriate biotype be employed.

While there has been general recognition of the effects upon the results of sensitivity tests of pH, temperature, and osmolarity, control of the inoculum is generally restricted to the broad phase of growth of the culture and the type of medium. In this respect a false assumption is often made, namely that growth of the cells in a chemically defined medium necessarily leads to defined populations. Factors such as growth rate, specific nutrient depletion, the cell cycle, and distribution of cells in biofilms and microcolonies are still neglected in spite of a large body of literature that demonstrates their profound influence (Brown, 1977; Al-Hiti and Gilbert, 1980; Costerton, Irwin and Cheng, 1981; Cozens and Brown, 1983). It would appear that the panels associated with designing protocols for preservative and disinfection testing, and perhaps also those engaged in screening for antimicrobial activity, have been slow to respond to the scientific advances of the last decade or two. It is now well established that in many infections the microorganism is characteristically growing slowly in an iron-restricted environment (Eudy and Burroughs, 1973; Weinberg, 1979). Such circumstances produce cell structures and associated sensitivities different from those obtained in conventional test protocols. Another neglected area concerns the cell density at which microbes are harvested and used for various tests of susceptibility to host defenses or to antimicrobial agents. It is now clear that many organisms produce density-dependent inducers of critical cell behavior (Jones et al., 1993, Williams et al., 1992). Furthermore, it is also well documented (Lam et al., 1980; Mayberry-Carson et al., 1984; Marrie and Costerton, 1985) that in many chronic infections microbes survive in both biofilms and, similarly, in the inanimate hospital environment (e.g., sinks and cooling towers). Again, such a mode of growth can give rise to properties very different from those obtained in typical test systems. Designers of sensitivity test protocols should pay attention to the properties of the microorganism as they are *in situ*.

PHENOTYPIC FLEXIBILITY

An important reason for lack of reproducibility and relevance is the plasticity of the microbial cell and especially the envelope. The cell envelope is remarkably flexible in both structure and composition and constantly interacts with its environment. This confers a great survival advantage in a changing environment, be it an infected host or an ecosystem in nature. Changes in envelope components such as phospholipid (Benns and Proulx, 1974; Gunter, Richter, and Smallbeck, 1975; Cozens and Brown, 1981), fatty acids (Minnikin, Abdulrahimzadeh, and Baddiley, 1971a, b), metal cations (Kenward, Brown, and Fryer, 1979), envelope-associated proteins and enzymes (Turnowsky et al., 1983), extracellular enzymes (Ombaka, Cozens, and Brown, 1983), and polysaccharides (Dean et al., 1977) accompany changes in growth rate and/or nutrient deprivation (Ellwood and Tempest, 1972; Tempest and Wouters, 1981). In turn, they influence susceptibility to chemical antimicrobial agents (Brown, 1977; Brown, Gilbert, and Klemperer, 1979; Brown and Williams, 1985b; Brown, Collier and Gilbert, 1990; Gilbert, Collier and Brown, 1990) and antibiotics (Finch and Brown, 1975; Boggis, Kenward, and Brown, 1979; Turnowsky et al., 1983) and host defenses (Brown and Williams, 1985a). Such influence can be either through change in permeability of the envelope towards an agent active at a point away from the envelope (Gilbert and Brown, 1978a, b, 1980; Nikaido, Rosenberg, and Foulds, 1983; Wright and Gilbert, 1987a, b) or through alteration in the predominance (Broxton, Woodcock, and Gilbert, 1984) or physiological importance (Nicas and Hancock, 1980) of that target. Thus, variation in the growth environment of a challenge inoculum will influence not only the reproducibility of a testing procedure but also its sensitivity and relevance. Variation may be inadvertent, through poor specification of the media, or it may be intrinsic to a particular technique, such as the use of cultures on solid media.

GROWTH RATE AND NUTRIENT DEPRIVATION

Microbiologists are organisms of habit. Where those habits are convenient, they have become particularly ingrained. One of these is to inoculate in the evening for an experiment next day. Thus, cultures as challenge inocula are commonly 16 h or 40 h old and often in the stationary phase of growth. Systems have evolved to suit this practice through "optimization" of the conditions of growth to give high yields of cells at this time. Such "optimization" is in reality "maximization" of the cells' growth potential and gives rise to cell populations totally unrepresentative of the world outside the test tube (*in vivo* or in a natural ecosystem), which is highly competitive and causes specialized adaptation for each circumstance. Under such conditions growth rates are likely to be in the order of 0.05/h rather than the 3.0/h typically obtained in standard laboratory media (Eudy and Burroughs, 1973; Brown, 1977; Weinberg, 1979; Gilbert, 1985). A considerable body of work has demonstrated the interrelation of specific growth rate and cellular physiology with sensitivity towards chemicals, antibiotics, and host defenses (Finch and Brown, 1975; Dean et al., 1977; Gilbert and Brown, 1978a, 1980; Gilbert and Wright, 1986; Tuomanen et al., 1986, Brown, Collier and Gilbert, 1990). In virtually all cases sensitivity to antimicrobial agents increases with increasing growth rate. The choice of rapidly grown inocula therefore gives an artificially high activity when testing disinfectant, preservative, and antibiotic containing systems, and may give a false impression of antibiotic sensitivity when used for routine sensitivity testing (Brown, 1977; Brown and Williams, 1985a, b). In this respect it is noteworthy that initial screening within industry for novel antimicrobial activity often employs rapidly grown cells. Dominance of a trend of increasing sensitivity with increasing growth rate could conceivably reflect this rather than representing an innate property of microorganisms. Screening with slowly grown cells, of an appropriate nutrient limitation, might select different compounds for development. Nevertheless there has been little published work specifically studying susceptibility properties of nongrowing, dormant vegetative microbes (Siegele and Kolter, 1992; Kaprelyants and Kell, 1993).

To be relevant rather than convenient, growth rates should be chosen to represent those likely to occur *in vivo*. At present the use of a chemostat would generally not be feasible, but growth systems can be so designed, for example, by utilizing a medium deficient in iron, that onset of the stationary phase is gradual and cells can be harvested reproducibly at a relatively slow growth rate, some time later.

Chemically defined media do not necessarily give biologically defined cells. Given that the performance characteristics of microorganisms vary dramatically according to the nature of the nutrient limiting growth, then it is important to re-create such conditions when attempting to increase the relevance of an *in vitro* test. In this respect, it is necessary that appropriate media be designed by controlled variation of each component such that one is restrictive and the remainder present to a controlled excess. The common approach, of defining the contents of the medium without necessarily knowing their influence upon growth, can lead to a culture being in stationary phase owing to a lack of more than one essential nutrient. Small variations of the incubation conditions or the constituents of the medium might then tilt the balance so that one essential nutrient is depleted first rather than another. This will inevitably contribute to a lack of interlaboratory reproducibility.

The effects of different nutrient limitations upon the USP preservative challenge test were determined by Al-Hiti and Gilbert (1980). They observed that the minimum concentrations required to satisfy the test, for three widely used preservatives, were some hundred times less than those recommended by the Pharmacopoeias. Also they showed that preserved broth systems might either pass or fail the test according to the nutrient limitation of the inocula.

While the nutritional status of microorganisms *in vivo* is difficult to assess and may vary from location to location, one may speculate about the nature of the growth limiting nutrient for various localized sites (Brown, 1977; Brown and Williams, 1985a, b). For example, it is now clear that iron restriction applies in many infections. Such a nutrient restriction ought to be employed when susceptibility to antibiotics or to host defenses is being evaluated. Similar criteria for disinfection testing and/or preservative testing are currently difficult to establish, but rigid definition and design of suitable media for realistically chosen test strains ought to increase markedly both the reproducibility and predictive effectiveness of such systems.

GROWTH AS ADHERENT BIOFILMS

There is now much evidence that many chronic infections involve organisms in adherent biofilms (pneumonia in cystic fibrosis: Lam et al., 1980; cystitis: Marrie and Costerton, 1983; endocarditis: Mills et al., 1984; osteomyelitis: Mayberry-Carson et al., 1984; Marrie and Costerton, 1985). Furthermore, infections of implants (Marrie and Costerton, 1984), various prosthetic devices (Nickel et al., 1985) and intrauterine contraceptive devices (Jacques, Marrie, and Costerton, 1986) commonly relate to the biofilm mode of growth. In some cases, the biofilms do not cause infection (Gristina, Salem, and Costerton, 1985). The antibiotic sensitivity of such organisms differs greatly from that shown after growth in suspension in conventional media (Costerton, Irwin, and Cheng, 1981; Marrie and Costerton, 1984; Nickel et al., 1985). With respect to environmental biocides, biofilms are the crucial problem (Costerton and Lashen, 1984).

Studies of biofilm susceptibility commonly use apparatus with no growth rate control and rich, complex media, and dubiously compare established biofilm populations with growing planktonic cultures (Brown, Allison and Gilbert, 1988; Gilbert, Collier and Brown, 1990). *In vitro* biofilm models have been reviewed (Brown and Gilbert, 1993) and are considered elsewhere in this book (Chapter 1.6).

SOLID OR LIQUID CULTURE FOR CONVENTIONAL SUSCEPTIBILITY TESTING

Biofilm microorganisms in the form of solid cultures have traditionally been used by microbiologists as a means of recognizing contaminant organisms and maintaining pure stock cultures. For the majority of regulatory tests the inocula are passaged several times on solid media to ensure purity and subsequently harvested by washing off the plates with a suitable medium.

Superficially, this procedure appears admirable. However, different workers will inoculate the plates to different colony densities, either by using different streak patterns or by being heavy- or light-handed during the primary inoculation. All microbiologists will have observed that in areas of plates where colony density is high, colony size is small. This relates to nutrient availability and/or production of inhibitory substances by individual colonies. An additional consequence of this will be that large colonies might well be oxygen-limited at the core yet depleted by some other nutrient at their periphery. Growth rate will also differ at different locations within the colony, a process contributing to colony morphology. Thus, different densities of colony on a plate will produce individual cells of different nutritional status and large colonies will be more heterogenous in this respect than small ones. Al-Hiti and Gilbert (1983) evaluated such effects with respect to *Pseudomonas aeruginosa* and *Escherichia coli* and their sensitivity to chlorhexidine diacetate. They observed that sensitivity decreased markedly as the colony density of the inoculum was increased towards confluence and that reproducibility determined from replicate experiments was greatest for liquid cultures and also increased with increasing colony density.

Similar, but smaller, changes in sensitivity can be observed for plate diffusion assays for susceptibility tests and microbiological assay. In the latter systems, however, workers generally utilize lawn cultures and the problem is minimized.

Deployment of liquid media and shaking incubators are therefore important if heterogeneity is to be reduced in the challenge inocula. Choice of appropriately designed liquid media can then determine the most appropriate nutrient depletion of growth for the application involved. This is currently not the case.

CONCLUSIONS

If conventional microbial challenge test procedures (and, indeed, susceptibility tests) are to be made more relevant, then attention must be paid to the nutrient environment from which the inocula are drawn. Shaken, liquid cultures will be less sensitive to antimicrobial agents in general and will also lead to a greater element of reproducibility than solid ones. Deployment of homogeneous suspensions depleted of single nutrients, possibly iron, and cultured at slow rates of growth to predetermined cell densities will increase relevance. For chronic infections involving surface growth, antibiotic screening using biofilm microorganisms would improve relevance. Similar considerations apply to biocide testing (e.g., sinks, cooling towers) and to studies on susceptibility to host defenses. There will be some objectives (including increase in understanding) that can be achieved only by using growth-rate-controlled, nutrient-limited cultures, planktonic or biofilm, and by studying cell properties through the cell cycle.

REFERENCES

Al-Hiti, M.M.A. and Gilbert, P. (1980). Changes in preservative sensitivity of the USP antimicrobial agents effectiveness-test microorganisms. *Journal of Applied Bacteriology* **49**, 119-26.

Al-Hiti, M.M.A. and Gilbert, P. (1983). A note on inoculum reproducibility: Solid versus liquid culture. *Journal of Applied Bacteriology* **55**, 173-6.

Benns, G. and Proulx, P. (1974). The effect of ATP and Mg^{2+} on the synthesis of phosphatidylglycerol in *Escherichia coli* preparations. *Biochimica et Biophysica Acta* **377**, 318-24.

Boggis, W., Kenward, M.A. and Brown, M.R.W. (1979). Effects of divalent metal cations in the growth medium upon the sensitivity of batch grown *Pseudomonas aeruginosa* to EDTA or polymyxin B. *Journal of Applied Bacteriology* **47**, 477-88.

Brown M.R.W. (1977). Nutrient depletion and antibiotic susceptibility. *Journal of Antimicrobial Chemotherapy* **3**, 198-201.

Brown, M.R.W., Allison, D.G. and Gilbert, P. (1988). Resistance of bacterial biofilms to antibiotics: a growth rate related effect? *Journal of Antimicrobial Chemotherapy* **22,** 777-780.

Brown, M.R.W., Collier, P.J. and Gilbert, P. (1990). Influence of growth rate on susceptibility to antimicrobial agents: modification of the cell envelope and batch and continuous culture studies. *Antimicrobial Agents and Chemotherapy* **34**, 1623-8.

Brown, M.R.W., Gilbert P. and Klemperer, R.M.M. (1979). Influence of the bacterial envelope on combined antibiotic action. **In:** *Antibiotic Interactions* (Williams J.D., Ed). pp. 69-86. Academic Press, London.

Brown, M.R.W. and Gilbert, P. (1993). Sensitivity of biofilms to antimicrobial agents. *Journal of Applied Bacteriology, Symposium Supplement* **74,** 87s-97s.

Brown, M.R.W. and Williams, P. (1985a). Influence of substrate limitation and growth phase on sensitivity to antimicrobial agents. *Journal of Antimicrobial Chemotherapy* **15**, Suppl. A, 7-14.

Brown, M.R.W. and Williams, P. (1985b). The influence of environment on envelope properties affecting survival of bacteria in infections. *Annual Reviews of Microbiology* **39**, 527-56.

Broxton, P., Woodcock, P.M. and Gilbert, P. (1984). Interaction of some polyhexamethylene biguanides and membrane phospholipids in *Escherichia coli*. *Journal of Applied Bacteriology* **57**, 115-24.

Costerton, J.W., Irwin, R.T. and Cheng, K.J. (1981). The bacterial glycocalyx in nature and disease. *Annual Reviews of Microbiology* **35**, 399-424.

Costerton, J.W. and Lashen, E.S. (1984). Influence of biofilm on efficacy of biocides on corrosion-causing bacteria. *Materials Performance* **23**, 34-7.

Cozens, R.M. and Brown, M.R.W. (1981). Chemical composition of outer membranes from nutrient depleted cultures of *Pseudomonas aeruginosa*. **In:** *Current Chemotherapy and Immunotherapy*. pp. 80-2. Proceedings of the 12th International Congress of Chemotherapy, Florence, Italy.

Cozens, R.M. and Brown, M.R.W. (1983). Effect of nutrient depletion on the sensitivity of *Pseudomonas cepacia* to antimicrobial agents. *Journal of Pharmaceutical Sciences* **72**, 1363-5.

Dean, A.C.R., Ellwood, D.C., Melling, J. and Robinson, A. (1977). The action of antibacterial agents on bacteria grown in continuous culture. **In:** *Continuous Culture,* Vol 6, *Applications and New Fields* (Dean, A.C.R., Ellwood, D.C., Evans, C.G.T. and Melling, J., Eds), pp. 69-86. Ellis Horwood, Chichester.

Ellwood, D.E. and Tempest, D.W. (1972). Effects of the environment on bacterial wall content and composition. *Advances in Microbial Physiology* **7**, 83-117.

Eudy, W.W. and Burroughs, S.E. (1973). Generation times of *Proteus mirabilis* and *Escherichia coli* in experimental infections. *Chemotherapy* **19**, 161-70.

Finch, J.E. and Brown, M.R.W. (1975). The influence of nutrient limitation in a chemostat on the sensitivity of *Pseudomonas aeruginosa* to polymyxin B and EDTA. *Journal of Antimicrobial Chemotherapy* **1**, 379-86.

Gilbert, P. (1985). The theory and relevance of continuous culture. *Journal of Antimicrobial Chemotherapy* **15**, Suppl. A, 1-6.

Gilbert, P. and Brown, M.R.W. (1978a). Influence of growth rate and nutrient limitation on the gross cellular composition of *Pseudomonas aeruginosa* and its resistance to 3- and 4-chlorophenol. *Journal of Bacteriology* **133**, 1066-72.

Gilbert, P. and Brown, M.R.W. (1978b). Effect of R-plasmid RP1 and nutrient depletion upon the gross cellular composition of *Escherichia coli* and its resistance to some uncoupling phenols. *Journal of Bacteriology* **133**, 1062-65.

Gilbert, P. and Brown, M.R.W. (1980). Cell-wall mediated changes in sensitivity of *Bacillus megaterium* to chlorhexidine and 2-phenoxyethanol associated with growth-rate and nutrient-limitation. *Journal of Applied Bacteriology* **48**, 223-30.

Gilbert, P., Collier, P.J. and Brown, M.R.W. (1990). Influence of growth rate on susceptibility to antimicrobial agents: biofilms, cell cycle, dormancy, and stringent response. *Antimicrobial Agents and Chemotherapy* **34**, 1865-8.

Gilbert, P. and Wright, N.E. (1986). Non-plasmidic resistance towards preservatives of pharmaceutical products. **In:** *Preservatives in the food, pharmaceutical and environmental industries* (Board, R.G. and Allwood, M.C., Eds), pp. 255-79, Academic Press, London.

Gristina, A.G., Salem, W. and Costerton, J.W. (1985). Bacterial adherence to biomaterials and tissue: the significance of its role in clinical sepsis. *Journal of Bone and Joint Surgery* **67**, 264-73.

Gunter, T., Richter, L. and Smallbeck, J. (1975). Phospholipids of *Escherichia coli* in magnesium deficiency. *Journal of General Microbiology* **86**, 191-3.

Jacques, M., Marrie, T.J. and Costerton, J.W. (1986). In-vitro quantitative adherence of microorganisms to intrauterine contraceptive devices. *Current Microbiology* **13**, 1337.

Jones, S., Yu, B., Bainton, N.J., Birdsall, M., Bycroft, B.W., Chabra, S.R., Cox, A.J.R., Golby, P., Reeves, P.J., Stephens, S., Winson, M.K., Salmond, G.P.C., Stewart, G.S.A.B. and Williams, P. (1993). The *lux* autoinducer regulates the production of exoenzyme virulence determinants in *Erwinia carotovora* and *Pseudomonas aeruginosa. The EMBO Journal*, **12**, 2477-82.

Kaprelyants, A.S. and Kell, D.B. (1993). Dormancy in stationary-phase cultures of *Micrococcus luteus*: flow cytometric analysis of starvation and resuscitation. *Applied and Environmental Microbiology* **59**, 3187-96.

Kenward, M.A., Brown, M.R.W. and Fryer, J.J. (1979). The influence of calcium or magnesium on the resistance to EDTA, polymyxin B or cold shock, and the composition of *Pseudomonas aeruginosa* in glucose or magnesium depleted culture. *Journal of Applied Bacteriology* **47**, 489-503.

Lam, J.S., Chan, R., Lam, K. and Costerton, J.W. (1980). Production of mucoid microcolonies by *Pseudomonas aeruginosa* within infected lungs in cystic fibrosis. *Infection and Immunity* **28**, 546-56.

Marrie, T.J. and Costerton, J.W. (1983). A scanning and transmission electron microscopic study of the surface of intrauterine devices. *American Journal of Obstetrics and Gynecology* **146**, 384-93.

Marrie, T.J. and Costerton, J.W. (1984). Scanning and transmission electron microscopy of *in situ* bacterial colonisation of intravenous and intraarterial catheters. *Journal of Clinical Microbiology* **19**, 687-93.

Marrie, T.J. and Costerton, J.W. (1985). Mode of growth of bacterial pathogens in chronic polymicrobial human osteomyelitis. *Journal of Clinical Microbiology* **22**, 924-33.

Mayberry-Carson, K.J., Tober-Meyer, B., Smith, J.K., Lambe, D.W. and Costerton, J.W. (1984). Bacterial adherence and glycocalyx formation in osteomyelitis experimentally induced with *Staphylococcus aureus*. *Infection and Immunity* **43,** 825-833.

Mills, J., Pulliam, L., Dall, L., Marzouk, J., Wilson, W. and Costerton, J.W. (1984). Exopolysaccharide production by various streptococci in experimental endocarditis. *Infection and Immunity* **43,** 359-367.

Minnikin, D.E., Abdulrahimzadeh, H. and Baddiley, J. (1971a). The interrelation of phosphatidylethanolamine and glycosyldiglycerides in bacterial membranes. *Biochemical Journal* **124,** 447-448.

Minnikin, D.E., Abdulrahimzadeh, H. and Baddiley, J. (1971b). The interrelation of polar lipids in bacterial membranes. *Biochimica et Biophysica Acta* **249,** 651-655.

Nikaido, H., Rosenberg, E.Y. and Foulds, J. (1983). Porin channels in *Escherichia coli:* studies with β-lactams in intact cells. *Journal of Bacteriology* **153,** 232-240.

Nicas, T.I. and Hancock, R.E.W. (1980). Outer membrane Protein H1 of *Pseudomonas aeruginosa:* involvement in adaptive and mutational resistance to ethylenediaminetetraacetate, polymyxin B and gentamicin. *Journal of Bacteriology* **143,** 872-878.

Nickel, J.C., Ruseska, I., Wright, J.B. and Costerton, J.W. (1985). Tobramycin resistance of *Pseudomonas aeruginosa* cells growing as a biofilm on urinary tract catheter materials. *Antimicrobial Agents and Chemotherapy* **27,** 619-624.

Ombaka, A., Cozens, R.M. and Brown, M.R.W. (1983). Influence of nutrient limitation of growth on stability and production of virulence factors of mucoid and non-mucoid strains of *Pseudomonas aeruginosa. Reviews in Infectious Diseases* **5,** 5880-5888.

Siegele, D.A. and Kolter, R. (1992). Life after log. *Journal of Bacteriology* **174** 345-8.

Smith, H. (1977). Microbial surfaces in relation to pathogenicity. *Bact. Rev.* **41,** 475-500.

Tempest, D.W. and Wouters, J.T.M. (1981). Properties and performance of microorganisms in chemostat culture. *Enzyme and Microbiological Technology* **3,** 283-290.

Tuomanen, E., Cozens, R., Tosch, W., Zak, O. and Tomasz, A. (1986). The rate of killing of *Escherichia coli* by β-lactam antibiotics is strictly proportional to the rate of bacterial growth. *Journal of General Microbiology* **132,** 1297-1304.

Turnowsky, F., Brown, M.R.W., Anwar, H. and Lambert, P.A. (1983). Effect if iron-limitation of growth rate on the binding of penicillin G to the penicillin binding protein of mucoid and non-mucoid *Pseudomonas aeruginosa. FEMS Microbiology Letters* **17,** 243-245.

Weinberg, E.D. (1979). Iron and infection. *Microbiological Reviews* **42,** 45-66.

Williams, P., Bainton, N.J., Swift, S., Chhabra, S.R., Winson, M.K., Stewart, G.S.A.B., Salmond, G.P.C. and Bycroft, B.W. (1992). Small molecule-mediated density-dependent control of gene expression in prokaryotes: Bioluminescence and the biosynthesis of carbapenem antibiotics. *FEMS Microbiology Letters*, **100**, 161-8.

Wright, N.E. and Gilbert, P. (1987a). Influence of specific growth rate and nutrient limitation upon the sensitivity of *Escherichia coli* towards Polymyxin B. *Journal of Antimicrobial Chemotherapy* **20,** 303-312.

Wright, N.E. and Gilbert, P. (1987b). Antimicrobial activity of N-alkyltrimethylammonium bromides: influence of specific growth rate and nutrient limitation. *Journal of Pharmacy and Pharmacology* **39,** 685-690.

1.2 Definition of Phenotype in Batch Culture

Michael R.W. Brown, Philip J. Collier, Rene J. Courcol, and Peter Gilbert

INTRODUCTION

Conventional batch culture, in liquid suspension or as biofilm/colonies on agar plates, has been outstandingly successful in providing convenient and inexpensive growth conditions for microbes. Nevertheless, many problems still remain regarding the relevance and reproducibility of batch grown inocula for various purposes. In the present context microbes are cultured to mimic the growth of pathogens or spoilage organisms *in vivo* or *in situ*, to screen for activity/novel compounds, for susceptibility/preservative efficacy testing, and to enhance understanding of microbial physiology and antimicrobial activity.

It is important to appreciate that there are numerous phenotypes depending on growth conditions. Thus, when culture conditions have been selected, subsequent cell properties have then also been predetermined to a large extent.

The influence on microbial physiology of temperature, pH, Eh, and osmolarity is long established and well recognized. Less widely recognized are the gross effects of specific nutrient deprivation and of growth rate, despite a vast literature (Brown and Williams, 1985b) and much recent work on the dormancy of nutrient-deprived, vegetative, non-spore-forming bacteria (Kaprelyants and Kell, 1993).

The growth rate per se of an invading organism will contribute to the outcome of an infection. *In vivo*, bacterial doubling times are typically slow, and the ability to acquire low levels of nutrient, in addition to generating a characteristic envelope, will influence growth rate and thus the attainment of a bacterial population sufficient to harm the host. Such considerations are important in selection. Low maximum specific growth rate (μ_{max}) and a high affinity (low K_s) for the substrate favors selection at low nutrient levels (Harder et al., 1977).

The purpose of this chapter is not to review this area again, but to exploit it to improve the definition of batch cultures grown in suspension. Thus, control of specific nutrient depletion can lead to increased reproducibility and relevance. The use of these principles in the production of bacterial spores is dealt with in Chapter 1.5.

0-8493-4752-1/95/$0.00+$.50

Cooper (1991) has proposed with respect to the division cycle that cell components could usefully be considered as three groups: the cytoplasm, the genome, and the cell surface. Cytoplasm is synthesized exponentially during the division cycle, DNA is synthesized as a series of linear rates of synthesis, and the surface of the cell increases to just enclose the cytoplasm and the genome. Batch-grown pure cultures contain subpopulations of microbes at different stages of the division cycle. For example, within limits of growth rates there is a constant period between termination of DNA synthesis and cell division (D period). Clearly, as growth rates vary, these periods will occupy varying proportions of the division time.

The role of the division cycle in producing synchronous cultures of predictable and reproducible properties has received virtually no attention from the point of view of susceptibility testing and will not be considered here. Nevertheless, it seems probable that studies of susceptibility through the cell cycle could lead to enhanced reproducibility of such tests if they were made at the center of a plateau of susceptibility.

SPECIFIC NUTRIENT DEPLETION IN BATCH CULTURE

There is a need for definition of terms (Brown and Williams, 1985a). We define "nutrient depletion" as an absolute lack of a nutrient in batch culture, and limitation as applicable to continuous culture. "Restriction" refers to a situation where the nutrient is present but relatively unavailable (e.g., iron *in vivo*). "Deprivation" is a generic term to cover all these possibilities.

The classical growth curve in batch culture is thought to include a lag phase in which adaptation and change to a new environment is involved, and a logarithmic phase in which the microbes are in balanced growth, with cells replicating exponentially and with reproducible properties along the log-linear line. This view of the logarithmic phase is a mistaken one and confuses demonstrable consistency of doubling time with assumed structural and physiological consistency. In fact, changes in envelope properties may take place several generations before the onset of stationary phase due to depletion of a specific nutrient. In iron-depleted media in batch culture, *Klebsiella pneumoniae* derepressed its high-affinity iron-uptake systems about three generations before the onset of stationary phase (Williams et al., 1984). Magnesium-depleted *Pseudomonas aeruginosa* in batch culture lost sensitivity to EDTA (Brown and Melling, 1969a), and to polymyxin B, depending on other metal cations in the medium. This sensitivity was fully restored only after about three generations in magnesium-plentiful medium (Brown and Melling, 1969b). Therefore, the use of logarithmic phase cells requires that they be harvested at least three generations both before the onset of stationary phase and after inoculation, otherwise they will demonstrate the envelope and associated susceptibility and other properties of nutrient-depleted stationary-phase cells or of the inoculum or of both. Thus, the existence of exponential replication is no assurance that envelope and associated properties are constant and reproducible.

The transition from late logarithmic to stationary phase may be immediate or gradual, depending on the nature of the depleted nutrient. Thus, limiting nutrients with low Ks and yield, such as carbon, will produce a rapid transition from logarithmic to stationary phase. In the case of elements such as calcium, magnesium, and iron, with high Ks and high yield, transitions produced will be slow and gradual. Slow transition into the stationary phase, characterized by metal cation limitation, allows changes to occur in the cells which relate to the efficiency of substrate utilization. Thus, taking magnesium limitation as an example, the substrate is used structurally as part of the cell wall, for ribosome stabilization and also as an enzyme cofactor. Slow-growing cells require less ribosomes and hence less magnesium than the fast-growing ones. Yield for slow-growing cells is therefore greater than for fast-growing ones. This has the effect of making the transition from the logarithmic to stationary phase even slower, as magnesium is redistributed in the cell, producing a much extended late-logarithmic phase. During this time the cells are able drastically to reorganize their structure/function to accommodate for the deficiencies in nutrient supply.

The concentration of growth limiting nutrient will affect both growth rate and/or the total extent of growth. For substrates, such as carbon, at relatively low concentrations, growth rate is reduced, while, at moderate and higher levels, growth rates are identical throughout most of the logarithmic phase but the biomass at which the cells enter stationary phase is restricted. The relationship between the concentration of the limiting nutrient and the stationary phase biomass is generally linear. The slope of this relationship represents the yield constant (Y). This is assuming that nothing else becomes limiting, such as oxygen or pH, nor that secondary and possibly toxic metabolites accumulate.

Furthermore, there is constant change throughout a batch culture. There is increasing nutrient utilization and production of extracellular products including population-dependent autoinducers (Jones et al., 1993). For reproducibility culture densities should be well above or below the critical concentration for intercell signaling (Williams et al., 1992). Thus, cells replicating towards the end of a batch growth "curve" are growing in an environment significantly different to that of forebears replicating several generations earlier at the beginning of a batch culture.

DEFINED NUTRIENT-DEPLETED MEDIA

An Empirical Approach

A simple method to define a nutrient-depleted medium is to lower the concentration of an appropriate essential nutrient until concentrations, which control and determine population density are obtained empirically. This approach has been widely used for media in chemostat studies.

Low-iron media have been used to mimic *in vivo* growth. Although iron is a ubiquitous media contaminant, there are various methods for reducing the available or absolute iron content of media. Waring and Werkman (1942) and

Donald et al. (1952) summarized the principles involved in the elimination of trace metals from growth media. These include the use of spent media, recrystallization, precipitation, absorption, chelation methods with synthetic chelating agents, and employment of biological agents as iron scavengers. The use of extraneous chemical or biological agents, however, is undesirable in some metabolic studies of bacteria because they may cause damage to bacterial cell membranes (Klebba et al., 1982). Iron chelators of biological origin such as transferrin, which has an association constant for iron of 10^{32}, have been used in several studies (Griffiths, 1983; Valenti et al., 1980), and are useful in that they more closely represent conditions *in vivo*. However, there are practical difficulties involved in the use of transferrins, namely, the need to dialyze out metal and citrate ions before use, and they are expensive. A synthetic iron chelator, desferroxamine (Desferal; Ciba-Geigy Ltd., Basel, Switzerland), has been reported to resemble closely iron chelators of biological origin and has been used clinically (van Asbeck et al., 1983).

Sometimes chemically defined salts media are inappropriate, e.g., for fastidious pathogens. Consequently, complex media are desirable. Recently, a rapid and simple method has been devised for the removal of iron and other cations from complex media (Domingue et al., 1990). The method involves adding Chelex 100 ion-exchange resin to tryptone soy or brain heart infusion broths (TSB, BHI) and stirring under controlled conditions. This removes iron and other cations from the complex media rapidly and efficiently. Dose-response kinetics were obtained for iron removal when broths were treated with differing amounts of resin. Satisfactory cation depletion resulted when TSB was extracted for between 0.75 to 1 h, and when BHI was extracted for 1 h, respectively. Iron depletion was assessed by growth characterization, as well as gel electrophoresis of *Pseudomonas aeruginosa* and *Pasteurella haemolytica* and immunoblotting of *Staphylococcus aureus* cultures. Atomic absorption spectroscopy programs were modified to ensure accuracy. The method is useful and economical, as large batches of depleted media can be prepared in a day, and repeated regeneration of the resin did not appear to affect its chelating capacity.

A Quantitative Approach

At constant conditions of pH and temperature, growth rate is dependent upon the concentration of nutrients (Monod, 1949). It can therefore be assumed, in the absence of accumulation of toxic products and, for aerobes, with excess oxygen that the onset of a slowing of the growth rate at the end of the exponential phase corresponds to the onset of the depletion of a particular nutrient (but see earlier about biochemical changes taking place generations earlier). When all other ingredients are in excess, then cell mass at this point is proportional to the initial concentration of that depleting nutrient. From a series of growth curves using different nutrient concentrations, a medium can be constructed which can be varied to permit depletion of any selected ingredient, with all others in known excess. Table 1 gives nutrient requirements for various microbes, based on this

TABLE 1
Nutrient Requirements for Various Gram-Negative Strains

Nutrient	Added Concentration (mmol/l) Required for Exponential Growth to Optical Density = 1.0 (420 nm) Organism			
	E. coli[a]	*E. coli*[b]	*P. mirabilis*[c]	*P. aeruginosa*[d]
Glucose	2.5	2.8	3.0	4.0
NH4+	1.85	1.75	5.0	4.0
K^+	0.026	0.042	0.035	0.062
Mg^{2+}	0.013	0.025	0.015	0.04
Fe^{2+}	0	0.0001	0	0.0006
SO^{2-}	0.026	0.019	0.05	0.052
PO^{3-}	0.17	0.56	17.0	0.32

Note: To determine requirement for each nutrient, other nutrients were present to excess, approximately × 10.

[a] *Escherichia coli* K12 W3110, buffered with 25 mmol/l 3-(N-morpholino)propane sulfonate (MOPS) pH 7.2.
[b] *Escherichia coli* K12 W3110, containing R-plasmid RP1, buffered as [a].
[c] *Proteus mirabilis* NCTC 5887, buffered with 25 mmol/l MOPS pH 7.0.
[d] *Pseudomonas aeruginosa* NCTC 6750 buffered with 60 mmol/l MOPS pH 7.4.

Adapted from Cozens, R.M., Klemperer, R.M.M., and Brown, M.R.W., in *Antibiotics: Assessment of Antimicrobial Activity and Resistance*, Academic Press, London, 1983, pp. 61–71.

approach. On approaching the stationary phase following depletion of an essential nutrient, at a *predetermined* population size, the cell will begin to make an envelope appropriate to that particular depletion. The change is gradual, and the longer the cells are "stationary", the greater may be the change in properties (Brown and Williams, 1985b). Media devised quantitatively in this way can easily have pH, oxygenation, and osmolarity defined.

USP Preservative Test Organisms

The quantitative approach has been used in which specific nutrient depletion was imposed to devise defined media for United States Pharmacopoeial preservative test organisms. (Al-Hiti and Gilbert, 1980). Initially the concentrations of either the carbon, nitrogen or phosphate sources were varied and the optical density of the cultures determined when they reached the stationary phase of growth. All remaining nutrients [$MgSO_4$, 0.5 m*M*; $FeNH_4(SO_4)_2$, 0.03

TABLE 2
Media Composition for the Growth of USP Test Strains *Escherichia coli*, *Pseudomonas aeruginosa*, *Staphylococcus aureus* and *Candida albicans*

Nutrient	Growth Limiting Nutrient Concentration (mM) to Give Stationary Phase at an Optical Density (E_{470}) of 1.0					Depletion
	Escherichia coli	*Staph. aureus*	*Ps. aeruginosa*		*Candida albicans*	
Glucose		3.0			6.0	Carbon-depleted[a]
Glycerol	8.5		12.5			Carbon-depleted[a]
Sodium citrate				9.0		Carbon-depleted[a]
K_2HPO_4	0.13	[b]	0.05	0.15	0.05	Phosphate-depleted[a]
$(NH_2)SO_4$	12	[c]	2.5	2.0	2.5	Nitrogen-depleted[a]

[a] If non-limiting added at five times these concentrations.
[b] No added phosphate for phosphate-depleted, otherwise 2.0 m*M*.
[c] No added $(NH_4)_2SO_4$ for nitrogen-depleted, otherwise 5.0 m*M*.

Note: Media also contained Mg SO_4, 0.5 mM;$FeNH_4(SO_4)_2$, 0.03 m*M*, and for *E. coli* and *Staph.aureus* KCl, 13.4m*M*. *S. aureus* cultures were supplemented with yeast extract (1.0 g/1), thiamine-HCl (1.0 mg/l) and biotin (0.15 mg/l) and that for *C. albicans* with biotin (0.15 mg/l)

Adapted from Al-Hiti, M.M.A. and Gilbert, P. *Journal of Applied Bacteriology*, 49, 119-126, 1980.

m*M;* and for *E. coli* and *Staphylococcus aureus* KCl, 13.4 m*M*] were present in excess. The media for *Staphylococcus aureus* were also supplemented with yeast extract (1.0 g/1), thiamine-HCl (1.0 mg/l) and biotin (0.15 mg/l) and those for *Candida albicans* with biotin (0.15 mg/l). The pH was measured before and after growth of the cultures to check the buffering capacity of the media (pH 7.2). Phosphate depleted and all the staphylococcal cultures were buffered using MOPS (200 m*M*); the remainder were buffered with phosphates (KH_2PO_4, 28 m*M;* K_2HPO_4, 72 m*M*). The concentrations of the carbon, nitrogen, and phosphate sources within the media causing the cultures to enter their stationary phase of growth at an optical density of 1.0 were used in the design of the simple-salts media and are listed in Table 2. The depleted nutrient was supplied at this concentration, and the remainder at five times this level.

For *Aspergillus niger* the USP Test specifies that the challenge inocula is a spore suspension. Solid media were therefore devised by Al-Hiti and Gilbert (1980) for the growth of this organism based on the simple-salts liquid media of Kobayashi et al. (1964) and solidified with 1% (w/v) bacteriological agar (Oxoid L11). Plates were inoculated centrally with an agar disk cut from a 7 d simple-salts plate culture of *A. niger*, using a 4 mm flamed cork borer. Rates of increase in colony size were determined directly by daily measurement over 7 d, and the density of spores within the colony was assessed daily for 7 d by flooding replicate plates with water, agitating with a glass spreader, and performing total spore counts on the final suspension. Concentrations of carbon, nitrogen, and phosphate sources were varied as with the liquid media.

TABLE 3
Media Composition for the Growth of *Aspergillus niger* ATCC 16404

Nutrient	Concentrations of Growth-Limiting Substrates (mmol/l) for *A. niger* in Bacteriological Agar (1% w/v)		
	Carbon-Limited	Phosphate-Limited	Nitrogen-Limited
Glucose	3.0	45.0	45.0
K_2HPO_4	72.0	0.05	72.0
$(NH_4)_2SO_4$	15.0	15.0	0.0

Note: Additives $MgSO_4$, 0.5 m*M*; $FeSO_4$, 0.03 m*M*; biotin, 0.15 mg/l; KH_2PO_4, 30 m*M* (carbon and nitrogen-limiting media only); MOPS buffer, 200 m*M* (phosphate limiting media only).

Adapted from Al-Hiti, M.M.A. and Gilbert, P. *Journal of Applied Bacteriology*, 49, 119–126, 1980.

Rates of increase in colony size altered linearly with respect to limiting nutrient concentration. At very low phosphate concentrations, however, although mycelial growth rate was limited by the phosphate concentration, the density of sporulation within the colony was very much reduced. Choices of limiting nutrient concentrations for the final media (Table 3) were therefore made subjectively, selecting those that restricted the rate of colony development but allowed a sufficient level of sporulation for harvesting and preparation of spore suspensions after 7 d incubation. For carbon-depleted and nitrogen-depleted cultures, these reduced the rate of increase in colony diameter by 50% from that on complete media. For phosphate-depleted cultures, however, where sporulation density varied with phosphate concentration, this was not possible, and that concentration giving a 50% reduction in spore density of the colony was selected.

REFERENCES

Al-Hiti, M.M.A. and Gilbert, P. 1980. Changes in preservative sensitivity for the USP antimicrobial agents effectiveness test micro-organisms. *Journal of Applied Bacteriology* **49:** 119-126.

Brown, M.R.W. and Melling, J. 1969a. Loss of sensitivity to EDTA by *Pseudomonas aeruginosa* grown under conditions of magnesium-limitation. *Journal of General Microbiology* **54:** 439-44.

Brown, M.R.W. and Melling, J. 1969b. Role of divalent cations in the action of polymyxin B and EDTA on *Pseudomonas aeruginosa*. *Journal of General Microbiology* **59:** 263-74.

Brown, M.R.W. and Williams, P. 1985a. The Influence of environment on envelope properties affecting survival of bacteria in infections. *Annual Reviews in Microbiology* **39:** 527-56.

Brown, M.R.W. and Williams, P. 1985b. Influence of substrate limitation and growth phase on sensitivity to antimicrobial agents. *Journal of Antimicrobial Chemotherapy* **15:** Suppl. A, 7-14.

Cooper, S. 1991. Bacterial growth and division. Academic Press, London.

Cozens, R.M., Klemperer, R.M.M. and Brown, M.R.W. 1983. The influence of cell envelope composition on antibiotic activity, pp. 61-71 in *Antibiotics: Assessment of Antimicrobial Activity and Resistance,* A.D. Russell and L.B. Quesnel (eds.), Academic Press, London.

Domingue, P.A.G., Mottle, B.M., Morck, D.W., Brown, M.R.W. and Costerton, J.W. 1990. A simplified rapid method for the removal of iron and other cations from complex media. *Journal of Microbiological Methods* **42:** 13-22.

Donald, C., Passey, B.I. and Swaby, R.J. 1952. A comparison of methods for removing trace metals from microbiological media. *Journal of General Microbiology* **7:** 211-220.

Griffiths, E. 1983. Availability of iron and survival of bacteria in infection, pp. 153-177 in *Medical Microbiology,* Vol. 3. C.S.F. Eason, J. Jeljasewicz, M.R.W. Brown, and P.A. Lambert (eds.), Academic Press, London.

Harder, W., Kuenen, J.G. and Matin, A. 1977. Microbial selection in continuous culture. *Journal of Applied Bacteriology* **43:** 1-24.

Jones, S., Yu, B., Bainton, N.J., Birdsall, M., Bycroft, B.W., Chhabra, S.R., Cox, A.J.R., Golby, P., Reeves, P.J., Stephens, S., Winson, M.K., Salmond, G.P.C., Stewart, G.S.A.B. and Williams, P. 1993. The *lux* autoinducer regulates the production of exoenzyme virulence determinants in *Erwinia carotovora* and *Pseudomonas aeruginosa, EMBO Journal* **12:** 2477-2482.

Kaprelyants A.S. and Kell D.B. 1993. Dormancy in stationary-phase cultures of *Micrococcus luteus*: flow cytometric analysis of starvation and resuscitation. *Applied and Environmental Microbiology* **59:** 3187-3196.

Klebba, P.E., McIntosh, M.A. and Neilands, J.B. 1982. Kinetics of biosynthesis of iron-regulated membrane proteins in *Escherichia coli. Journal of Bacteriology* **149:** 880-888.

Kobayashi, G.S., Friedman, L. and Kofroth, J.F. 1964. Some cytological and pathogenic properties of spheroplasts of *Candida albicans. Journal of Bacteriology* **88:** 795-801.

Monod, J. 1949. The growth of bacterial cultures. *Annual Reviews of Microbiology* **3:** 371-394.

Valenti, P., De Stasio, A., Seganti, L., Mastromarino, P., Sinibaldi, L. and Orsi, N. 1980. Capacity of staphylococci to grow in the presence of ovotransferrin or $CrCl_3$ as a character of potential pathogenicity. *Journal of Clinical Microbiology* **11:** 445-447.

Van Asbeck, B.S., Marcelis, J.H., Marx, J.J.M., Stuyvenberg, A., van Kats, J.H. and Verhoef, J. 1983. Inhibition of bacterial multiplication by the iron chelator desferroxamine: potentiating effect of ascorbic acid. *European Journal of Clinical Microbiology* **2:** 426-431.

Waring, W.S. and Werkman, C.H. 1942. Growth of bacteria in an iron-free medium. *Archives of Biochemistry and Biophysics* **1:** 303-310.

Williams, P., Brown, M.R.W. and Lambert, P.A. 1984. Effect of iron deprivation on the production of siderphores and outer membrane proteins in *Klebsiella aerogenes. Journal of General Microbiology* **130:** 2357-365.

Williams, P., Bainton, N.J., Swift, S., Chhabra, S., Winson, M.K., Stewart, G.S.A.B., Salmond, G.P.C. and Bycroft, B.W. 1992. Small molecule-mediated density-dependent control of gene expression in prokaryotes: bioluminescence and the biosynthesis of carbapenem antibiotics. *FEMS Microbiology Letters* **100,** 161-168.

1.3 Changes in Bacterial Cell Properties in Going from Exponential Growth to Stationary Phase

Maria M. Zambrano and Roberto Kolter

INTRODUCTION

Bacteria are remarkable in their capacity to grow under a variety of adverse conditions and to inhabit a wide range of ecosystems. In order to do so, these organisms must be capable of responding quickly to changes in their environment. More importantly, they must also be able to withstand long periods of nutritional deprivation while within environments that are often restricted in the availability of one or more essential nutrients (Morita, 1988). Indeed, the very efficiency with which bacteria are capable of utilizing available nutrient and energy sources ensures that these will be quickly depleted. Bacteria may therefore seldom encounter conditions that sustain prolonged balanced growth, and most probably alternate between prolonged periods of starvation and sporadic growth. The ability of bacteria to respond to the extreme conditions of nutritional insufficiency is therefore crucial to ensure the species' long-term survival. Understanding the strategies that enable microorganisms to remain viable during periods of prolonged starvation is therefore of central importance in microbial ecology and, because of the possible persistence of microbial pathogens, is of concern in the area of public health (Roszak and Colwell, 1987).

Bacterial genera, such as *Bacillus* and *Myxococcus*, have developed elaborate mechanisms to maintain viability during starvation. Accordingly, in response to nutrient deprivation, highly differentiated spores are formed which are more resistant to environmental stresses than vegetative cells (Kaiser,

0-8493-4752-1/95/$0.00+$.50

1984; Losick and Youngman, 1984). Many bacterial species, however, respond to starvation without undergoing cellular differentiation such as this. Through the application of molecular genetic and biochemical techniques, we have begun to understand some of the mechanisms employed by nondifferentiating bacteria such as *Escherichia*, *Salmonella*, and *Vibrio* which enable them to survive starvation. As with the spore-forming organisms, these organisms also enter a program which results in metabolically less active yet more resistant cells as a response to nutrient deprivation (Kjelleberg et al., 1987; Matin et al., 1989; Siegele and Kolter, 1992).

Many of the recent studies concerning the starvation response of non-differentiating bacteria have been conducted on *Escherichia coli*. In order to survive both in the intestinal tract of mammals and in the more highly variable environments outside the host, *E. coli* must be capable of adapting to nutrient insufficiency and fluctuations in physical conditions, such as pH, temperature, and osmolarity. Laboratory conditions under which cells are studied are generally optimized for growth and offer plentiful nutrient supplies, and may not accurately reflect what *E. coli* encounters in nature. Under these optimal conditions, cells grow rapidly until they start to exhaust the least available essential nutrient. At this point, the growth rate decreases and populations stop their exponential increase in biomass. This constitutes a period of transition between exponential growth and the onset of what has become known as the stationary phase. "Stationary phase" is a descriptive term used to define the period during batch culture in which there is no further increase in cell number (Kolter, Siegele and Tormo, 1993). Interest in this aspect of *E. coli* growth in closed environments has increased greatly over the past few years because it provides a model system for the study of molecular mechanisms of bacterial survival.

CHANGES THAT RESULT FROM STARVATION

Starved *E. coli* cells undergo a series of morphological and physiological changes that distinguish them from cells taken from exponentially growing cultures. Microscopic examination has shown that such cells become much smaller and spherical in stationary phase (Lange and Hengge-Aronis, 1991a; Zambrano et al., 1993). The onset of starvation also triggers changes in membrane fatty acid composition (Cronan, 1968), in cell wall structure (Tuomanen and Cozens, 1987) and in the superhelical density of reporter plasmid DNA (Balke and Gralla, 1987). At low temperatures and low osmolarity, starved *E. coli* can also produce a fibronectin-binding filament named curli (Olsen, Jonsson and Normark, 1989) proposed to convey an advantage in natural environments. More importantly, cells adjust to nutrient insufficiency by decreasing their overall metabolic rate. The observed increases in the rate of protein degradation, early in stationary phase, and the rapid degradation of RNA (Mandelstam, 1960) have led to the suggestion that such cellular constituents might provide the supplies needed to maintain low endogenous metabolic activity (Siegele and Kolter, 1992). Cells can also accumulate storage polymers such as glycogen (Preiss and Romeo, 1989; Romeo and Preiss, 1989) in stationary phase and produce compounds such as trehalose, which is an important protectant against osmotic stress and high temperature (Csonka and Hanson, 1991; Hengge-Aronis et al., 1991). Analogous to the resistance of spores to a variety of environmental assaults, the starvation response in *E. coli* generally results in increased resistance to a variety of inimical agents. In comparison to exponentially growing cells, such stationary phase-induced resistance provides greater protection against heat, oxidative stress and osmotic challenge (Jenkins, Chaisson and Matin, 1988; Jenkins, Shultz and Matin, 1988; Matin, 1991).

Entry into stationary phase is characterized by an overall decrease in protein synthesis (Davis, Luger and Tai, 1986; Groat et al., 1986), together with the expression of specific sets of proteins. These newly synthesized proteins, important for maintaining prolonged viability (Reeve, Amy and Matin, 1984), can vary depending on the nature of the exhausted nutrient that has led to growth arrest (Groat et al., 1986; Chapter 1.2). A core set of approximately 15 proteins appears always to be induced, independent of the particular starvation conditions (Groat et al., 1986; Schultz, Latter and Matin, 1988). It has been suggested that these proteins, designated as *Pex* (*P*ost-*ex*ponential), may play a role in the development of stress resistance in stationary phase cells (Matin, 1991). Thus, regulatory mechanisms operating during the entry into stationary phase are of critical importance as growing cells shift their metabolism from one that promotes rapid growth to one that can ensure starvation survival. Although the nature of the signal that triggers the starvation response in *E. coli* is unknown, recent findings are beginning to elucidate some of the regulatory pathways involved.

REGULATION OF GENE EXPRESSION IN STATIONARY PHASE

The expression of many genes in stationary phase is regulated by the product of the *rpoS* gene. This regulatory gene was independently identified by several groups due to its role in different cells functions. Mutations in *appR*, which affect acid phosphatase production (Tuoati, Dassa and Boquet, 1986), *nur*, involved in resistance to near ultraviolet light (Tuveson, 1981), and *katF*, required for production of the stationary phase–specific catalase (Loewen and Triggs, 1984), all mapped to the same region on the *E. coli* chromosome (minute 59). It was later recognized that these all represented mutations in the same locus (Lange and Hengge-Aronis, 1991b). Sequence analysis of the *katF* gene indicated that its predicted protein product had strong similarity with the major sigma factor in *E. coli*, σ^{70} (Mulvey and Loewen, 1989). Based on sequence predictions, as well as on the regulatory role played in stationary phase cells, the designation of *rpoS* was proposed (Lange and Hengge-Aronis, 1991b). The fact that *rpoS* mutants have reduced viability and fail to develop the increase in resistance that is characteristic of starved cells further emphasized this gene's central role in stationary phase physiology (Lange and Hengge-Aronis, 1991b; McCann, Kidwell and Matin, 1991). Evidence that the *rpoS* gene product does in fact function as a sigma factor *in vitro* has only recently been provided (Tanaka et al., 1993). To reflect its role as an alternative sigma subunit of RNA polymerase in stationary phase cells, the gene has been designated as *rpoS* and its protein product as σ^S.

Two-dimensional gel electrophoretic analysis of protein synthesis has revealed the induction of at least 30 proteins to be dependent on *rpoS* following carbon starvation (Lange and Hengge-Aronis, 1991a, b; McCann, Kidwell and Matin, 1991). Several members of the *rpoS* regulon have already been identified. Some of the σ^S-regulated genes involved in stress protection include *katE*, which encodes the stationary phase catalase HPII (Loewen, Switala and Triggs-Raine, 1985; Loewen and Triggs, 1984), *xthA*, encoding exonuclease III involved in DNA repair (Sak, Eisenstark and Touati, 1989) and *dps*, the gene for a novel histone-like protein which protects cells against H_2O_2 (Almiron et al., 1992). The osmotically regulated genes *osmB* (Jung et al., 1990) and *osmY*, of unknown function (Yim and Villarejo, 1992), and *treA* and *otsAB*, involved in trehalose breakdown and synthesis, respectively (Hengge-Aronis et al., 1991), are also expressed in stationary phase in a σ^S-dependent manner. Additional members of the σ^S regulon include *glgS*, which is involved in glycogen synthesis (Hengge-Aronis and Fischer, 1992), *appA*, the gene for acid phosphatase (Touati, Dassa and Boquet, 1986), the *mcc* genes for microcin C7 production (Diaz-Guerra, Moreno and San Millan, 1989), *bolA*, involved in cell morphology (Bohannon et al., 1991; Lange and Hengge-Aronis, 1991b), the *cyxAB* genes, encoding a potential third cytochrome oxidase (Dassa et al., 1992), and *csgA*, the curli filament structural gene (Olsen et al., 1993). In addition to its role as a positive regulator, *rpoS* can also repress expression of several proteins in stationary phase (Lange and Hengge-Aronis, 1991b; McCann, Kidwell and Matin, 1991).

The mechanism by which *rpoS* mediates expression of these genes is still unknown. The lack of consensus sequences in the promoter regions suggests either a regulatory cascade of gene expression or regulation with the aid of accessory factors to enable efficient recognition of promoters (Hengge-Aronis, 1993; Kolter, Siegele and Tormo, 1993). Another important aspect of the stationary phase response involves understanding regulation of *rpoS* itself. Given the importance of σ^S-dependent gene expression in stationary phase it will be interesting to determine how temporal regulation of *rpoS*, and of σ^S-regulated genes, is achieved. Evidence obtained so far, through the use of *lacZ* fusions to *rpoS* and some of its regulated genes, suggests that there is transcriptional and post-transcriptional regulation of σ^S activity (Loewen et al., 1993; McCann, Fraley and Matin, 1993).

The changes in gene expression and protein synthesis that occur in stationary phase involve additional regulatory factors besides σ^S. For example, several of the proteins expressed upon carbon starvation require cAMP but not *rpoS* (Matin, 1991; McCann, Kidwell and Matin, 1991). The stationary phase–induced synthesis of three heat-shock proteins — DnaK, GroEL, and HtpG — is dependent on the alternative heat-shock sigma factor σ^{32} (Jenkins, Auger and Matin, 1991).

The induction of microcin B17 production at the onset of stationary phase also occurs independently of *rpoS*. In fact, transcription of the microcin B17 structural gene *mcbA*, has been shown to require σ^{70} *in vitro* (Bohannon et al., 1991). Ultimately, it will be of interest to identify the signal or signals which trigger the starvation response in stationary phase *E. coli*. Although several candidates have been suggested (Siegele and Kolter, 1992; Young, Alvarez and Bernlohr, 1990; Young and Bernlohr, 1991), the nature of the signal and how it mediates induction of gene expression is still unknown. Regardless of whether these signals differ depending upon the cause of growth arrest or whether a common signal exists, each must be translated somehow into regulation of expression via σ^S-dependent and σ^S-independent pathways.

POPULATION CHANGES DURING STARVATION

Cells respond to conditions of nutrient exhaustion by carefully controlling gene expression and protein synthesis. The starvation response in *E. coli* helps to ensure prolonged survival of the cells by reducing endogenous metabolism and preparing them for possible adverse conditions. When fresh nutrient supplies become available, cells must also be able to respond rapidly by exiting stationary phase and resuming growth. These physiological responses to changing environmental conditions are well-regulated and reversible processes. Cells, however, can also respond to changes in their external environment by undergoing genetic changes. In this respect, the high population numbers and rapid growth rates of bacteria have in fact been exploited for selection of mutants with new metabolic capabilities (Mortlock, 1982).

Continued exposure to different growth conditions can select for mutants that have growth capabilities that are enhanced over the parental strain. Population changes have been known to take place for a long time in *E. coli* cultures that are kept growing continually in chemostats or by serial transfer techniques (Dykhuizen and Hartl, 1983). These conditions select for favorable mutations which confer competitive advantages over the parental strain. The successive appearance of such mutants in a continuous culture has been termed "nonselective" or "periodic" selection (Atwood, Schneider and Ryan, 1951; McDonald, 1955; Norvick and Szilard, 1950). More recently it has been shown that prolonged growth at elevated temperature selects *E. coli* cells better able to grow under high temperature conditions than the wild type (Bennet, Dao and Lenski, 1990). It has even been possible to select *E. coli* cells that can grow and form colonies at temperatures that are lethal for wild-type cells (54°C) (Droffner and Yamamoto, 1991).

More unusual, perhaps, has been the reported high mutation rates in starved cells (Cairns, Overbaugh and Miller, 1988; Hall, 1990, 1991). These mutations, presumed to occur in nongrowing populations, apparently arise after a particular selective pressure has been imposed on the starved cells. The fact that postselection mutations confering advantages on the cells seem to occur at much higher rates than nonadvantageous mutations has led to extensive discussions regarding the classical view that mutations arise independently of selection (Lenski and Mittler, 1993; Stahl, 1992). Although it is now largely accepted that mutations can arise in stationary phase cells, the molecular mechanisms involved are still unknown. The issue of whether these constitute "directed" mutations is therefore still unresolved (Foster, 1992). The identification of genetic defects which specifically affect the postselection mutation rate opens up new possibilities for understanding the mechanisms involved (Boe, 1990; Rebeck and Samson, 1991).

E. coli cells can elicit a variety of physiological responses to ensure long-term viability upon starvation. In addition, genetic changes can also enhance a cell's capacity to compete or to survive in nutrient-restricted conditions. Recently, we have identified and characterized mutant cells with a growth advantage in stationary phase cultures (Zambrano et al., 1993). We discovered that prolonged incubation under conditions of starvation selects for mutants with a growth advantage over the original parental strain. This phenotype is made evident when cells from a several-day-old (aged) culture of *E. coli* are mixed as a minority with a one-day-old (young) culture. In these mixed cultures, cells from the aged culture can grow and cause the death of the cells from the young culture. Some of the mutations that can confer this phenotype have indeed been found to be in *rpoS*, the structural gene for the stationary phase-specific sigma factor, σ^s. This indicates that stationary phase *E. coli* populations are dynamic and that population changes take place as mutant cells, with a growth advantage, take over the population. It is therefore clear that microorganisms have evolved diverse strategies that allow them to adapt to changing nutritional conditions throughout all phases of their life cycle.

ACKNOWLEDGMENTS

This chapter was adapted from the introduction to the doctoral thesis of M.M. Zambrano. Work on this topic in the laboratory is supported by a grant from the National Science Foundation (DMB-9207323) to R. Kolter. M.M. Zambrano was the recipient of a Ryan Predoctoral Fellowship. R. Kolter received an American Cancer Society Faculty Research Award.

REFERENCES

Almiron, M., A. Link, D. Furlong, and R. Kolter. 1992. A novel DNA-binding protein with regulatory and protective roles in starved *Escherichia coli*. *Genes and Development* **6**, 2646-2654.

Atwood, K. C., L. K. Schneider, and F. J. Ryan. 1951. Periodic selection in *Escherichia coli*. *Proceedings of the National Academy of Sciences U.S.A.* **37**, 146-155.

Balke, V. L., and J. D. Gralla. 1987. Changes in the linking number of supercoiled DNA accompany growth transitions. *Journal of Bacteriology* **169**, 4499-4506.

Bennet, A. F., K. M. Dao, and R. E. Lenski. 1990. Rapid evolution in response to high temperature selection. *Nature* **346**, 79-81.

Boe, L. 1990. Mechanism for induction of adaptive mutations in *Escherichia coli*. *Molecular Microbiology* **4**, 597-601.

Bohannon, D. E., N. Connell, J. Keener, A. Tormo, M. Espinosa-Urgel, M. M. Zambrano, and R. Kolter. 1991. Stationary-phase-inducible "gearbox" promoters: differential effects of *katF* mutations and the role of σ^s. *Journal of Bacteriology* **173**, 4482-4492.

Cairns, J., J. Overbaugh, and S. Miller. 1988. The origin of mutants. *Nature* **335**, 142-145.

Cronan, J. E. 1968. Phospholipid alterations during growth of *Escherichia coli*. *Journal of Bacteriology* **95**, 2054-2061.

Csonka, L. N., and A. D. Hanson. 1991. Prokaryotic osmoregulation: Genetics and physiology. *Annual Reviews of Microbiology* **45**, 569-606.

Dassa, J., H. Fsihi, C. Marck, M. Dion, M. Kieffer-Bontemps, and P. L. Boquet. 1992. A new oxygen-regulated operon in *Escherichia coli* comprises the genes for a putative third cytochrome oxidase and for pH 2.5 acid phosphatase *(appA)*. *Molecular and General Genetics* **229**, 342-352.

Davis, B. D., S. J. Luger, and P. C. Tai. 1986. Role of ribosome degradation in the death of starved *E. coli* cells. *Journal of Bacteriology* **166**, 439-445.

Diaz-Guerra, L., F. Moreno, and J. L. San Millan. 1989. *appR* gene product activates transcription of Microcin C7 plasmid genes. *Journal of Bacteriology* **171**, 2906-2908.

Droffner, M. L., and N. Yamamoto. 1991. Prolonged environmental stress via a two step process selects mutants of *Escherichia*, *Salmonella* and *Pseudomonas* that grow at 54°C. *Archives of Microbiology* **156**, 307-311.

Dykhuizen, D. E., and D. L. Hartl. 1983. Selection in chemostats. *Microbiology Reviews* **47**, 150-168.

Foster, P. L. 1992. Directed mutation: between unicorns and goats. *Journal of Bacteriology* **174**, 1711-1716.

Groat, R. G., J. E. Schultz, E. Zychlinsky, A. Bockman, and A. Matin. 1986. Starvation proteins in *Escherichia coli*: Kinetics of synthesis and role in starvation survival. *Journal of Bacteriology* **168**, 486-493.

Hall, B. G. 1990. Spontaneous point mutations that occur more often when they are advantageous than when they are neutral. *Genetics* **126**, 5-16.

Hall, B. G. 1991. Is the occurrence of some spontaneous mutations directed by environmental challenges? *The New Biology* **3**, 729-733.

Hengge-Aronis, R. 1993. Survival of hunger and stress: the role of *rpoS* in early stationary phase gene regulation in *Escherichia coli*. *Cell* **72**, 165-168.

Hengge-Aronis, R., and D. Fischer. 1992. Identification and molecular analysis of *glgS*, a novel growth-phase-regulated and *rpoS*-dependent gene involved in glycogen synthesis in *Escherichia coli*. *Molecular Microbiology* **6**, 1877-1886.

Hengge-Aronis, R., W. Klein, R. Lange, M. Rimmele, and W. Boos. 1991. Trehalose synthesis genes are controlled by the putative sigma factor encoded by *rpoS* and are involved in stationary phase thermotolerance in *Escherichia coli*. *Journal of Bacteriology* **173**, 7918-7924.

Jenkins, D. E., E. A. Auger, and A. Matin. 1991. Role of RpoH, a heat shock regulator protein, in *Escherichia coli* carbon starvation protein synthesis and survival. *Journal of Bacteriology* **173**, 1992-1996.

Jenkins, D. E., S. A. Chaisson, and A. Matin. 1990. Starvation-induced cross protection against osmotic challenge in *Escherichia coli*. *Journal of Bacteriology* **172**, 2779-2781.

Jenkins, D. E., J. E. Schultz, and A. Matin. 1988. Starvation-induced cross protection against heat or H_2O_2 challenge in *Escherichia coli*. *Journal of Bacteriology* **170**, 3910-3914.

Jung, J. U., C. Gutierrez, F. Martin, M. Ardourel, and M. Villarejo. 1990. Transcription of *osmB*, a gene encoding an *Escherichia coli* lipoprotein, is regulated by dual signals. *Journal of Biological Chemistry* **265**, 10574-10581.

Kaiser, D. 1984. Regulation of multicellular development in *Myxobacteria*. pp. 197-218 *in* R. Losick and L. Shapiro (eds.), *Microbial Development*. Cold Spring Harbor Laboratory, Cold Spring Harbor, N.Y.

Kjelleberg, S., M. Hermansson, P. Marden, and G. W. Jones. 1987. The transient phase between growth and nongrowth of heterotrophic bacteria, with emphasis on the marine environment. *Annual Reviews of Microbiology* **41**, 25-49.

Kolter, R., D. A. Siegele, and A. Tormo. 1993. The stationary phase of the bacterial life cycle. *Annual Reviews of Microbiology* **47**, 855-874.

Lange, R., and R. Hengge-Aronis. 1991a. Growth phase–regulated expression of *bolA* and morphology of stationary phase *Escherichia coli* cells is controlled by the novel sigma factor σ^S (*rpoS*). *Journal of Bacteriology* **173**, 4474-4481.

Lange, R., and R. Hengge-Aronis. 1991b. Identification of a central regulator of stationary phase gene expression in *Escherichia coli*. *Molecular Microbiology* **5**, 49-59.

Lenski, R. E., and J. E. Mittler. 1993. The directed mutation controversy and neo-Darwinism. *Science* **259**, 188-194.

Loewen, P. C., J. Switala, and B. L. Triggs-Raine. 1985. Catalase HPI and HPII in *Escherichia coli* are induced independently. *Archives of Biochemistry and Biophysics* **243**, 144-149.

Loewen, P. C., and B. L. Triggs. 1984. Genetic mapping of *katF*, a locus that with *katE* affects the synthesis of a second catalase species in *Escherichia coli*. *Journal of Bacteriology* **160**, 668-675.

Loewen, P. C., I. von Ossowski, J. Switala, and M. R. Mulvey. 1993. KatF (σ^S) synthesis in *Escherichia coli* is subject to posttranscriptional regulation. *Journal of Bacteriology* **175**, 2150-2153.

Losick, R., and P. Youngman. 1984. Endospore formation in *Bacillus*, pp 63-88 *in* R. Losick and L. Shapiro (eds.), *Microbial Development*. Cold Spring Harbor Laboratory, Cold Spring Harbor, N.Y.

Mandelstam, J. 1960. The intracellular turnover of protein and nucleic acids and its role in biochemical differentiation. *Bacteriology Reviews* **24**, 289-308.

Matin, A. 1991. The molecular basis of carbon-starvation-induced general resistance in *Escherichia coli. Molecular Microbiology* **5**, 3-10.

Matin, A., E. A. Auger, P. H. Blum, and J. E. Schultz. 1989. Genetic basis of starvation survival in nondifferentiating bacteria. *Annual Reviews of Microbiology* **43**, 293-316.

McCann, M. P., C. D. Fraley, and A. Matin. 1993. The putative σ factor KatF is regulated post-transcriptionally during carbon starvation. *Journal of Bacteriology* **175**, 2143-2149.

McCann, M. P., J. P. Kidwell, and A. Matin. 1991. The putative σ factor KatF has a central role in development of starvation-mediated general resistance in *Escherichia coli. Journal of Bacteriology* **173**, 4188-4194.

McDonald, D. J. 1955. Segregation of the selective advantage obtained through orthoselection in *Escherichia coli. Genetics* **40**, 937-950.

Morita, R. Y. 1988. Bioavailability of energy and its relationship to growth and starvation survival in nature. *Canadian Journal of Microbiology* **34**, 436-441.

Mortlock, R. P. 1982. Metabolic acquisitions through laboratory selection. *Annual Reviews of Microbiology* **36**, 259-284.

Mulvey, M. R., and P. C. Loewen. 1989. Nucleotide sequence of *KatF* of *Escherichia coli* suggests *KatF* protein is a novel σ transcription factor. *Nucleic Acids Research* **17**, 9979-9991.

Novick, A., and L. Szilard. 1950. Experiments with the chemostat on spontaneous mutations of bacteria. *Proceedings of the National Academy of Science, U.S.A.* **36**, 708-719.

Olsen, A., A. Arnqvist, M. Hammar, S. Sukupolvi, and S. Normark. 1993. The RpoS sigma factor relieves H-NS mediated transcriptional repression of *csgA*, the subunit gene of fibronectin binding curli in *Escherichia coli. Molecular Microbiology* **7**, 523-536.

Olsen, A., A. Jonsson, and S. Normark. 1989. Fibronectin binding mediated by a novel class of surface organelles on *Escherichia coli. Nature* **338**, 652-655.

Preiss, J., and T. Romeo. 1989. Physiology, biochemistry and genetics of bacterial glycogen synthesis, pp 183-233 *in* A. H. Rose and D. Tempest (eds.), *Advances in Microbial Physiology.* Academic Press, London.

Rebeck, G. W., and L. Samson. 1991. Increased spontaneous mutation and alkylation sensitivity of *Escherichia coli* strains lacking the *ogt* O^6-methylguanine DNA repair methyltransferase. *Journal of Bacteriology* **173**, 2068-2076.

Reeve, C. A., P. S. Amy, and A. Matin. 1984. Role of protein synthesis in the survival of carbon-starved *Escherichia coli* K-12. *Journal of Bacteriology* **160**, 1041-1046.

Romeo, T., and J. Preiss. 1989. Genetic regulation of glycogen biosynthesis in *Escherichia coli*: *In vitro* effects of cyclic AMP and guanosine 5′-diphosphate 3′-diphosphate and analysis of *in vivo* transcripts. *Journal of Bacteriology* **171**, 2773-2782.

Roszak, D. B., and R. R. Colwell. 1987. Survival strategies of bacteria in the natural environment. *Microbiology Reviews* **51**, 365-379.

Sak, B. D., A. Eisenstark, and D. Touati. 1989. Exonuclease III and the catalase hydroperoxidase II in *Escherichia coli* are both regulated by the *katF* gene product. *Proceedings of the National Academy of Science, U.S.A.* **86**, 3271-3275.

Schultz, J. E., G. I. Latter, and A. Matin. 1988. Differential regulation by cyclic AMP of starvation protein synthesis in *Escherichia coli. Journal of Bacteriology* **170**, 3903-3909.

Siegele, D. A., and R. Kolter. 1992. Life After Log. *Journal of Bacteriology* **174**, 345-348.

Stahl, F. W. 1992. Unicorns Revisited. *Genetics* **132**, 865-867.

Tanaka, K., Y. Takayanagi, N. Fujita, A. Ishihama, and H. Takahashi. 1993. Heterogeneity of the principal sigma factor in *Escherichia coli*: the *rpoS* gene product, σ^{38}, is a principal sigma factor of RNA polymerase in stationary-phase *Escherichia coli*. *Proceedings of the National Academy of Science, U.S.A.* **90**, 3511-3515.

Touati, E., E. Dassa, and P. L. Boquet. 1986. Pleiotropic mutations in *appR* reduce pH 2.5 acid phosphatase expression and restore succinate utilization in CRP-deficient strains of *Escherichia coli*. *Molecular and General Genetics* **202**, 257-264.

Tuomanen, E., and R. Cozens. 1987. Changes in peptidoglycan composition and penicillin-binding proteins in slowly growing *Escherichia coli*. *Journal of Bacteriology* **169**, 5308-5310.

Tuveson, R. W. 1981. The interaction of a gene *(nur)* controlling near-UV sensitivity and the *polA1* gene in strains of *E. coli* K-12. *Photochemistry and Photobiology* **33**, 919-923.

Yim, H. H., and M. Villarejo. 1992. *osmY*, a new hyperosmocally inducible gene, encodes a periplasmic protein in *Escherichia coli*. *Journal of Bacteriology* **174**, 3637-3644.

Young, C. C., and R. W. Bernlohr. 1991. Elongation factor Tu is methylated in response to nutrient deprivation in *Escherichia coli*. *Journal of Bacteriology* **173**, 3096-3100.

Young, C. C., J. D. Alvarez, and R. W. Bernlohr. 1990. Nutrient-dependent methylation of a membrane-associated protein of *Escherichia coli*. *Journal of Bacteriology* **172**, 5147-5153.

Zambrano, M. M., D. A. Siegele, M. Almiron, A. Tormo, and R. Kolter. 1993. Microbial competition: *Escherichia coli* mutants that take over stationary phase cultures. *Science* **259**, 1757-1760.

1.4 Continuous Culture of Microorganisms in Liquid Media: Effects of Growth Rate and Nutrient Limitation

Peter Gilbert and Michael R.W. Brown

INTRODUCTION

When microorganisms are grown in batch culture their environment is in continual flux. As nutrients are consumed, secondary metabolites are exported to the medium, biomass increases, and availability of oxygen drops (Chapters 1.2 and 3.1). The cumulative effects of this flux are that, even within a single flask of medium, the physiology of the cells is difficult to predict at any given time, and simple standardization of the timing of harvest for the preparation of inocula is often inadequate for ensuring that a reproducible phenotype is collected on successive occasions. In batch culture, in the absence of an inhibitor, growth ceases through lack of one or more essential nutrients (oxygen, carbon source, etc.). The nature of this depletion, to a significant extent, defines the physiological status of the cells in stationary phase (see Chapters 1.2, 1.3, and 3.1). In addition, phenotype is affected by the rate of division of the cells in mid- and late-logarithmic phases of growth. This may be affected/regulated, not only through changing the pH of the medium and/or the incubation temperature, but also by changing the source from which the critical nutrients are derived. Thus, growth with sodium succinate as a sole carbon source will often give shorter generation times than growth with glucose or glycerol as the carbon source, and, for prototrophic strains, supplies of amino acids as the nitrogen source will give faster overall growth than ammonium salts. Batch cultures are not, therefore, amenable to the study of the effects of pH, temperature, or C/N source upon physiology, since changing any of these

0-8493-4752-1/95/$0.00+$.50

will also change the growth rate. The effects of the manipulated physico-chemical parameter and that of changed growth rate are impossible to separate.

Continuous culture (Monod, 1949; Herbert, Ellsworth and Telling, 1956), on the other hand, is an example of an open *in vitro* culture system which allows for the study of microorganisms growing under steady-state conditions which are relevant to their natural habitats. Open cultures provide a constant input of substrate, and continual removal of spent medium, waste products, and biomass. Bacterial growth occurs at a constant rate in a fixed physico-chemical environment and can therefore be controlled by the imput of medium. Nutrient concentration, pH, metabolic products, and oxygen tension, all of which change during batch culture, can be maintained at a steady state in continuous culture, while controlling the growth rate, up to the maximum (μ_{max}), for the chosen medium. If different media are to be compared, then growth rates may be chosen which are beneath μ_{max} for each of them. Alternatively, comparisons are often made at a fixed μ_{REL} (i.e., $\mu_{REL} = \mu/\mu_{max}$ for each medium) (Tempest, 1976). The effects of growth rate can therefore be distinguished from those due to phenotypic adaptation of the cells to the environment. Conversely physico-chemical parameters, such as pH and temperature, can be altered with the growth rate fixed to a constant value. From the point of view of experimental control, this situation is ideal. It must be borne in mind, however, that continuous cultures are highly selective environments, and will often, in the long term, select for a different genotype/phenotype from that which dominated the original inoculum.

CONTINUOUS CULTURE: THEORETICAL CONSIDERATIONS

At any given time during batch culture, the specific growth rate (μ) of the cells, defined as the rate of increase in biomass per unit of organism concentration, is related to the concentration of an essential nutrient up to a saturating concentration, at which growth rate is maximized (μ_{max}). This dependence, analogous to Michaelis-Menten kinetics, was described mathematically by Monod (1949)

$$\mu = \mu_{max}\left(\frac{S}{K_S + S}\right) \tag{1}$$

where S = the substrate concentration in the culture

K_S = a saturation constant, analogous to K_M and equal to the substrate concentration at which $\mu = {}^1/_2\ \mu_{max}$

In batch culture this gives growth at μ_{max} throughout most of the logarithmic phase; nutrient limitation of growth rate only becomes significant late in the culture, and affects the pattern of onset of stationary phase (Chapter 1.2).

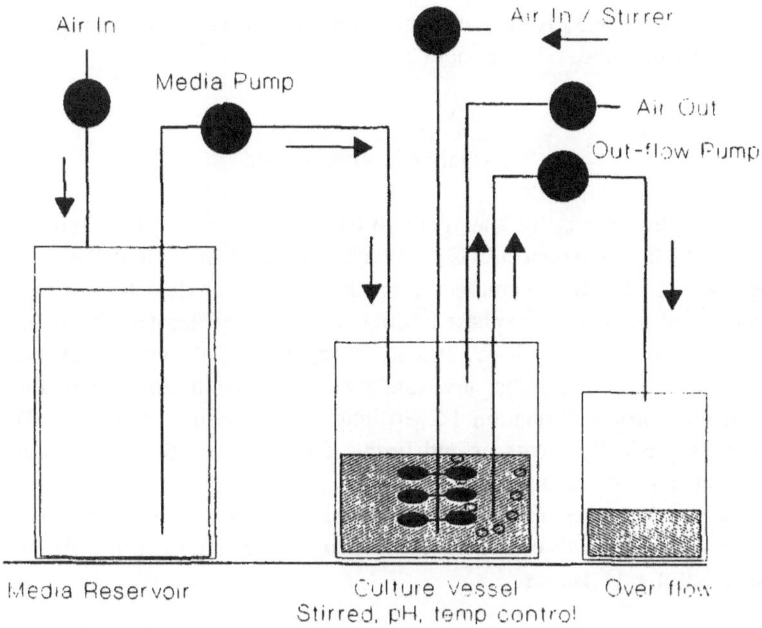

FIGURE 1 Diagrammatic representation of chemostat.

Nutrient limitation of growth rate in batch cultures is only short lived, since substrate concentrations are constantly changed by consumption of nutrient. In continuous cultures, however, steady states are achieved, at which rate-limiting concentrations of chosen substrates may be perpetuated (Herbert, Ellsworth and Telling, 1956).

A diagrammatic representation of a chemostat is given in Figure 1. The apparatus consists of a culture vessel (volume v ml) fitted with a constant level device. The vessel is aerated, thoroughly mixed by impellors, and maintained at constant temperature and pH. The culture vessel is continually replenished with fresh medium, and spent medium and biomass are removed, at a single fixed flow rate (f ml/h), maintaining the culture volume (v) constant. Thorough agitation of the vessel contents and aeration ensure that the added medium is rapidly and completely distributed throughout an oxygenated culture. The residence time for any individual bacterial cell within the culture vessel is a function of the dilution rate (D), defined as

$$D = \frac{f}{v} \tag{2}$$

When the fermenter is set up and inoculated the culture grows, as in batch, at a rate equivalent to μ_{max} (h^{-1}). Provided that μ_{max} is greater than D (h^{-1}), the biomass in the culture increases at a rate equal to $\mu_{max} - D$, and at any given biomass the concentration of each substrate in the culture (S) is equal to

$S_R - X \bullet Y$ (where S_R is the incoming substrate concentration, X is the biomass in the culture, and Y is the yield constant).

$$Y = \frac{\text{weight of bacteria formed}}{\text{weight of substrate used}} \tag{3}$$

This situation is strictly analogous to batch culture where D = zero. In both culture systems the concentration of each substrate remaining in the culture (S) decreases as biomass increases. In batch culture, such depletion brings about an onset of the stationary phase through the loss of the least available nutrient. In continuous culture, where nutrients are replenished, the growth rate becomes reduced to $< \mu_{max}$ by the now rate limiting concentration (S) of the least available nutrient (Equation 1.). So long as μ remains greater than D, then biomass gradually increases and further reduces the substrate concentration remaining in the culture. If μ becomes less than D, then the biomass in the fermenter decreases. Eventually (in practice after five volume changes have taken place), growth of the biomass balances the rate of inflowing medium such that $\mu = D$. Hence,

$$\mu = D = \frac{\text{In } 2}{\text{generation time}} \tag{4}$$

and

$$S = K_S\left(\frac{D}{(\mu_{max} - D)}\right) \tag{5}$$

At steady state the equilibrium biomass in the culture (X) is given by

$$\overline{X} = Y(S_R - S) \tag{6}$$

If Equation 5 is substituted into Equation 6, then

$$\overline{X} = Y\left(S_R - K_S\left(\frac{D}{\mu_{max} - D}\right)\right) \tag{7}$$

Since μ_{max}, K_S, and Y are constants, for a given organism in a given medium, then the steady-state biomass and substrate concentration (S) in the vessel depend solely on the values of S_R and D (Equations 5 and 7). Varying the substrate concentration in the incoming medium therefore affects the steady-state biomass (Figure 2), whereas the dilution rate directly controls the specific growth rate. If the dilution rate is set to a value greater than μ_{max}, then steady state will be lost and the culture will be washed out of the fermenter at a rate $D - \mu_{max}$. The critical dilution rate at which this occurs is termed D_{MAX}.

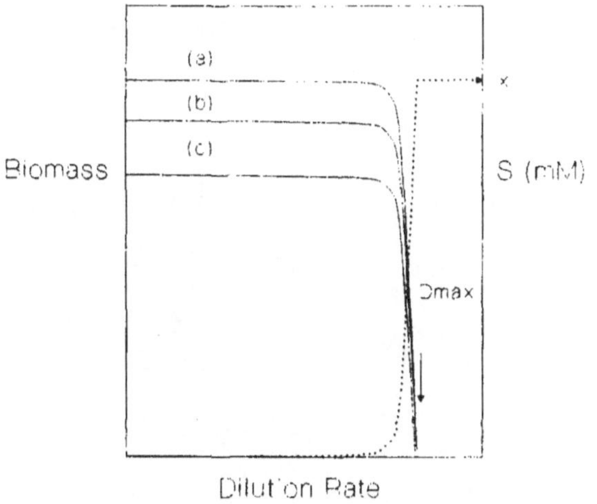

FIGURE 2 The effect of limiting substrate concentration in the incoming medium (S_R) upon the steady-state biomass in an ideal chemostat over a range of dilution rates. K_R line (a) > K_R line (b) > K_R line (c). The dotted line indicates the residual substrate concentration (S) within the culture, which is the same in all three instances.

Major differences exist, therefore, between the variable parameters available for control in batch and continuous culture. In batch culture the nature of the medium determines μ_{max}, which is expressed during logarithmic phase. During logarithmic phase the nature of the rate-limiting nutrient and hence μ might change since different substrates are consumed at different rates, a reflection of Y (Equation 3) (Chapter 3.1). The concentration of the least available nutrient determines the biomass at onset of stationary phase (nutrient depletion), and the physiology of the stationary phase cells (Chapter 1.2). The nutrient causing onset of stationary phase need not be the same as that limiting growth in logarithmic phase. In chemostats, the least available nutrient will, at steady state, determine the growth rate (nutrient limitation), and its concentration will determine the steady-state biomass. Other nutrients, present in excess, will establish μ_{max} for the system.

The relationship between steady-state biomass and dilution rate, illustrated in Figure 2, represents a substrate for which there is no maintainance requirement for endogenous metabolism, and for which the yield constant (Y) is unchanged with the growth rate of the cells. Such a situation exists for nitrogen limitation (Figure 2) but not when the growth-limiting substrate is such as carbon, magnesium, iron, or oxygen (Figure 3).

Carbon Limitation

Cells will endogenously catabolize carbon substrate in order to maintain viability. Carbon substrate catabolized in this manner will not contribute towards the generated biomass. Rates of endogenous metabolism are more or less unaffected by the growth rate of the cells; thus, the proportion of the

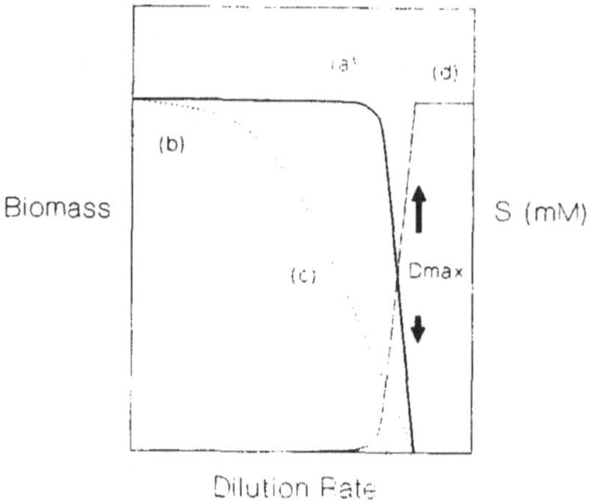

FIGURE 3 The effect of different rate-limiting nutrients upon the steady-state biomasses achieved in a chemostat at various dilution rates. (a) is an ideal substrate with no endogenous catabolism and constant Y, (b) is a carbon substrate where endogenous catabolism becomes significant at low dilution rate, and (c) is a substrate such as iron or magnesium where Y depends upon growth rate or oxygen where the rate of supply is independent of the dilution rate; (d) indicates the residual substrate concentration (S) within the culture, which follows a similar pattern in all three instances.

available carbon substrate used endogenously by the cells becomes significant, in terms of the generated biomass, at slow growth rates. The biomass of carbon-limited cultures, therefore, decreases with decreasing dilution rate below some critical value, usually ca. $\mu = <0.075h^{-1}$ (Figure 3, line b). At slow growth rates, such as these, it is often the case that a proportion of the cells become dormant and are washed out of the fermenter (Pirt, 1972). In order to maintain reproducibility of an inoculum taken from the chemostat it is therefore advisable that, for carbon-limited populations, dilution rates not go below this value.

Magnesium and Iron Limitation

Substrates such as magnesium and iron differ from carbon sources in that while they are not lost by catabolism, the requirements of the cell increase with increasing growth rate. Thus, a major functional role for magnesium is ribosome stabilization. Since fast-growing cells must contain a greater number of ribosomes than slow-growing ones, the yield constant for magnesium is decreased accordingly. Similarly, a major destination for iron is cytochromes, and the requirement for these will be increased in rapidly metabolizing cells. The effect on steady-state biomass of the yield constant for the growth-limiting substrate (Y) decreasing with increasing growth rate is shown in Figure 3 (line c). Biomass in the fermenter will decrease progressively with increasing dilution rate up to washout (D_{MAX}). In spite of the changed biomass the concentration of the rate-limiting substrate in the culture vessel will be as before (line d, Figure 3).

Oxygen Limitation

The situation for oxygen limitation is more complicated. The major complication is that oxygen is usually supplied to the fermenter independently of the medium. Oxygen supply is therefore independent of the dilution rate. Two fermenter configurations are commonly employed, those where oxygen/air is supplied at a constant rate and those where oxygen/air is supplied on demand, sensed through an oxygen electrode. In the first situation, with the air supply fixed, then with all other nutrients present in excess and at a fixed dilution rate, the biomass will increase so long as $\mu > D$, until the biological oxygen demand of the culture (BOD) exceeds the rate of oxygen input. A steady state will eventually be established where once again $\mu = D$. If the flow rate of the culture is increased, then the metabolic rate of the cells, and hence their BOD, will increase. Since BOD will now exceed the rate of oxygen supply, μ will become less than D and biomass will be washed out. A new steady state will establish with μ equal to the new dilution rate. The biomass in the culture will therefore decrease, as for magnesium and iron as substrates, as an inverse function of dilution rate (Figure 3, line c). In this instance the rate of dissolution of oxygen into the culture is analogous to K_R and determines biomass at steady state. Where oxygen tension in the fermentation vessel can be controlled, it is possible to obtain oxygen limitation at various dilution rates while maintaining a constant biomass, and thereby constant physico-chemical environment (pH, excess nutrients etc.).

Turbidostats

Turbidostats are continuous cultures which have been designed to maximize the biomass generated by the fermenter. In the turbidostat the rate of flow of fresh medium is governed by the turbidity of the culture. If the turbidity of the culture decreases (i.e., $\mu < D_{MAX}$), then the flow of medium stops. When the turbidity is maximal or rising, then medium is added. The flow of medium is therefore discontinuous and ensures that at steady state the growth rate of the culture is equivalent to μ_{max}.

CONTINUOUS CULTURE: PRACTICAL CONSIDERATIONS

Basic assumptions made of any continuous culture fermenter are that fresh medium is completely and instantaneously mixed throughout a homogeneous culture such that nutrient is available to all cells equally, and that air/oxygen supply does not limit the biomass generated (unless oxygen limitation or anaerobic growth is sought).

Culture Homogeneity

If incoming nutrient is not equally available to all of the cells within the culture, then the basic assumption made previously about dilution rate controlling growth rate no longer applies. Two factors commonly affect culture homogeneity in continuous culture systems: formation of biofilms/floccules and foaming.

Foaming

While foaming of the culture will affect nutrient distribution and possibly provide a focus for the concentration of cells, it is readily combated either by changing the medium or by the addition of antifoaming agents. It must be borne in mind, however, that the presence of antifoam will markedly affect the surface properties of the cultured cells.

Floccules

Organisms may become flocculated, or they may attach to the walls of the vessels and grow as biofilms (Chapter 1.6). In such instances the availability of nutrient is greater to the unattached, planktonic cells, and to those cells on the outside of a biofilm or floccule, than it is to those cells located deeply within the cell mass. A distinction must be drawn between floccules and attached biofilms in that the former may be washed out of the culture vessel whereas biofilms, attached to the fermenter wall, are retained.

The residence time of a floccule within the fermentation vessel is given, as for single cells, by Equation 2. While the arguments made previously for homogeneous cultures still apply, at steady state the net growth rate, rather than the growth rate of individual cells, is governed by D. Since the cells on the exterior of each floccule will be growing at a faster rate than those within it, then each floccule will contain some cells with $\mu > D$ and others with $\mu < D$. While the width of the distribution of μ about D will increase with the size of the floccules, the biomass will remain constant. Such systems find application as fluidized bed reactors (Bryers, 1993) but are not amenable for the reproducible generation of inocula for performance tests.

Biofilms

If cells attach, as biofilms, to the interior of the fermentation vessel, then their residence time is significantly increased over those cells which are growing planktonically. Such cells might even become dormant, or sporulate, without being washed from the culture. Provided that the biofilms do not detach and contaminate the planktonic culture, then for the planktonic, single cells $\mu = D$. Steady state will, however, have been destroyed since, with time, the suspended biomass will become decreased through nutrient loss to the biofilm, and the total biomass in the vessel (planktonics plus biofilm) will increase. Growth of the attached population will eventually lead to the development of thick biofilms with marked concentration gradients of the limiting nutrient established across their depth (Wimpenny, Peters and Scourfield, 1989). Cells deep within the biofilm, adjacent to the fermenter wall, will be growing, if at all, at rates significantly reduced from $\mu = D$. If the biofilms become sloughed-off, then they may, to all intents and purposes, be considered as floccules (above), but, since the bulk of the planktonic cells will be at $\mu = D$, they will have a net growth rate less than $\mu = D$. Wall growth within a chemostat, such as this, has generally been regarded as an unfortunate problem. Recently, however, chemostat cultures have been widely employed for the long-term culture of biofilms (West et al., 1989; Anwar et al., 1989; Keevil, Dowsett and Rogers, 1993; Chapters 1.6, 3.6, and 3.7). In such techniques the biofilms are deliberately formed on test pieces suspended within the culture. Use of a chemostat in such instances should not, however, imply any form of control over the physiology and growth rate of the attached population (Chapter 3.1).

Continuous Culture Fermenter Design

Continuous culture systems vary in complexity, from simple stirred flasks with provision for the addition of buffered media and the removal of culture to sophisticated fermenters, computer-interfaced to achieve control and maintainance of pH, Eh, and oxygen tension, together with a means of detecting and combatting foam. Sizes of fermentation vessels vary between 50 ml and 25 l, for experimental systems, up to 5000 l for industrial plants. As the size of the fermenter is increased, or as greater steady-state biomasses are sought, then achieving adequate oxygen/air supplies together with complete mixing of fresh medium with culture becomes problematic. In order to overcome these problems the fermenters are often baffled and vigorously agitated by powerful impellors.

As aeration rates are increased in order to accommodate high biomasses, evaporation losses can become significant. Loss of water from a fermenter by evaporation will, over an extended time, greatly affect the tonicity of the culture. In many fermenters condensers, associated with the air outflow, return water to the culture. A simple, small-scale, all-glass fermenter suitable for generating susceptibility test inocula has been described by Gilbert and Stuart (1977); larger, more sophisticated fermenters are available from the major laboratory suppliers.

MODULATION OF ANTIMICROBIAL SUSCEPTIBILITY IN CONTINUOUS CULTURE

An effect of different growth-limiting nutrients, in the chemostat, is to give rise to cells with radically different cell envelopes (Holme, 1972; Ellwood and Tempest, 1972; Brown and Melling, 1969a, b; Brown, 1975). The general physiology of the cells becomes adapted in a number of ways:

1. Usage of the growth-limiting nutrient is rationalized through the use of alternative substrates, modification of cell composition, and/or reduction in the amounts of macromolecules containing these substances.
2. The cell surface becomes adapted in order to increase its affinity for the growth-limiting material, thus making uptake into the cytosol more efficient and competitive. Similarly, production of extracellular enzymes (Evans et al., 1994) and chelators (Brown and Williams, 1985) might increase availability of the limiting nutrient.
3. The cellular growth rate will become reduced to the maximum permissible, given (1) and (2).

With gram-negative bacteria changes in the cell envelope are widely reported (Ellwood and Tempest, 1972; Nikaido and Nakae, 1979; Lugtenberg and Van Alphen, 1983). Altered phospholipid content (Pechey et al., 1974; Teuber and Bader, 1976; Ikeda et al., 1984), porin protein composition (Harder et al., 1981; Williams et al., 1984), lipopolysaccharide content (Tamaki et al., 1971) and cation contents (Brown and Melling 1969a, b; Gilleland et al., 1974; Melling et al., 1974; Nichols et al., 1989; Boggis et al., 1979; Shand et al., 1985) of the cell envelopes have all been variously reported to influence antimicrobial susceptibility towards preservatives and disinfectant molecules (Brown, 1977; Gilbert and Wright, 1986; Brown et al., 1990; Gilbert et al., 1990) and antibiotics (De La Rosa et al., 1982; Brown and Williams, 1985; Tuomanen et al., 1986) for a wide range of bacterial species (Tempest et al., 1968; Meers and Tempest, 1970; Minnikin et al., 1971; Dean, 1972; Dean et al., 1976; Holme, 1972; Broxton et al., 1984). The action of antimicrobial agents is thought to be modulated, under such circumstances, through:

1. Changes in the relative abundance of the target molecule/material.
2. Alteration of surface charge and thereby initial drug adsorption.
3. Alteration of the fluidity and lipophilicity of the various envelope compartments, thereby affecting optimal drug lipophilicity (log P_o).
4. Alteration of porin-protein expression and content, thereby affecting permeation of small hydrophilic agents.

A number of groups have used continuous culture in order to evaluate the effects of growth rate upon the sensitivity of cells towards antibiotics, disinfectants, and preservatives. A general conclusion to be drawn from such studies is that slowly growing cells are particularly recalcitrant to inactivation (Finch and Brown, 1975; Gilbert and Brown, 1978; Tuomanen et al., 1986; Brown et al., 1990; Gilbert et al., 1990).

Susceptibility Towards Chemical Biocides

Finch and Brown (1975) and Gilbert and Brown (1978) observed that *P. aeruginosa* became particularly sensitive towards ethylenediaminetetraacetic acid and chlorinated phenols as the growth rate was increased. In one of these studies (Gilbert and Brown 1978), the increased sensitivity, towards phenolic biocides, was linked with marked reductions in the total lipopolysaccharide content of the envelopes at the faster growth rates. Since increases in lipopolysaccharide also resulted in decreased levels of drug uptake by the cells, it was suggested that lipopolysaccharide formed a barrier in the outer membrane against adsorption of these agents.

Wright and Gilbert (1987a) investigated the interrelation of the chlorhexidine sensitivity of *E. coli* and the growth rate, for four nutrient limitations. Nitrogen- and carbon-limited cultures showed an overall increase in sensitivity as growth rate increased while magnesium- and phosphorus-limited cultures showed an opposite trend of increased resistance. At the extremes of growth rate tested, different orders of sensitivity were observed between nutrient limitations. When $\mu < 0.08\ h^{-1}$, sensitivity was seen to decrease with different nutrient limitations in the sequence carbon limitation > phosphorus-limitation > magnesium-limitation > nitrogen-limitation, while as faster growth rates ($\mu > 0.4\ h^{-1}$), the sequence was altered to carbon limitation > nitrogen limitation > phosphorus limitation > magnesium limitation. Overall, carbon-limited cultures were most sensitive towards chlorhexidine, with this limitation showing the least dependency upon growth rate.

In a related study with the same organism, Wright and Gilbert (1987b) investigated the effects of growth rate and nutrient limitation upon the activity of a homologous series of *n*-alkyltrimethylammonium bromides (C*n*TABs) against *E. coli*. Growth inhibitory and bactericidal activities of these compounds are parabolically related to the *n*-alkyl chain length of these compounds, and thereby to compound lipophilicity (log P)(Al-Taae et al., 1986). The chain length at which optimal activity is demonstrated varies between different cell types and reflects the lipophilicity and barrier properties of the cell envelopes (Hansch and Clayton, 1973). Wright and Gilbert (1987b) argued that alteration in envelope lipophilicity through changes in growth rate and nutrient limitation might be expected to produce changes in log P_0 and also in the degree of activity demonstrated by the optimally active compound. Figure 4 shows the effect of *n*-alkyl chain length and growth rate upon the sensitivity of cells prepared under four nutrient limitations, and Figure 5 shows the effects of nutrient limitation and growth rate upon the activity of one compound, cetrimide (C_{16}TAB). In all cases resistance maximized at growth rates between $0.1\ h^{-1}$ and $0.23\ h^{-1}$ and decreased markedly at the faster growth rates. The results of this study support the hypothesis that growth rate and nutrient

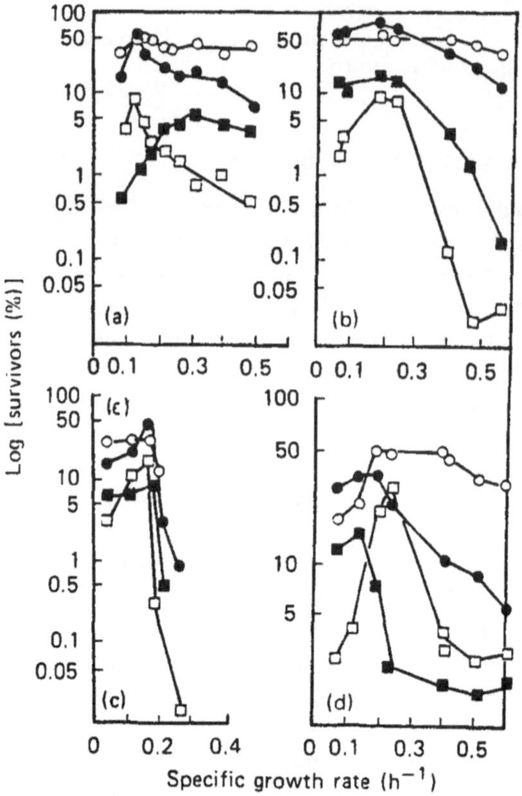

FIGURE 4 Steady-state viability (30min) following exposure to C_{12}TAB (26 m*M*, ○), C_{14}TAB (12 m*M*, ●), C_{16}TAB (4.1 m*M*, □) and C_{18}TAB (3.8 m*M*, ■) of *Escherichia coli*, previously grown in a chemostat under (a) magnesium limitation, (b) nitrogen limitation, (c) phosphate limitation, and (d) carbon limitation at a variety of growth rates. (From Wright, N.E., and Gilbert, P., *Journal of Pharmacy and Pharmacology*, 39, 685-690, 1987. With permission.)

limitation alter the overall lipophilicity of the cell envelope and thereby influence the optimal value of log P required by compounds in order to traverse it.

Susceptibility Towards Antibiotics

Through control of growth rate and the application of particular nutrient limitation, continuous culture techniques have been widely used to model infections. Using such techniques, growth rate and nutrient deprivation have been identified as fundamental modulators of antibiotic activity. In this respect it has become apparent that the antibiotics ceftoxidine and ceftriaxone have little or no activity against slowly growing cultures of *Escherichia coli* (Cozens et al., 1986; Tuomanen et al., 1986). In contrast, the β-lactam CGP 17520 is particularly effective against slow-growing cultures with activity directed against

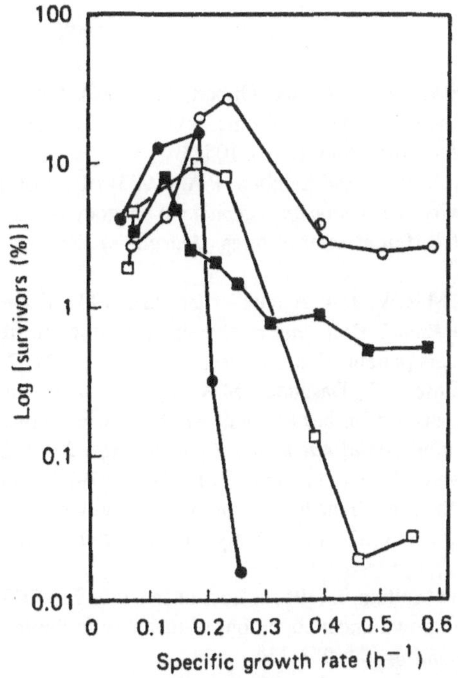

FIGURE 5 Steady-state viability (30 min) following exposure to C_{16}TAB (4.1 m*M*) of *Escherichia coli*, previously grown in a chemostat under (■) magnesium limitation, (□) nitrogen limitation, (●) phosphate limitation and (○) carbon limitation at a variety of growth rates. (From Wright, N.E., and Gilbert, P., *Journal of Pharmacy and Pharmacology*, 39, 685-690, 1987. With permission.)

PBP 7 (Cozens et al., 1986; Tuomanen and Schwartz, 1987). Since expression of PBPs is highly growth-rate dependent, this has consequences for β-lactam antibiotic susceptibility (Turnowsky et al., 1983, Tuomanen et al., 1986).

Various studies have related changes in gross cell-envelope composition to alterations in polymyxin susceptibility (Finch and Brown, 1975; Melling, Robinson and Ellwood, 1974), and the activity of polymyxin shown to vary up to 10-fold through changes in nutrient limitation and specific growth rate (Dorrer and Teuber, 1977; Wright and Gilbert, 1987c). The aminoglycoside antibiotics tobramycin and streptomycin are also growth-rate dependent in their action (Muir, Van Heeswick and Wallace, 1984; Raulston and Montie, 1989; Evans et al., 1990; Duguid et al., 1992a) as are the newer quinolone agents (Evans et al., 1991; Duguid et al., 1992b). Such effects are not restricted to antimicrobial susceptibility and have also been reported to influence profoundly immunogenicity of microbes (Anwar et al., 1985; Brown and Williams, 1985) as well as susceptibility to host defenses (Anwar, Brown and Lambert, 1983) and extracellular virulence factor production (Ombaka, Cozens and Brown, 1983; Evans, Brown and Gilbert, 1994).

REFERENCES

Al-Taae, A.N., Dickinson, N.A. and Gilbert, P. (1986) Antimicrobial activity and physico-chemical properties of some alkyltrimethyl ammonium bromides. *Letters in Applied Microbiology* **1**: 101-105.

Anwar, H., Brown, M.R.W. and Lambert, P.A. (1983) Effect of nutrient depletion on the sensitivity of *Pseudomonas cepacia* to phagocytosis and serum bactericidal activity at different temperatures. *Journal of General Microbiology* **129:** 2021-2027.

Anwar, H., Brown, M.R.W., Day, A. and Weller, P.H. (1985) Outer membrane antigens of mucoid *Pseudomonas aeruginosa* isolated directly from the sputum of a cystic fibrosis patient. *FEMS Microbiology Letters* **24:** 235-239.

Anwar, H., van Biesen, T., Dasgupta, M.K., Lam, K. and Costerton, J.W. (1989) Interaction of biofilm bacteria with antibiotics in a novel in vitro chemostat system. *Antimicrobial Agents and Chemotherapy* **33:** 1824-1826.

Boggis, W., Kenward, M.A. and Brown, M.R.W. (1979) Effects of divalent metal cations in the growth medium upon the sensitivity of batch grown *Pseudomonas aeruginosa* to EDTA or Polymyxin A. *Journal of Applied Bacteriology* **47:** 477-488.

Brown, M.R.W. and Melling, J. (1969a) Loss of sensitivity to EDTA by *Pseudomonas aeruginosa* grown under conditions of magnesium limitation. *Journal of General Microbiology* **54**: 263-274.

Brown, M.R.W. and Melling, J. (1969b) Role of divalent cations in the action of polymyxin B and EDTA on *Pseudomonas aeruginosa. Journal of General Microbiology* **59:** 263-274.

Brown, M.R.W. (1975) The role of the envelope in resistance. In *Resistance of* Pseudomonas aeruginosa (Brown, M.R.W., Ed.) John Wiley and Sons, London, pp 71-107.

Brown, M.R.W. (1977) Nutrient depletion and antibiotic susceptibility. *Journal of Antimicrobial Chemotherapy* **3:** 198-201.

Brown, M.R.W. and Williams, P. (1985) The influence of environment on envelope properties affecting survival of bacteria in infections. *Annual Reviews in Microbiology* **39:** 527-556.

Brown, M.R.W., Collier, P.J. and Gilbert, P. (1990) Influence of growth rate on the susceptibility to antimicrobial agents: modification of the cell envelope and batch and continuous culture studies. *Antimicrobial Agents and Chemotherapy* **34**: 1623-1628.

Broxton, P., Woodcock, P.M. and Gilbert, P. (1984) Interaction of some polyhexamethylene biguanides and membrane phospholipids in *Escherichia coli. Journal of Applied Bacteriology* **57**: 115-124.

Bryers, J.D. (1993) The biotechnology of interfaces. In *Microbial Cell Envelopes: Interactions and Biofilms* (Quesnel, L.B., Handley, P.S. and Gilbert, P. Eds) Blackwell Scientific Press, London, pp 98-109.

Cozens, R.M., Tuomanen, E., Tosch, W., Zak, O., Suter, J. and Tomasz, A. (1986) Evaluation of the bactericidal activity of β-lactam antibiotics upon slowly growing bacteria cultured in the chemostat. *Antimicrobial Agents and Chemotherapy* **29:** 797-802.

Dean, A.C.R. (1972) Influence of environment on the control of enzyme synthesis. *Journal of Applied Chemistry and Biotechnology* **22**: 245-259.

Dean, A.C.R., Ellwood, J., Melling, J. and Robinson, A. (1976) The action of antibacterial agents on bacteria grown in continuous culture, In *Continuous Culture 6: Applications and New Fields* (Dean, A.C.R., Ellwood, D.C., Evans, C.G.T. and Melling, J., Eds) Chichester, Ellis Horwood, pp 251-261.

De La Rosa, E.J., De Pedro, M.A. and Vasquez, D. (1982) Modification of penicillin binding proteins of *Escherichia coli* associated with changes in the state of growth of the cells. *FEMS Microbiology Letters* **14**: 91-94.

Dorrer, E. and Teuber, M. (1977) Induction of polymyxin resistance in *Pseudomonas fluorescens* by phosphate limitation. *Archives of Microbiology* **114:** 87-89.

Duguid, I.G., Evans E., Brown, M.R.W. and Gilbert, P. (1992a) Effect of biofilm culture upon the susceptibility of *Staphylococcus epidermidis* to tobramycin. *Journal of Antimicrobial Chemotherapy*, **30**: 803-810.

Duguid, I.G., Evans, E., Brown, M.R.W. and Gilbert, P. (1992b) Growth-rate-independent killing by ciprofloxacin of biofilm-derived *Staphylococcus epidermidis:* evidence for cell-cycle dependency. *Journal of Antimicrobial Chemotherapy* **30:** 791-802.

Ellwood, D.C. and Tempest, D.W. (1972) Effects of environment on bacterial wall content and composition. *Advances in Microbial Physiology* **7:** 83-117.

Eudy, W.W. and Burroughs, S.E. (1973) Generation times of *Proteus mirabilis* and *Escherichia coli* in experimental infections. *Chemotherapy* **1:** 161-170.

Evans, D.J., Brown, M.R.W., Allison, D.G. and Gilbert, P. (1990) Susceptibility of bacterial biofilms to tobramycin: role of specific growth rate and phase in the division cycle. *Journal of Antimicrobial Chemotherapy* **25:** 585-591.

Evans, D.J., Allison, D.G., Brown, M.R.W. and Gilbert, P. (1991) Susceptibility of *Pseudomonas aeruginosa* and *Escherichia coli* biofilms towards ciprofloxacin: effect of specific growth rate. *Journal of Antimicrobial Chemotherapy* **27:** 177-184.

Evans, E., Brown, M.R.W. and Gilbert, P. (1994) Iron cheletor, exopolysaccharide and protease production on *Staphylococcus epidermidis*: a comparative study of the effects of specific growth rate in biofilm and planktonic culture. *Microbiology* **140:** 153-157.

Finch, J.E. and Brown, M.R.W. (1975) The influence of nutrient-limitation in a chemostat on the sensitivity of *Pseudomonas aeruginosa* to polymyxin and EDTA. *Journal of Antimicrobial Chemotherapy* **1:** 379-386.

Gilbert, P. and Stuart, A. (1977) Small-scale chemostat for the growth of mesophilic and thermophilic microorganisms. *Laboratory Practice* **26:** 627-628.

Gilbert, P. and Brown, M.R.W. (1978) Influence of growth rate and nutrient limitation on the gross cellular composition of *Pseudomonas aeruginosa* and its resistance to 3- and 4-chlorophenol. *Journal of Bacteriology* **133:** 1066-1072.

Gilbert, P. and Wright, N.E. (1986) Non-plasmidic resistance towards preservatives in pharmaceutical and cosmetic products, In *Preservatives in the Food, Pharmaceutical and Cosmetics Industries* (Board, R.G. and Allwood, M.C., Eds) Society for Applied Bacteriology Technical Series, London, Academic Press, pp 255-279.

Gilbert, P., Brown, M.R.W. and Costerton, J.W. (1987) Inocula for antimicrobial sensitivity testing: a critical review. *Journal of Antimicrobial Chemotherapy* **20**: 147-154.

Gilbert, P., Collier, P.J. and Brown, M.R.W. (1990) Influence of growth rate on susceptibility to antimicrobial agents: biofilms, cell cycle, dormancy and stringent response. *Antimicrobial Agents and Chemotherapy* **34:** 1865-1868.

Gilleland, H.E., Jr., Stinnett, J.D. and Eagon, R.G. (1974) Ultrastructural and chemical alteration of the cell envelope of *Pseudomonas aeruginosa* associated with resistance to EDTA resulting from growth in a Mg^{2+}-deficient medium. *Journal of Bacteriology* **117:** 302-311.

Hansch, C. and Clayton, J.M. (1973) Lipophilic character and biological activity of drugs: the parabolic case. *Journal of Pharmaceutical Science* **62:** 1-32.

Harder, K.J., Nikaido, H. and Matsuhashi, M. (1981) Mutants of *Escherichia coli* that are resistant to certain β-lactam compounds lack the OMP-F porin. *Antimicrobial Agents and Chemotherapy* **20:** 549-552.

Herbert, D, Ellsworth, R. and Telling, R.C. (1956) The continuous culture of bacteria: a theoretical and experimental study. *Journal of General Microbiology* **14:** 601-622.

Holme, T. (1972) Influence of environment on the content and composition of bacterial envelopes. *Journal of Applied Chemistry and Biotechnology* **22:** 391-399.

Ikeda, T., Ledwith, A., Bamford, C.H. and Hann, R.A. (1984) Interaction of a polymeric biguanide with phospholipid membranes. *Biochimica et Biophysica Acta* **769**: 57-66.

Keevil, C.W., Dowsett, A.B. and Rogers, J. (1993) *Legionella* biofilms and their control. In *Microbial Biofilms: Formation and Control* (Denyer, S.P., Gorman, S.P. and Sussman, M. Eds.) Blackwell Scientific Press, London, pp 201-216.

Lugtenberg, B. and Van Alphen, L. (1983) Molecular architecture and functioning of the outer membrane of *Escherichia coli* and other Gram-negative bacteria. *Biochimica et Biophysica Acta* **737:** 51-115.

Meers, J.L. and Tempest, D.W. (1970) The influence of growth limiting substrate and medium NaCl concentration on the synthesis of magnesium binding sites in the walls of *Bacillus subtilis* var *niger*. *Journal of General Microbiology* **63:** 325-331.

Melling, J., Robinson, A. and Ellwood, D.C. (1974) Effect of growth environment in a chemostat on the sensitivity of *Pseudomonas aeruginosa* to polymyxin B sulphate. *Proceedings of the Society for General Microbiology* **1:** 61.

Minnikin, D.E., Abdulrahimzadeh, H. and Baddiley, J. (1971) The interrelation of phosphatidylethanolamine and glycosyldiglycerides in bacterial membranes. *Biochemical Journal* **124:** 447-448.

Monod, J. (1949) The growth of bacterial cultures. *Annual Reviews of Microbiology* **3:** 371-394.

Muir, M.E., Van Heeswick, R.S. and Wallace, B.J. (1984) Effect of growth rate on streptomycin accumulation by *Escherichia coli* and *Bacillus megaterium*. *Journal of General Microbiology* **130:** 2015-2022.

Nichols, W.W., Evans, M.J., Slack, M.P.E. and Walmsley, H.L. (1989) The penetration of antibiotics into aggregates of mucoid and non-mucoid *Pseudomonas aeruginosa*. *Journal of General Microbiology* **135:** 1291-1303.

Nikaido, H. and Nakae, T. (1979) The outer-membrane of Gram-negative bacteria. *Advances in Microbial Physiology* **20**: 163-250.

Ombaka, A., Cozens, R.M. and Brown, M.R.W. (1983) Influence of nutrient limitation of growth on stability and production of virulence factors of mucoid and non-mucoid strains of Pseudomonas aeruginosa. *Review in Infectious Diseases* **5:** 5880-5888.

Pechey, D.T., Yau, A.O.P. and James, A.M. (1974) Total and surface lipids of cells of *Pseudomonas aeruginosa* and their relationship to gentamycin resistance. *MICROBIOS* **11**: 77-86.

Pirt, J. (1972) Prospects and problems in continuous flow culture of micro-organisms. *Journal of Applied Chemistry and Biotechnology* **22:** 55-64.

Raulston, J.E. and Montie, T.C. (1989) Early cell envelope alterations by tobramycin associated with its lethal action on *Pseudomonas aeruginosa. Journal of General Microbiology* **135:** 3023-3034.

Shand, G.H., Anwar, H., Kadurugamuwa, J., Brown, M.R.W., Silverman, S.H. and Melling, J. (1985) *In-vivo* evidence that bacteria in urinary tract infection grow under iron restricted conditions. *Infection and Immunity* **48:** 35-39.

Tamaki, S., Sato, T. and Matsuhashi, M. (1971) Role of lipopolysaccharide in antibiotic resistance and bacteriophage adsorption of *Escherichia coli* K12. *Journal of Bacteriology* **105:** 968-975.

Tempest, D.W., Dicks, J.W. and Ellwood, D.C. (1968) Influence of growth conditions on the concentration of potassium in *Bacillus subtilis* var *niger* and its positive relationship to cellular ribonucleic acid, teichoic acid and teichuronic acid. *Biochemical Journal* **106:** 237-243.

Tempest, D.W. (1976) The concept of relative growth rate: its theoretical basis and practical application. In *Continuous Culture 6: Applications and New Fields* (Dean, A.C.R., Ellwood, D.C., Evans, C.G.T. and Melling, J. Eds) Ellis Horwood, Chichester, pp 349-53.

Teuber, M. and Bader, J. (1976) Action of polymyxin B on bacterial membranes: Phosphatidylglycerol and cardiolipin induced susceptibility to polymyxin B in *Achloplasma laidlawii. Antimicrobial Agents and Chemotherapy* **9:** 26-35.

Tuomanen, E., Cozens, R., Tosch, W., Zak, O. and Tomasz, A. (1986) The rate of killing of *Escherichia coli* by β-lactam antibiotics is strictly proportional to the rate of bacterial growth. *Journal of General Microbiology* **132:** 1297-1304.

Tuomanen, E. and Schwartz, J. (1987) Penicillin binding protein 7 and its relationship to lysis of non-growing *Escherichia coli. Journal of Bacteriology* **169:** 4912-4915.

Turnowsky, F., Brown, M.R.W., Anwar, H. and Lambert, P.A. (1983) Effect of iron-limitation of growth rate on the binding of penicillin G to the penicillin binding protein of mucoid and non-mucoid *Pseudomonas aeruginosa. FEMS Microbiology Letters* **17:** 243-245.

West, A.A., Araujo, R., Dennis, P.J.L., Lee, J.V. and Keevil, C.W. (1989) Chemostat models of *Legionella pneumophila.* In *Airborne Deteriogens and Pathogens* (Flannagan, B. Ed), Biodeterioration Society, London, pp. 107-116.

Williams, P., Brown, M.R.W. and Lambert, P.A. (1984) Effect of iron deprivation on the production of siderophores and outer-membrane proteins in *Klebsiella aerogenes. Journal of General Microbiology* **130:** 2357-2365.

Wimpenny, J.W.T., Peters, A. and Scourfield, M. (1989) Modelling spatial gradients. In "Structure and function of biofilms" (Charackalis, W.G. and Wilderer, P.A., Eds.) pp 111-127. John Wiley, Chichester, England.

Wright, N.E. and Gilbert, P. (1987a) Influence of specific growth rate and nutrient limitation upon the sensitivity of *Escherichia coli* towards chlorhexidine diacetate. *Journal of Applied Bacteriology* **82:** 309-314.

Wright, N.E. and Gilbert, P. (1987b) Antimicrobial activity of n-alkyl trimethyl ammonium bromides: Influence of specific growth rate and nutrient limitation. *Journal of Pharmacy and Pharmacology* **39:** 685-690.

Wright, N.E. and Gilbert, P. (1987c) Influence of specific growth rate upon the sensitivity of *Escherichia coli* towards polymyxin B. *Journal of Antimicrobial Chemotherapy* **20:** 303-312.

1.5 The Influence of Nutrient Environment During Vegetative Growth on the Properties of Bacterial Endospores

Peter Gilbert, Michael, R.W. Brown, Ibrahim S.I. Al-Adham, and Norman Hodges

INTRODUCTION

Various genera of bacteria are able to survive adverse conditions through the formation of metabolically dormant endospores. Since the physical properties and characteristics of endospores have been well reviewed in the literature (Gould and Hurst, 1969; Gould and Dring, 1974; Russell, 1982), it is not the purpose of the current contribution to reiterate these. Rather it is our intention to examine how the nature of the growth medium in which the vegetative cells have grown, up to and including sporogenesis, affects the nature and performance properties of the spores.

Spores are important to sterilization and disinfection processes since their low metabolic rate, relative impermeability to hydrophilic agents, and relative dehydration render them among the most heat- and biocide-resistant forms of microorganisms in the natural world. Chemical and physical sterilization processes must therefore be able to contend with the presence of such entities, which are often, therefore, included in validation procedures as biological indicators (Chapter 3.4).

The nature of the complex nutrient medium used to grow the sporulating culture has been reported, by a variety of workers in the early part of this century, to have a bearing on spore properties, but when such claims were

0-8493-4752-1/95/$0.00+$.50

investigated in one of the first well-designed and statistically analyzed comparative studies, no significant difference was found among three such media (El-Bisi and Ordal, 1956). This has, however, not subsequently been a consistent finding; for example, while it was reported that ethylene oxide resistance was influenced by the nature of the complex medium (Dadd, McCormick and Daley, 1983), the same investigators reported that resistance is unaffected by synthetic rather than complex media (Dadd, Stewart and Town, 1983).

As with the growth of all vegetative microorganisms, the nature of the medium influences many of the cellular properties (see Chapter 1.2). During the logarithmic phase, growth rates are determined by the nature of the ingredients (e.g., glucose, glycerol, succinate, or glutamate as carbon sources) in the medium, the incubation temperature, and the pH. Each of these factors will have some influence on the properties of endospores produced. While all substrates will initially be present to excess in a culture, one of them will be growth-rate limiting either in its utilization by the cell or in its uptake. Growth rates in the logarithmic phase will therefore vary among media. Sporogenesis will generally not occur while the cells are in the logarithmic phase of development. As growth progresses, however, and each nutrient is consumed, concentration of some nutrients becomes critical and reduces the rate of cell growth and division to less than μ_{max}. The cells have entered their stationary phase of development. Progression of the growth of the culture from logarithmic phase to stationary phase can be abrupt, where the depleted nutrient is essential and converted to cellular mass with low efficiency (e.g., carbon source), or it might be extended over many hours (e.g., oxygen and cation depletion; see Chapters 1.2 and 3.1). Slowing of cellular growth rate initiates sporogenesis in endospore-forming microorganims and can trigger alternative dormancy states in non-sporeformers. This will occur at a critical concentration of the limiting nutrient and is independent of the cell density. Thus, if sporogenesis follows onset of stationary phase induced by depletion of magnesium ions at cell densities of 10^6 and 10^7 per ml, then the relative amounts of magnesium available to each pre-sporulating cell will differ greatly. Since sporogenesis commences at a finite point in the cell division cycle (i.e., with DNA replication), then each cell in the population will commence sporulation at a different time after critical conditions are met. Thus, some cells will initiate sporogenesis immediately, whereas others will have to complete cell division and reenter a new division cycle before this is possible. Each cell will consequently have access to different levels of the depleted nutrient during sporogenesis. These levels might be insufficient to allow sporogenesis to go to completion. Thus, the proportion of cells in a population able to sporulate to produce heat-resistant endospores varies with the nature of the depleting nutrient (Table 3). Such conditions produce spore crops which are highly heterogenous with respect to their structure and properties. This ensures, in nature, that maturation rate, dormancy, and subsequently ease of germination differ markedly within the population and reduces the likelihood of simultaneous germination.

Faced with depletion of certain nutrients, notably magnesium and manganese, sporogenesis is able to go to completion for only a few of the cells. For those cells where sporogenesis is completed, the composition and properties of the endospores will clearly be affected by the relative amounts of critical nutrients available at initiation, and whether or not sufficient carbon substrates had remained in order to support endogenous metabolism. Three factors additional to growth — temperature, pH, and sources of critical nutrients — will therefore affect the eventual properties of the endospores. These are the nature of the depleted nutrient (carbon, nitrogen, phosphorus, magnesium source, etc.; see Hodges and Brown, 1975; Lee and Brown, 1975), the cell density at the time of onset to stationary phase (Al-Adham, 1989), and the relative concentration of inessential nutrients and cations. Thus, when *Bacillus stearothermophilus* sporulation was induced by exhaustion of the nitrogen, the resulting spores exhibited a moist heat D_{110} of 459 min, but this was reduced to 196, 171, and 108 min by exhaustion of phosphate, sulfate, and glucose, respectively (Lee and Brown, 1975). By altering the availability of excess nutrients, sporulation in high concentrations of calcium (Amaha and Ordal, 1957; Friesen and Anderson, 1974 and references therein) and manganese (Friesen and Anderson, 1974) were shown to increase heat resistance, whereas elevating the phosphate concentration has been shown to have the opposite effect (Friesen and Anderson, 1974; and citations in Russell, 1982).

Whereas many of the studies on the influence of metal ions on resistance were undertaken some years ago using *Bacillus* species, there is little recent information on *Clostridia,* except for the observation that elevated concentrations of transition metals in the sporulation medium are reflected in elevated concentrations in the spores of *Clostridium botulinum* and high spore contents of iron and copper promote heat sensitivity (Kihm et al., 1990).

SPORULATION MEDIUM AND DEFINITION OF SPORE PHENOTYPE

Grelet (1957) devised sporulation media in which each of the components in a defined medium was in turn made the growth-limiting factor while other components were present in excess. He found sporulation to occur when the growth-limiting nutrient for *Bacillus megaterium* was glucose, nitrate, phosphate, iron, or zinc, but not to occur when potassium, magnesium, or manganese were the limiting factors.

Using this approach with *Bacillus megaterium*, Brown and Hodges (1974) developed and quantified the composition of defined media for various specific nutrient depletions (Table 1). In essence, the growth of batch cultures was separately limited by the availability of glucose, ammonium, sulfate, potassium, phosphate, manganese, and magnesium. Maximum population density (E_{420}) for graded concentrations of each limiting nutrient was plotted against nutrient concentration, and a linear plot was obtained below a critical concentration. The concentration of each of these nutrients could then be selected which caused onset of stationary phase and sporogenesis at

TABLE 1
Composition of Media Used for the Production of Variously Depleted Endospores of *Bacillus megaterium*

Medium Constituent	Nature of Nutrient Depletion: Glucose	NH₄	SO₄	PO₄	K	Mn	Mg	Glucose plus Mg[b]
	Concentration (Molar)							
$Na_2HPO_4 \cdot 12H_2O$	0.0539	0.0539	0.0539	[a]	2×10^{-4}	0.0539	0.0539	0.0539
KH_2PO_4	0.0128	0.0128	0.0128			0.0128	0.0128	0.0128
NH_4Cl	0.007	[a]	0.007	0.007	0.007	0.007	0.007	0.00635
KCl		0.007		0.0128	[a]			
Na_2SO_4	2.5×10^{-5}	2.5×10^{-5}	[a]	2.5×10^{-5}	2.5×10^{-5}	2.5×10^{-5}	6.3×10^{-5}	8.7×10^{-5}
$MgSO_4 \cdot 7H^2O$	4×10^{-5}	4×10^{-5}		4×10^{-5}	4×10^{-5}	4×10^{-5}	[a]	5×10^{-6}
$FeSO_4 \cdot 7H_2O$	2×10^{-6}	2×10^{-6}		2×10^{-6}	2×10^{-6}	2×10^{-6}	2×10^{-6}	2×10^{-6}
$MnSO_4 4H_2O$	4×10^{-7}	4×10^{-7}		4×10^{-7}	4×10^{-7}	[a]	4×10^{-7}	4×10^{-7}
Glucose	[a]	0.01	0.01	0.01	0.01	0.02	0.01	0.003
HEPES Buffer				0.025	0.025			
$MgCl_2 \cdot 6H_2O$			4×10^{-5}					
$FeCl_2 \cdot 4H_2O$			2×10^{-6}					
$MnCl_2 \cdot 4H_2O$			4×10^{-7}					

Note: All concentrations are expressed as molarities and all media are adjusted to pH 7.4.

[a] Limiting constituent concentration varied.

[b] Reformulated medium used to induce simultaneous glucose and magnesium depletion (Ca content 1 µg/ml).

Data taken from Hodges, N.A. and Brown, M.R.W., in *Spores VI*, American Society for Microbiology, Washington, D.C., 1975.

TABLE 2
Concentrations of Critical Nutrients Required to Support the Growth of *Bacillus stearothermophilus* to an Optical Density (E_{420} nm) of 1.0

	Strains			
	Auxotrophs NCIB 8157, 8919, 8920; NCTC 10003		Prototrophs NCTC 10003	
Media Constituent	Limiting Concn. (M)	Non-limiting Concn. (M)	Limiting Concn. (M)	Non-limiting Concn. (M)
L-Glutamate (no NH4 added)[a]	1.2×10^{-3}	6×10^{-3}	1.2×10^{-3}	6.0×10^{-3}
L-Glutamate (no glucose added)[a]	4.0×10^{-3}	4.0×10^{-3}	20×10^{-3}	20×10^{-3}
D-Glucose	3.3×10^{-3}	16.5×10^{-3}	3.3×10^{-3}	16.5×10^{-3}
Ammonium	2.2×10^{-3}	11.0×10^{-3}	2.2×10^{-3}	11.0×10^{-3}
DL-Methionine	4.5×10^{-5}	2.2×10^{-4}	—	—
Sulfate	—	—	5.5×10^{-5}	2.8×10^{-4}
Magnesium (Mg^{++})	1.8×10^{-5}	9.0×10^{-5}	1.8×10^{-5}	9.0×10^{-5}
Manganese (Mn^{++})	5×10^{-6}	2.5×10^{-5}	5×10^{-6}	2.5×10^{-5}
Nicotinic Acid	6.0×10^{-7}	3.0×10^{-6}	—	—
Thiamine	2.5×10^{-8}	1.3×10^{-7}	—	—
Biotin	8.0×10^{-10}	4.0×10^{-9}	—	—
Calcium (Ca^{++})	Trace	Trace	Trace	Trace
Iron (Fe^{++})	Trace	Trace	Trace	Trace

Note: Buffer: Na_2HPO_4 (1.76×10^{-2} mol/l) and KH_2PO_4 (7.3×10^{-3} mol/l) pH 7.0–7.2.

[a] L-glutamate may be used as an alternative carbon/nitrogen source.

Data taken from Lee, Y.H., Brown, M.R.W., and Cheung, H.-Y., *Journal of Applied Bacteriology*, 53, 179–187, 1982.

predetermined cell densities. Under conditions of magnesium depletion for this organism (Brown and Hodges, 1974) and also for *B. stearo-thermophilus* (Al-Adham, 1989), two phases of growth occurred separated by a plateau, indicating a dual efficiency of conversion of magnesium into cellular materials. Sporulation occurred in all cultures except those limited by potassium, manganese, or magnesium. Spores were produced in magnesium-limited cultures provided that glucose was simultaneously depleted. Spores produced under different conditions of nutrient depletion varied in germination characteristics, heat resistance and spore volume. Hodges et al. (1980) reported that spore crops of nutrient-depleted cultures had more reproducible resistance characteristics than occurred with complex media.

Quantitatively defined minimal media for culture (60°C) of prototrophic and auxotrophic strains of *B. stearothermophilus* were devised by Lee, Brown, and Cheung (1982). The composition of these media are given in Table 2. Spores produced from media nutrient depleted in different ways had composition

TABLE 3
Properties and Composition of *Bacillus stearothermophilus* Endospores Produced under Conditions of Various Nutrient Depletion (Table 2)

	Property		Composition ($\mu g/10^8$ spores)			Composition ($\mu g/10^7$ spores)	
Nutrient Depletion	**Sporulation (%)**	**D_{110} (min)**	**Mn^{2+}**	**Mg^{2+}**	**Ca^{2+}**	**DAP**	**Hexosamine**
Carbon depletion	24.8	379	1.41	0.25	5.2	0.54	1.19
Nitrogen depletion	35.8	44	1.29	0.74	3.8	0.59	1.35
Carbon/Mg depletion	12.9	232	3.24	0.26	5.4	0.67	0.67
Sulfate depletion	72.2	228	1.27	0.36	3.9	0.47	1.18
Phosphate depletion	34.8	92	1.10	5.70	4.0	0.53	2.02

After Lee, Y.H., Ph.D. Thesis, University of Aston, U.K., 1976.

and germination properties which varied according to the depleting nutrient (Table 3; Lee, 1976; Cheung, Vitkovic and Brown, 1982). Al-Adham (1989) examined the effect of stationary phase cell density at onset of sporulation upon spore properties. Media were designed which restricted the growth of *B. stearothermophilus* NCTC 10003 to optical densities (E_{470} nm) of 0.3, 0.5, and 1.0 through depletion of magnesium, phosphate, or carbon source. Significant differences were noted not only between the properties of differently depleted cells but also according to the stationary phase cell density (Table 4).

In addition to the nature of the growth limiting nutrients addition of chemicals other than growth substrates can affect spore properties. Thus, the presence of increasing concentrations of ethanol in the sporulation medium up to 5% v/v has been shown (Figure 1) progressively to reduce the thermal resistance of spores of *B. megaterium, B.subtilis*, and *B. stearothermophilus* (Khoury, Lombardi and Slepecky, 1987). This sensitization was attributed to possible changes in membrane lipid composition and fluidity.

SPORULATION TEMPERATURE

Numerous workers have reported the effects of sporulation temperature on the resistance of the harvested spores without unanimity in their conclusions. A minority of reports indicate that temperature of sporulation has little or no influence on resistance (Dadd, McCormick and Daley, 1983; Alcock and Brown, 1985), but far more commonly enhanced resistance of *B. stearothermophilus* (Friesen and Anderson, 1974; Khoury, Lombardi and Slepecky, 1987), *Bacillus subtilis* (Khoury, Lombardi and Slepecky, 1987; Condon, Bayarte and Sala, 1992) and *B. megaterium* (Khoury, Lombardi and Slepecky, 1987) has been associated with increased sporulation temperature. This relationship, however, is not necessarily linear, and plots of incubation temperature against resistance

TABLE 4
Effect of Nutrient Depletion and Stationary Phase Cell Density on the Composition and Properties of *Bacillus stearothermophilus* NCTC 10003 Endospores.

	Property				Composition μg/10^7 spores			
Nutrient Depletion	Stationary Phase OD_{470} nm	Sporulation	D_{110}	Germination	DAP	Mn^{2+}	Mg^{2+}	Ca^{2+}
Glycerol (2.2 m*M*)	1.0	23.0%	15 min	47%	9.0	0.07	0.04	2.62
Glycerol (0.7 m*M*)	0.3	nd	15 min	55%	1.8	0.06	0.06	0.56
Magnesium (0.15 m*M*)	1.0	11.8%	2.5 min	6.3%	25.0	3.27	0.05	41.5
Magnesium (0.05 m*M*)	0.3	nd	4.0 min	<1.0%	2.1	1.91	0.05	6.26
Phosphate (0.98 m*M*)	1.0	25.6%	0.5 min	5.6%	35.0	0.08	0.07	8.01
Phosphate (0.49 m*M*)	0.5	nil	nd	nd	nd	nd	nd	nd

After Al-Adham, I.S.I., Ph.D. Thesis, University of Manchester, U.K., 1989.

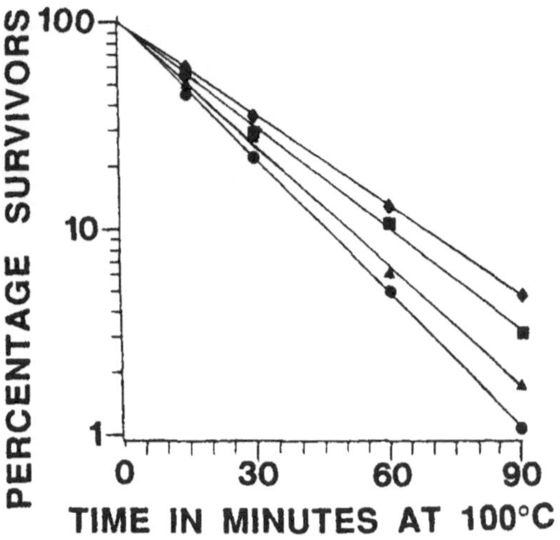

FIGURE 1 Heat resistance of *Bacillus stearothermophilus* spores after growth and sporulation in ethanol. Ethanol concentration, (♦), 0%; (■), 1%; (▲), 2%; (●), 3%. (Adapted from Khoury, P.H., Lombardi, S.J., and Slepecky, R.A., *Current Microbiology,* 15, 15–19, 1987.)

often show that the latter reaches a plateau or even begins to fall when sporulation occurs at or near the maximum permissible growth temperature (Condon, Bayarte and Sala, 1992; Friesen and Anderson, 1974).

The correlation between resistance and sporulation temperature which is observed within a single species, is also seen when a variety of species are examined. In general the spores of thermophils are more resistant to heat than those of mesophils and psychrophils, so that the species at the top of the "league table" of heat resistance are invariably thermophilic strains, e.g., *B. stearothermophilus, Bacillus coagulans* and *Clostridium thermosaccharolyticum.* Despite the differences that might have been expected from the employment of several different sporulation media, Beaman and Gerhardt (1986) found a firm linear correlation between the optimum sporulation temperature and the degree of heat resistance for the spores of 13 strains of seven *Bacillus* species (Figure 2).

GROWTH AND SPORULATION AT VARIOUS pH

The effect of pH during sporulation on the properties of the spores which are subsequently harvested is often difficult to discern because the pH value may change progressively during growth and sporulation depending upon the strength of the buffer and the initial concentration of sugar available to generate organic acids. The pH profile which is frequently observed is a steady fall during vegetative growth with a minimum just after the end of the log phase,

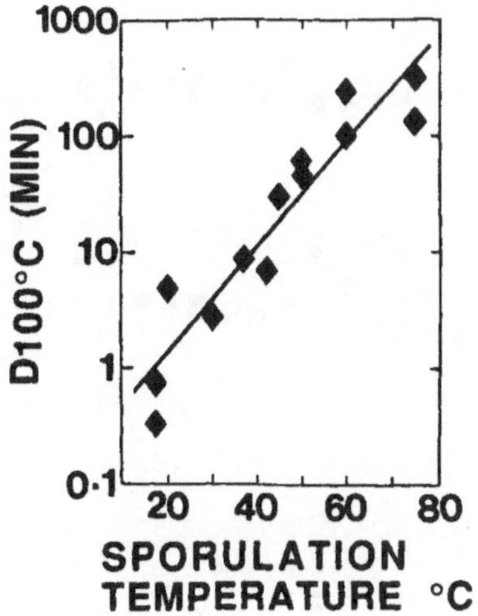

FIGURE 2 Heat resistance in relation to sporulation temperature for lysozyme-sensitive spores of 13 strains from seven *Bacillus* species. (Adapted from Beaman, T.C. and Gerhardt, P., *Environmental Microbiology*, 52, 1242–1246, 1986. With permission.)

which may be followed, particularly in *Bacillus* species (see for example Friesen and Anderson, 1974), by a rise but not a full return to the initial pH value as sporulation proceeds. Alternatively, pH may remain static, as is more commonly the case in *Clostridia* (Craven, 1990; Pang, Carroad and Wilson, 1983). A further complication is that media which support high cell densities of *Bacillus* species may create conditions under which the cells sporulate with suboptimal oxygen concentrations so that any effects of this might also be superimposed upon the properties of the spores.

The influence of pH on *Bacillus* spore heat resistance has received little attention (Russell, 1982), but the indications available in the literature suggest that it has no marked effect (Friesen and Anderson, 1974 and references therein). In the case of *Clostridia*, pH has occasionally (Craven, 1990; Pang, Carroad and Wilson, 1983), but not invariably (Alcock and Brown, 1985) been stated to have a definite effect, but such reports are relatively isolated and it would appear that pH is not likely to be a major factor in the determination of spore properties.

CULTURE AGE PRIOR TO HARVESTING

There is a limited amount of evidence that the resistance of spores may, in some cases, be increased by prolonged incubation in the sporulation medium. The heat resistance of PA 3679 (Alcock and Brown, 1985) and resistance to heat, hydrogen peroxide, and peracetic acid of a strain of *B. subtilis* (Figure 3; Leaper, 1987), were reported to increase when the incubation period between

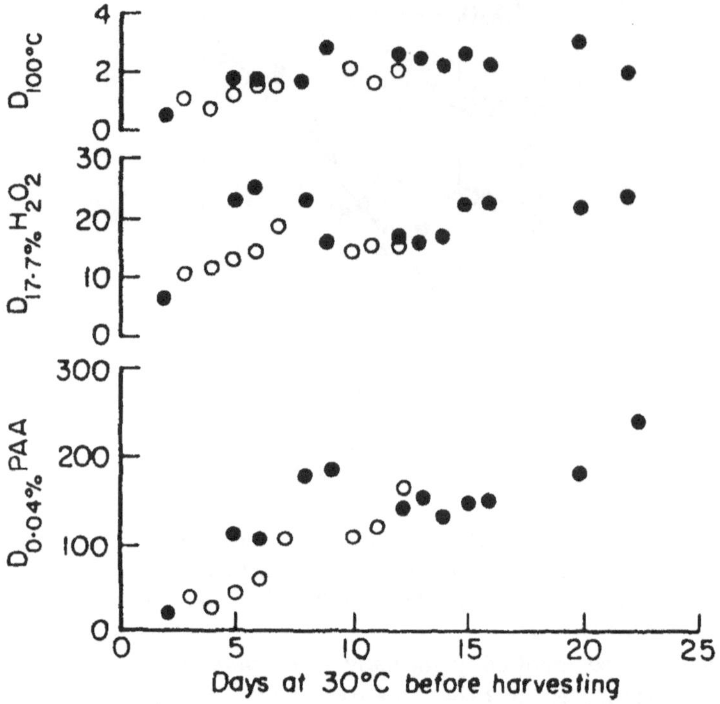

FIGURE 3 Influence of interval between sporulation and harvesting of spores of *Bacillus subtilis* on the D-value for the resistance to 0.04% peracetic acid at 20°C ($D_{0.04\%}$ PAA); 17.7% hydrogen peroxide at 20°C ($D_{17.7\%}$ H_2O_2) and heating at 100°C (D100°C for (●), batch one, and (○), a second batch of spores. (Adapted from Leaper, S., *Letters in Applied Microbiology*, 4, 55–57, 1987. With permission.)

sporulation and harvesting was extended substantially beyond the time required for liberation of free spores from sporangia. In the case of PA 3679 spores were found to be most resistant after 7 days incubation, and heat resistance declined when incubation was extended beyond that time (Alcock and Brown, 1985).

REFERENCES

Al-Adham, I.S.I. 1989. Growth and sporogenesis of *Bacillus stearothermophilus* under conditions of nutrient depletion. PhD Thesis, University of Manchester, UK.

Alcock, S.J. and K.L. Brown. 1985. Heat resistance of P A 3679 (NCIB 8053) and other isolates of *Clostridium sporogenes*. In G.J. Dring, D.J. Ellar and G.W. Gould (eds.), *Fundamental and Applied Aspects of Bacterial Spores*. Academic Press, London, p. 261.

Amaha, M. and Z.J. Ordal. 1957. Effect of divalent cations in the sporulation medium on the thermal death rate of *Bacillus coagulans* var. *thermoacidurans*. *Journal of Bacteriology* **74**:596-604.

Beaman, T.C. and P. Gerhardt. 1986. Heat resistance of bacterial spores correlated with protoplast dehydration mineralisation and thermal adaptation. *Applied and Environmental Microbiology* **52:**1242-1246.

Brown, M.R.W. and N.A. Hodges. 1974. Growth and sporulation characteristics of *Bacillus megaterium* under different conditions of nutrient limitation. *Journal of Pharmacy and Pharmacology* **26:**217-227.

Cheung, H.-Y, L. Vitkovic and M.R.W. Brown. 1982. Dependence of *Bacillus stearothermophilus* spore germination on nutrient depletion and manganese. *Journal of General Microbiology* **128:**2403-2409.

Condon, S., M. Bayarte and F.J. Sala. 1992. Influence of the sporulation temperature on the heat resistance of *Bacillus subtilis. Journal of Applied Bacteriology* **73:**251-256.

Craven, S. E. 1990. The effect of the pH of the sporulation environment on the heat resistance of *Clostridium perfringens* spores. *Current Microbiology* **22:**233-237.

Dadd, A.H., C.M. Stewart and M.M. Town. 1983. A standardised monitor for the control of ethylene oxide sterilization cycles. *Journal of Hygiene, Cambridge* **91:**93-100.

Dadd, A.H., K.E. McCormick and G.M. Daley. 1983. Factors influencing the resistance of biological monitors to ethylene oxide. *Journal of Applied Bacteriology* **55:**39-48.

El-Bisi, H.M. and Z.J. Ordal. 1956. The effect of certain sporulation conditions on the thermal death rate of *Bacillus coagulans* var. *thermoacidurans. Journal of Bacteriology* **71:**1-16.

Friesen, W.T. and R.A. Anderson. 1974. Effects of sporulation conditions and cation-exchange treatment on the thermal resistance of *Bacillus stearothermophilus* spores. *Canadian Journal of Pharmaceutical Science* **9:**50-53.

Gould, G.W. and A. Hurst. (Eds.) 1969. *The Bacterial Spore*, Academic Press, London.

Gould, G.W. and G.J. Dring. 1974. Mechanisms of heat resistance. *Advances in Microbial Physiology* **11:**137-164.

Grelet, N. 1957. Growth limitation and sporulation. *Journal of Applied Bacteriology* **20:**315-324.

Hodges, N.A. and M.R.W. Brown. 1975. Properties of *Bacillus megaterium* spores formed under conditions of nutrient limitation. In P. Gerhardt, H.L. Sadoff, and R. N. Costilow (eds.), *Spores VI*. American Society for Microbiology, Washington.

Hodges, N.A., J. Melling and S.J. Parker. 1980. A comparison of chemically defined and complex media for the production of *Bacillus subtilis* spores having reproducible resistance and germination characteristics. *Journal of Pharmacy and Pharmacology* **32:**126-130.

Khoury, P.H., S.J. Lombardi and R.A. Slepecky. 1987. Perturbation of the heat resistance of bacterial spores by sporulation temperature and ethanol. *Current Microbiology* **15:**15-19.

Kihm, D.J., M.T. Hutton, J.H. Hanlin, and E.A. Johnson. 1990. Influence of transition metals added during sporulation on heat resistance of *Clostridium botulinum* spores. *Applied and Environmental Microbiology* **56:**681-685.

Leaper, S. 1987. A note on the effect of sporulation conditions on the resistance of *Bacillus* spores to heat and chemicals. *Letters in Applied Microbiology* **4:**55-57.

Lee, Y.H. and M.R.W. Brown. 1975. Effect of nutrient limitation on sporulation of *Bacillus stearothermophilus. Journal of Pharmacy and Pharmacology*, Suppl. **27:**22P.

Lee, Y.H. 1976. Growth, sporulation and spore properties of *Bacillus stearothermophilus* in chemically defined media. PhD Thesis, University of Aston, UK.

Lee, Y.H., M.R.W. Brown and H.-Y. Cheung. 1982. Defined minimal media for the growth of prototrophic and auxotrophic strains of *Bacillus stearothermophilus*, *Journal of Applied Bacteriology* **53:**179-187.

Pang, K.A., P.A. Carroad and A.W. Wilson. 1983. Effect of culture pH on D value, cell growth and sporulation rates of P.A. 3679 spores produced in an anaerobic fermentor. *Journal of Food Science* **48:**467-470.

Russell, A.D. 1982. Inactivation of bacterial spores by thermal processes (moist heat), In *The Destruction of Bacterial Spores*. Academic Press, London.

Russell, A.D. 1982. *The Destruction of Bacterial Spores*. Academic Press, London.

1.6 Influence of the Environment on the Properties of Microorganisms Grown in Association with Surfaces

Peter Gilbert, Anne E. Hodgson, and Michael R.W. Brown

INTRODUCTION

In nature, free-living or planktonic populations of microorganisms are unusual; rather, they grow in an association with surfaces to form functional consortia known as biofilms (Costerton et al., 1987). The physiological properties of the organisms, including susceptibility towards antibiotics and biocides, is markedly affected by such associations. Surface-associated growth is therefore of considerable importance when selecting phenotypes which are representative of various natural habitats (Gilbert, Collier and Brown, 1990). In this respect, biofilms have been recognized and studied from a diverse range of surface types and environments.

In aquatic environments, microbial consortia are often found in association with sediments and stones or at the air/water interface as "mats" (Paerl, 1975). Additionally, microorganisms will often flocculate and form aggregates in nutrient-rich liquid environments such as waste treatment plants. The properties of these aggregates are affected as much by the cell-cell contact provided within the floccule as by the supplies of nutrients to the solution phase. In industrial plants, biofouling of surfaces is generally problematic, and has been observed almost universally wherever organisms survive and grow. Sites for biofilm formation of particular interest are with respect to the design and testing of biocides and biocidal/noncolonizable surfaces. These include

0-8493-4752-1/95/$0.00+$.50

pipework, the hulls of ships, heat exchangers and cooling towers (Lappin-Scott and Costerton, 1989; Lee and West, 1991; Shaw et al., 1985).

Autochthonous and pathogenic colonizers of animals and plants also exhibit preferential surface growth — as, for example, the natural flora of the skin and gut and dental plaque (Gibbons and van Houte, 1975; Cheng et al., 1981; Dazzo, 1984; Costerton et al., 1987). Of increasing concern in the clinical context has been the colonization of implanted biomaterials where such growth is often associated with a recalcitrance to antibiotic treatment (Brown and Gilbert, 1993). Such resistance often means that the device must be surgically removed from the patient while antibiotic treatment is continued (Elliott, 1988; Dickinson and Bisno, 1989a, b). Failure to do this commonly leads to a reoccurrence of the infection once treatment has ceased. Organisms grown upon such surfaces should therefore form an important part of the screens for antimicrobial development, and possibly part of the routine clinical assessement of antibiotic susceptibility (Anwar et al., 1992).

Biofilms are collections of microorganisms associated with a surface and organized within an extracellular polymer matrix. While growths as biofilms associated with medical prostheses and within tissues are frequently monocultures, in other sites mixed populations are more often encountered. These might include communities of mixed bacterial species, together with fungi and protozoa. In these the community as a whole often has a metabolic capability unrepresented by any single member species, and the interspecies relationships might be synergistic. Such microbial consortia range in thickness from coated monolayers of cells a few micrometers thick to extensive films several centimeters deep. Since in a thick biofilm the majority of the population of cells will be in association with other cells rather than the colonized surface, then some of the properties of biofilms will be shared by flocculated and clumped organisms.

Given the widespread occurrence of surface growth, it is perhaps surprising that relative to the study of planktonic populations, sessile growth of microorganisms has received little direct experimental study. This is presumably because the systematic study of bacterial biofilms requires the development of complex *in vitro* culture systems. Such methods of culture have only recently come into widespread use. Nevertheless, it is now clear that the structure and physiology of attached cells and of cells organized into biofilms differs substantially from that of planktonic cells grown within the same nutrient medium (Fletcher, 1984; Williams, 1988; van Loosdrecht et al., 1990). There is now considerable evidence, from the laboratory and also from field experience, that sessile cells are significantly more resistant to the effects of many biocides and antimicrobial drugs, and are less responsive to immune defenses (Brown et al., 1988; Hoyle et al., 1990).

Current screening by the pharmaceuticals and chemicals industries for novel antibiotics, preservatives, and biocides involves predominately the use of inocula grown planktonically in nutrient-rich media. New agents are therefore often optimized for their activity against planktonic cells before they are "field-tested" in natural situations involving attached cells and biofilms. The sessile mode of growth must be studied and understood if appropriate model systems are to be developed which duplicate, *in vitro*, the targets for antimicrobial strategies. The development and application of *in vitro* models of attached growth to the primary screening process is essential if anti-biofilm agents are to be developed.

SURFACE COLONIZATION AND GROWTH

The process of microbial colonization of surfaces is sequential and involves a number of distinct stages. Each of these must be studied and understood if antiadhesive strategies are to developed and assessed on a rational basis. First, cells must access the surface from the bulk phase, which is usually liquid. This can be accommodated in three ways: diffusion, convective transport, and active motility. On arrival at the surface, adhesion is dependent upon the physico-chemical properties of the cell surface relative to the substratum. This initial adhesion is weak, mediated through electrostatic interactions, and largely reversible (Marshall et al., 1971; van Loosdrecht et al., 1990). Reversibly bound cells are held in a secondary minimum where the electrostatic repulsion of the two negatively charged surfaces are balanced by the van der Waals forces of attraction. Cells may subsequently become firmly attached to the surface through the presence of structures, on their surfaces, such as fibrils and pili. These are able to overcome the repulsive forces that exist between the two negatively charged surfaces, through their shape and rigidity, and bind irreversibly at the primary energy minimum where van der Waals forces overcome the electrostatic repulsion (Gilbert, Evans and Brown, 1993). Even in the absence of surface appendages, exopolymers produced by the organisms while reversibly bound to the surface diffuse away from the cells to create polymer bridges which irreversibly hold the organisms to the substratum (Costerton et al., 1981). Such exopolymers are produced in abundance by the attached cells. These form the glycocalyx, which further "glues" the cells to the surface, and eventually envelops them. Division of the attached cells within the glycocalyx results in the formation of microcolonies, which may coalesce to form a more extensive biofilm. Growth of the biofilm occurs by internal replication and by recruitment of cells from the bulk phase, culminating in a biofilm consisting of microcolonies and single cells embedded within a highly hydrated glycocalyx which occludes the surface.

Established biofilms may serve as reservoirs of microorganisms, facilitating the dispersal of cells into the bulk phase. This is a necessary aspect of the sessile mode of growth without which the organisms would be incapable of colonizing new sites. In infections, cells dispersed from biofilms, associated with prosthetic devices, generate bacteremias and may go on to produce secondary infections such as endocarditis. A bacteremia is often the first indication of infection. While the bacteremias often respond readily to antibiotic therapy, the sessile focus of infection does not. Thus, initial improvements in the patient's condition are all too often followed by relapses after treatment.

RESISTANCE PROPERTIES OF BACTERIAL BIOFILMS

The resistance of bacterial biofilms and attached cells towards treatments with biocides and antibiotics are of particular interest in the context of microbiological challenge testing and screening for novel antimicrobial activity. Understanding of the mechanisms of such resistance inputs to the choice of relevent inocula which will generate predictive outcomes. Such resistance might be related solely to reduced growth rate and imposition of specific nutrient limitations within the depths of an established biofilm (Section 1.2). Other equally attractive hypotheses, however, attribute resistance to functional characteristics of biofilms. These generally refer either to action of the glycocalyx as a diffusion barrier to drug penetration or to the induction of attachment-specific, drug-resistant physiologies.

Central to all of these hypotheses is the principle that attached cells within a biofilm differ significantly from their planktonic, "domesticated", laboratory counterparts. Most antimicrobial systems have been developed and optimized in laboratories employing planktonic broth cultures as inocula, at least in the initial screening stage. Consequently the possibility must not be overlooked that resistance in biofilms reflects inappropriate selection of compounds for development during the screening process rather than innate recalcitrance of biofilms to antimicrobial agents. While explanations for the observed lack of sensitivity towards today's antimicrobial agents may well be explicable in physiological terms, such propositions cannot be disregarded until primary screening for antimicrobial activity routinely involves realistic, relevant challenge systems.

INFLUENCE OF SURFACE ATTACHMENT AND GROWTH UPON MICROBIAL ACTIVITY

Microorganisms are remarkably flexible in their physiology and respond to changing environmental conditions (Brown and Williams, 1985; Williams, 1988; Robertson and Meyer, 1992). Failure to adapt their physiology in an appropriate manner would place individual organisms at a disadvantage within the environment. In the "real world" microorganisms are generally slow- or nongrowing (Gilbert et al., 1990) as a result of a particular nutrient scarcity. They will experience other than ideal pH, osmolarity, and temperature, and may in addition be exposed to inimical agents such as antibiotics, biocides, antibodies, serum factors, and phagocytes. While the influence of various nutritional and environmental factors upon microbial properties has been discussed earlier (Chapter 1.2), their contribution towards the observation of an attached phenotype must be considered. Thus, growth rate, nutrient limitation, and nutrient availability will affect the properties of cells grown at surfaces. Furthermore, some of the physiological processes associated with growth on surfaces, such as glycocalyx deposition, while possibly providing a degree of homeostasis, will place additional constraints upon the availability of nutrients for individual organisms. Within a complex and thick biofilm, nutrient, pH, and gaseous gradients will be established. Accordingly, cells variously located within the film will experience different environmental stresses and adapt their phenotype accordingly. In planktonic cultures all members of the population will experience essentially the same physico-chemical environment and express similar phenotypes. Biofilm populations, on the other hand, will contain a very heterogenous mixture of phenotypes for each member species. Survival of biofilms under adverse conditions and during chemical treatments might therefore represent survival of the fittest phenotype and death of the remainder. Because of their heterogenous nature, biofilms will always contain, somewhere within them, the extremes of phenotype. This is in addition to the expression, by cells, of attachment- and biofilm-specific physiologies.

Role of the Glycocalyx

Electron microscopy of antibody-stabilized preparations of biofilm together with X-ray crystallographic studies show the glycocalyx to be an ordered array of fine fibers providing a thick, continuous, hydrated, polyanionic matrix around the cells (Costerton et al., 1981). Such exopolymer matrices are termed "glycocalyx" and are generally composed of fibrillar polysaccharide materials but may also include globular glycoproteins (Sutherland, 1977). Biofilm-associated cells may have an improved nutritional status through the localized capture of charged substrates by the polyanionic glycocalyx (Sutherland, 1977) and through concentration of extracellular products. Growth within a microcolony or biofilm also provides some homeostasis with respect to the environment; community members are thus shielded from the environmental stresses experienced by planktonic organisms. Evidence exists of gradients of pH and oxygen tension within established biofilms (Wimpenny et al., 1989). It is therefore envisaged that within mature biofilms there will exist many different microniches, each particularly suited to a different community member. For example, deep within thick bacterial biofilms there may be anoxic regions resulting from oxygen utilization by organisms nearer the biofilm surface, where gaseous diffusion is more efficient. Such conditions may favor the growth of anaerobes. The close proximity of cells within biofilms facilitates the exchange of genetic information (Bale et al., 1988).

The nature of the glycocalyx surrounding attached cells suggests that extracellular products may be held in close association with the cells rather than released into the bulk phase, as would be the case in planktonic culture. This is often advantageous to component cells. Colonizers of metabolizable substrates (e.g., cellulose fibers in the bovine rumen) release digestive enzymes which remain focused at the surface, producing efficient substrate digestion. Similarly, β-lactamases produced by pathogenic bacteria growing as biofilms remain trapped by the glycocalyx matrix, ideally located to protect cells from attack by β-lactam antibiotics (Giwercman et al., 1991). Perhaps the most significant advantage of sessile growth is the substantial protection it affords from a huge range of hostile agents. In infections, such agents may include phagocytes, antibodies, surfactants, complement and other serum factors, and antibiotics. The recalcitrance of surface-associated microorganisms towards antibiotic treatment and immune clearance contributes to the chronic nature of many of these infections (Costerton et al., 1987). Correspondingly, surface-associated microorganisms in natural and industrial environments are largely protected from the action of grazing protozoa, bacteriophages, and biocides.

The Glycocalyx as a Moderator of Drug/Biocide Resistance

Possession of such an envelope will undoubtedly influence profoundly the access of molecules and ions, including protons (Costerton et al., 1981), to the cell wall and membrane. It is tempting, therefore, to suggest that the glycocalyx physically prevents access of antimicrobials to the cell surface and that the recalcitrance of biofilms is purely and simply a matter of exclusion (Costerton, 1977; Slack and Nichols, 1981, 1982; Costerton et al., 1987). Such explanations have been refuted (Gordon et al., 1988; Nichols et al., 1988, 1989) on the basis that reductions in diffusion coefficient for antibiotics such as tobramycin and cefsulodin within biofilms or microcolonies are insufficient to account for the observed changes in susceptibility of the contained cells. In this light, Gristina et al. (1989) examined susceptibilities towards antibiotics of biofilms of slime-producing and non-slime producing strains of *S. epidermidis* and saw no difference, suggesting the slime not to be of significance in antibiotic penetration. Alternative suggestions as to the barrier role of the glycocalyx come from the proposition that in addition to acting as a diffusion barrier, binding of the agent by the exopolymers and cells on the periphery of the microcolony will reduce availability of the drug. Clearly, whether or not the glycocalyx constitutes a physical barrier to antibiotic penetration depends greatly upon the nature of the antibiotic, the binding capacity of the glycocalyx towards it, the levels of agent used therapeutically, and the rate of growth of the microcolony relative to antibiotic diffusion rate. For antibiotics such as tobramycin and cefsoludin, therefore, such effects are suggested to be minimal (Nichols et al., 1988, 1989). Evans et al. (1991a), however, assessed susceptibility towards the quinolone antimicrobial ciprofloxacin, of mucoid and nonmucoid strains of *Pseudomonas aeruginosa* grown as biofilms. In this study possession of the mucoid phenotype was clearly associated with decreases in susceptibility. In a similar vein, activities of chemically, highly reactive biocides such as iodine and iodine-polyvinylpyrollidone complexes (Favero et al., 1983) are substantially reduced by the presence of protective exopolymers. In such instances not only will the polymers act as adsorption sites, but they will also react chemically with, and neutralize, biocides. Similarly, the relatively large accumulation of drug-inactivating enzymes such as β-lactamases (Giwercman et al., 1991) within the glycocalyx will create marked concentration gradients of the antibiotic across them and protect, to some extent, the underlying cells.

In summary, reductions in diffusion coefficients across the glycocalyx, relative to liquid media, are alone insufficient to account for antibiotic recalcitrance of biofilms since at equilibrium, concentrations at the cell surfaces will be the same as those in the bathing medium. With losses either through chemical or enzymic inactivation of the agent occurring in the biofilm, the reduction in diffusion coefficient might, however, facilitate resistance to the agent. Such protection might also apply where the adsorptive capacity of the glycocalyx is very high with respect to biocide. Thus, in some instances possession of an exopolymer glycocalyx will modify susceptibility, but in other cases its effects will be negligible.

Attachment-Specific Physiologies

Surface-induced stimulation of bacterial activity has often been described in the literature (Zobell, 1943; Fletcher, 1986). Dagostino et al. (1991) have attempted to identify such activity with the induction of particular genes. They utilized transposon mutagenesis to insert, randomly, into the chromosome of *E. coli* a marker gene which lacked its own promoter element. They then went on to isolate mutants which did not express the gene when grown on agar or in liquid media, but which did so when the cells were attached to a polystyrene surface (Dagostino et al., 1991). There is still, however, much debate as to whether surface-induced stimulation of metabolic activity reflects derepression/induction of specific operons/genes or are purely manifestations of the physico-chemical presence of the surface on the surroundings of the cell (van Loosdrecht et al., 1990). Indirect effects of surfaces might include the accumulation of many substrates at surfaces which will therefore be available in increased abundance for attached organisms (Power and Marshall, 1988). Some cells within biofilms may grow at rates reduced from that of planktonic cells in the same medium, hence many of the genetic responses to stationary phase seen in liquid batch cultures will also be apparent (Chapter 1.4). Whether or not bacteria possess touch receptors remains to be proven. If they do, then they will clearly be of immense importance not only in the initiation and formation of biofilms but also in their expressed physiology. Candidate physiologies for touch-induction are now appearing regularly in the literature, but, with the exception of cell-density responsive transcriptional activation (below), molecular genetic bases for the phenomena have not been proposed.

Degradation of the substrate nitrilotriacetate, which does not adsorb to surfaces, is enhanced when the degradative organisms are attached to inert surfaces (McFeters et al., 1990). This suggests increased production of degradative enzmes by attached cells. Similarly, gliding bacteria lack extracellular polymer biosynthesis when grown in suspension culture (Humphrey et al., 1979; Abbanat et al., 1988), but rapidly initiate/increase such synthesis following irreversible adhesion to a surface. In this respect Evans et al. (1994) cultured biofilms of *Staphylococcus epidermidis* at various controlled growth rates and examined production of a number of extracellular virulence factors. Comparisons of such data to planktonic cultures grown in the chemostat showed significant enhancement of extracellular protease, siderophores, and exopolymers by the attached biofilms (Figure 1). Significantly, these differences became greater with increasing growth rate for extracellular proteases and siderophores, yet smaller for the exopolymer. Clearly, exopolymer production is enhanced not only by attachment but also through reduction in cellular growth rate. Work on the regulation of lateral flagella gene transcription in *Vibrio parahaemolyticus* showed it to produce a single polar flagellum in liquid, and numerous lateral, unsheathed flagella on solid culture media (Belas et al., 1984, 1986; McCarter et al., 1988). They recognized that this reflected increased viscosity experienced by the organism that restricted movement of the polar flagellum. The net effect, however, was a switching on of the *laf* genes through contact with a surface. Lee and Falkow (1990) recognized that reduced oxygen tensions such as might be experienced by cells enveloped within a biofilm or in association with a surface caused the triggering of expression of *Salmonella* invasins.

Quorum Sensing

Transcriptional Activation

It has long been appreciated that the properties of some groups of microorganisms — for example, feeding and sporulation in myxobacters and actinomycetes, and swarming motility in *Vibrio parahaemolyticus* and *Proteus mirabilis* (Allison and Hughes, 1992) — exhibit cooperative behavioral patterns. Intercellular communication has likewise been recognized for several years in slime molds (Kaiser and Losick, 1993). In recent years properties of cells and genetic responses of populations have been identified that are responsive to population density (Williams et al., 1992; Jones et al., 1993). The first such observations concerned the luminescence of the marine bacterium *Vibrio fischerii*. This organism does not luminesce when grown as dilute planktonic cultures but does so when the organism is either concentrated within the light organs of fish or when, in batch culture, cell density exceeds a critical point. Early experimenters noted that when cell supernatants were transferred from dense cell populations to less dense ones then luminescence was induced (Eberhard, 1972; Nealson, 1977). Eventually such phenomena were tracked down to the production of an autoinducer (Eberhard et al., 1986) which accumulates in batch culture and which will switch on luminescence when its concentration is sufficiently high. Clearly, concentrations of the autoinducer will become greater not only as cell density is increased through growth but also if organization of the cells restricts dilution of the autoinducer by diffusion. Growth of cells in association with surfaces and within biofilms offers the

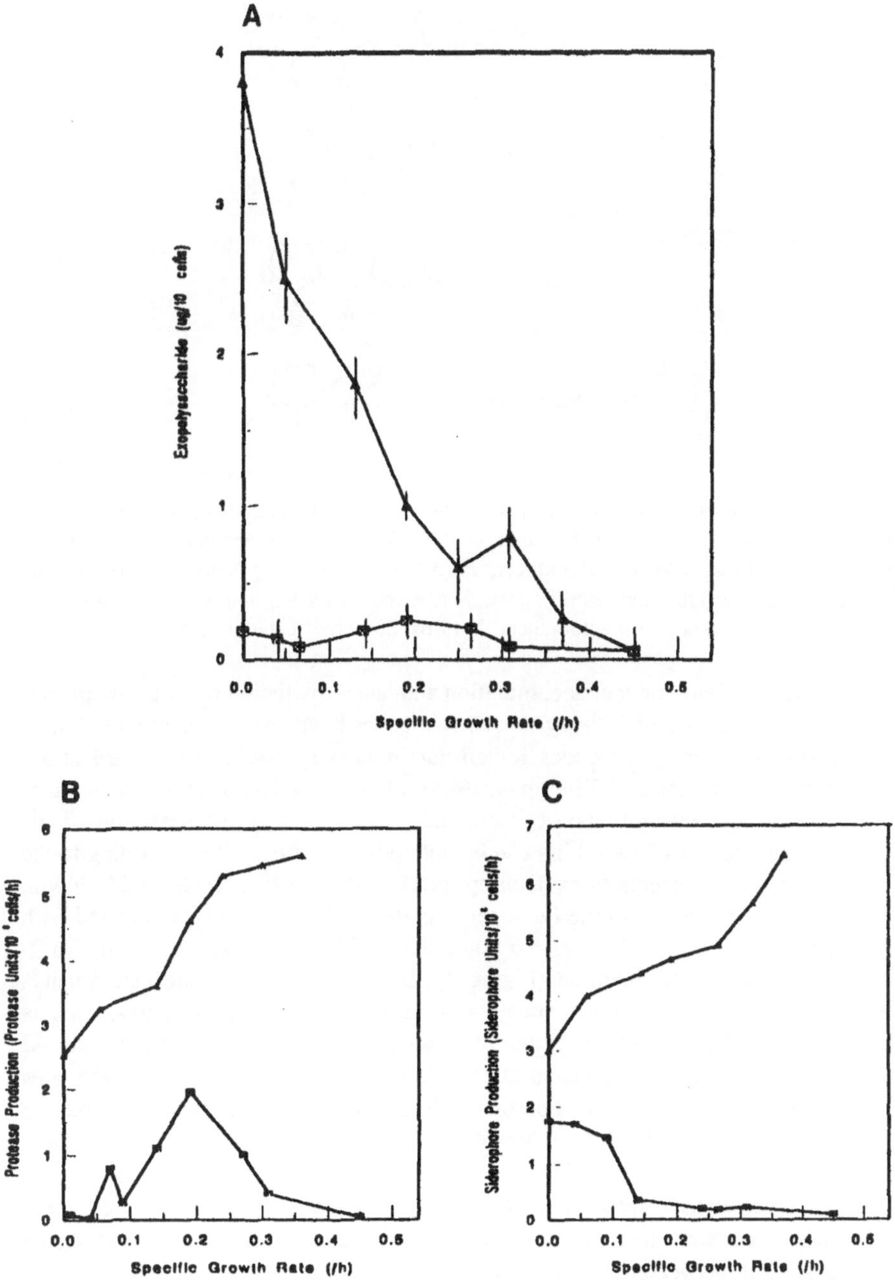

Figure 1 Production of exopolysaccharide (A), extracellular protease (B), and siderophore (C) by biofilms (▲) and planktonic (■) populations of *Staphylococcus epidermidis* cultured at various specific growth rates. (From Evans, E., et al., *Microbiology*, 140, 153-157, 1994. With permission.)

A

B

C

D

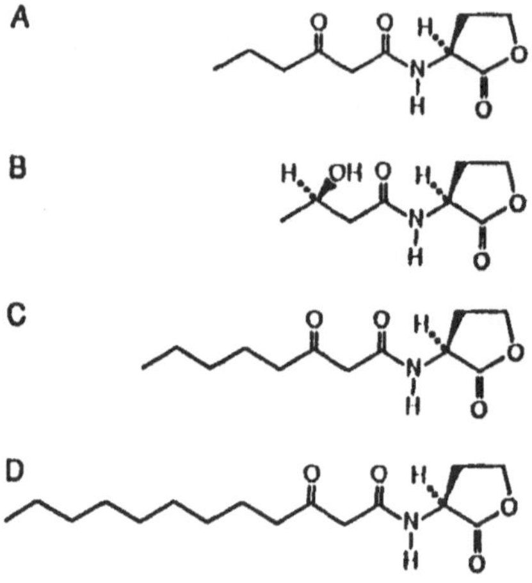

Figure 2 Structures of *Vibrio fischerii* autoinducer (VAI) and related autoinducers for cell-density responsive transcriptional activation. (A), VAI, N-(3-oxohexanoyl)-L-homoserine lactone; (B) *Vibrio harveyi* autoinducer, N-(3-hydroxybutanoyl)-L-homoserine lactone; (C) *Agrobacter tumefasciens* autoinducer, N-(3-oxooctanoyl)-L-homoserine lactone and (D) *Pseudomonas aeruginosa* autoinducer, N-(3-oxodecanoyl)-L-homoserine lactone.

prospect of autoinducer accumulation and hence switching on of a new phenotype. In the case of *Vibrio fischerii* two genes have been isolated and cloned, *LuxR* and *LuxI. LuxI* codes for an autoinducer synthetase (Eberhard et al., 1986; Engebrecht and Silverman, 1984), whereas *LuxR* codes for a membrane-associated transcriptional activator that is switched by the detection of the autoinducer. *LuxR* switching can be both positive or negative according to the autoinducer concentrations (Dunlop and Greenberg, 1988; Shadel and Baldwin, 1991). Other genes in the *Lux* operon code for the machinery associated with luminescence. The *V. fischerii* autoinducer (VAI) is 3-oxo-N-(tetrahydro-2-oxo-3-furanyl)hexanamide (Figure 2). *LuxR* and *LuxI* mediate cell-density-dependent control of *lux* gene transcription. At low cell densities, *LuxI* is transcribed at a basal level and VAI accumulates slowly in the medium. At sufficiently high VAI concentrations, VAI interacts with *LuxR*, which activates transcription of the *Lux* operon leading to luminescence and the positive autoregulation of *LuxI*. A review of these regulatory elements has recently been published (Meighan and Dunlop, 1993).

If cell-density-responsive transcriptional activation was not restricted solely to luminescence, then its implications in biofilms is enormous. Within the past 2 years it has become apparent that several species of bacteria contain regulatory systems analogous to *LuxR* and *LuxI*. These systems regulate conjugal transfer in *Agrobacterium tumefasciens* Ti plasmids (Piper et al., 1993) and extracellular virulence factor production in *Pseudomonas aeruginosa* (Gambello and Iglewski, 1991; Williams et al., 1992; Jones et al., 1993; Gambello, Kaye, and Iglewski, 1993). There is also evidence for homologous systems in *Erwinia, Rhizobia,* and *Escherichia*. These regulatory systems are now known as the

LuxR superfamily and recognize a number of chemically related autoinducers (Figure 2). Their role in the properties of attached bacterial populations remains to be determined.

IN VITRO MODELS OF BIOFILMS

Observations *in vivo* and *in situ* repeatedly demonstrate the recalcitrance of bacterial biofilms towards inimical treatments which are of demonstrable efficacy towards planktonic populations. In order to evaluate fully the various proposals concerning the mechanisms of such recalcitrance, *in vitro* models are required which differentiate between adherence and the associated influences of glycocalyx, nutrient status, and growth rate. Biofilms may be described simply as functional consortia of microbial cells within extracellular polymer matrices and associated with surfaces. Nevertheless, their physiology, metabolism, and organization are greatly dependent upon the nature of those surfaces and also upon the prevailing physico-chemical environment. Thus, biofilms associated with various ecological niches differ significantly from one another. Accordingly, numerous laboratory models have been developed in attempts to reproduce them. These models vary in complexity from growth on solidified nutrient media to complex biofermenters. Choice of an appropriate model system must therefore be tempered not only by technical constraints and understanding of the *in vivo/in situ* organization of the biofilm, but also by a consideration of all the physiological processes which might influence the end result. In attempts closely to duplicate the conditions prevailing in the natural biofilms many *in vitro* studies lack experimental control of the physiologies of the cells. The results which such studies generate, therefore, add little to the observations which have been made in the field. This is particularly the case for studies involving antimicrobial strategies. The main approaches used for *in vitro* studies of biofilm physiology will be reviewed here and reference made to their ability to distinguish between the various hypotheses concerning biofilm recalcitrance. Detailed accounts of the individual appoaches to biofilm culture will be given as appropriate in the applications section of this volume (Chapters 3.2 and 3.7).

CLOSED GROWTH MODELS

Solid:Air Interfaces

The simplest *in vitro* method of producing a biofilm population is to inoculate the surface of an agar plate to produce a confluent lawn culture. Such cultures possibly mimic those bacteria isolated from soft tissue infections, and, while not fully duplicating conditions of growth *in vivo*, model the close proximity of individual cells to one another and the various gradients (Lorian, 1989). In this respect, colonies grown on agar may be representative of biofilms at solid:air interfaces and provide samples for biochemical analysis. If agar cultures are to be deployed, then it is essential that they be prepared as confluent lawns. Individual colonies on solid media show size variation according to the colony density, presumably mediated through the effects of nutrient availability and oxygen tension change (Shapiro, 1987). Al-Hiti and Gilbert (1983) assessed the susceptibility of suspensions prepared from variously seeded nutrient agar plates and noted not only differences towards a number of pharmaceutical preservatives, but also increased heterogeneity of response within the larger colonies. While such heterogeneity is likely to reflect that found within dense biofilms, it is likely to detract from the reproducibility of response of inocula generated in this fashion.

Strains of *Pseudomonas aeruginosa*, isolated from the lungs of cystic fibrosis patients, showed significant increases in mucous production when grown on agar compared to liquid-grown cultures (Chan et al., 1984). Variation in the expression of cell envelope proteins, and therefore potential antigenicity, of cells grown in liquid media versus agar-grown cells has been demonstrated in a number of organisms including *Esherichia coli*, *Pseudomonas aeruginosa* (Critchley and Basker, 1988), and *Staphylococcus aureus* (Cheung and Fischetti, 1988).

Agar polysaccharide is polyanionic in nature and may therefore reduce the bioavailability of cations such as iron and magnesium, which needs to be considered when interpreting results. This also has a bearing on antimicrobial susceptibility testing, as interaction with cationic agents, such as aminoglycosides or quaternary ammonium compounds, could lead to misleading results. Comparisons of susceptibility made between agar-grown organisms must therefore involve harvested cells, free from agar residuals. While this negates any protective effects of the glycocalyx, significant increases in MIC and MBC have been reported for agar-grown *Haemophilus influenza* (Bergeron et al., 1987), *Pseudomonas aeruginosa* (DeMatteo et al., 1981), *E. coli* (Al-Hiti and Gilbert, 1983; Hohl and Felber, 1988) and *S. epidermidis* (Kurian and Lorian, 1980) relative to broth cultures. Ideally, cells should be removed from the agar surface prior to susceptibility testing yet be maintained as intact biofilm. Millward and Wilson (1989) and Nichols et al. (1989) attempted to overcome these difficulties by growing cultures, not on the agar itself, but upon cellulose nitrate membrane filters (0.45 μm porosity) placed in direct contact with the agar surface. In this procedure, the filter is subsequently removed from the agar and the culture exposed to the antibacterial agent as an intact biofilm. By comparing the susceptibility of intact biofilms determined in this way with those dispersed prior to antibacterial treatment, it is possible to assess the relative contribution of biofilm organization to antibacterial resistance.

Solid:Liquid Interfaces

Solid:liquid interfaces are frequently encountered at sites of biofilm formation. At such interfaces the solid surface might be nutritionally inert, such as that presented by the wall of a pipe or the surface of an implanted medical device, or it might provide nutrients/cations for microbial utilization. The liquid phase might be static or it might impose shear stresses upon the developing biofilm, causing cells to be shed and possibly influencing exopolymer deposition (Deretic et al., 1989). Many of these biofilm attributes may be modeled in closed growth systems, enabling studies of antimicrobial susceptibility. Thus, Gristina et al. (1987) placed sample disks directly into inoculated media while Prosser et al. (1987) pre-exposed disks to bacterial suspensions before transferring them to sterile broth. In both instances, the disks were monitored for the development of attached populations and subjected to susceptibility testing.

Using such approaches the susceptibility of a variety of gram-negative bacteria towards amidinocillin, cephaloridine, cefamandole and chlorhexidine has been shown to be substantially reduced (Prosser et al., 1987; Dix et al., 1988; Stickland et al., 1989; Stickler and Hewitt, 1991) and the tobramycin resistance of *P. aeruginosa* and *S. epidermidis* and vancomycin resistance of *S. epidermidis* to be increased 20–1000-fold (Gristina et al., 1987) for biofilms relative to planktonic populations. As noted by Brown et al. (1988), however, these experimental systems are uncontrolled with respect to growth rate and often make dubious comparisons between slow-growing biofilm cells and relatively fast growing planktonic cells. As such, the effects of growth rate cannot be distinguished from those associated with adhesion. Nevertheless, biofilms constructed in this manner can be used as a primary screen (Anwar et al., 1989a, b) when there are many antibiotics to be tested.

Open Growth Models (Non-Steady-State)

Chemostat-Based Models

Closed (batch) cultures do not model adequately many natural ecosystems which are continuously perfused with fresh supplies of nutrients. Natural populations therefore have the potential to grow over extended periods of time, often at very much reduced rates. Continuous culture systems also provide for a continuous supply of fresh nutrients and are often employed in the laboratory for planktonic populations in order to control and regulate growth rate under defined nutrient conditions. An obvious extension to this is to include submerged test pieces in a chemostat and to monitor the attached population (Keevil et al., 1987; Anwar et al., 1989a, b) or to flow medium continuously over a substratum. In such approaches, small, flat tiles of the test material are suspended by silk or cotton thread into steady-state chemostat cultures (Keevil et al., 1987). Cells from the liquid phase attach to the tiles, grow and divide, and biofilm formation progresses, which can be monitored by periodic removal of tiles for analysis.

Growth-rate control in continuous culture fermenters depends upon the establishment of a steady-state biomass which is perfectly mixed with and in equilibrium with the extracellular medium (Herbert et al., 1956). When biofilms are generated within continuous fermenters, although the rate of supply of nutrients may be constant, the biofilms cannot be regarded as being in steady-state since the adherent population and the total biomass in the fermenter will increase as a function of time and, by the definition of biofilms, there will be incomplete mixing.

While such models make an effective long-term model of environmental and medical biofilms, use of continuous culture should not infer growth-rate control. Indeed, as the biofilms develop, initially through sequestration of planktonic cells to the substratum and secondarily through growth of the attached cells, growth rates will decrease. Thus, the properties of the formed biofilm are likely to change substantially with incubation time in the chemostat. Indeed, Anwar et al. (1990, 1992) report substantial differences in antibiotic susceptibility of "old" (7 days) and "young" (2 days) biofilms generated in this manner. The approach has been used with success to model the formation of dental plaque on methacrylate dental resins with mixed inocula representing organisms of the oral cavity (Keevil et al., 1987), growth of *Legionella* biofilms in cooling systems where the medium deployed was treated, sterilized tap water (West et al., 1989, 1990), and also for a variety of investigations of antibiotic susceptibility (Anwar et al., 1989a, b). In these latter studies antibiotic susceptibilities of staphylococcal and pseudomonad biofilms increased for the adherent population, grown on suspended tiles within a chemostat as they are in the environment. Nevertheless, little understanding of the mechanisms involved can be gained from these studies.

The Robbins Device

Probably the most widely used of all systems for the study of adherent microbial populations, the Robbins device (Figure 5, Chapter 3.7, McCoy et al., 1981) is comprised of a plastic/metal tube section into which pieces of test material can be inserted such that the material forms part of the internal tube wall. As one end of the tube is continually fed with a suitable batch or chemostat culture of microorganisms, attachment to the internal walls and biofilm formation will occur, which can be sampled at intervals by removal of test pieces. This approach is an effective model for biofouling in industrial tubular flow systems and has been used to assess the effectiveness of biocides and other treatments at removing biomass (Ruseska et al., 1982; Costerton and Lashen, 1984). The device may usefully be incorporated into existing industrial water systems to allow the monitoring of biofouling and the effectiveness of removal strategies *in situ*. The Robbins device represents an open system in that there is a continuous supply of fresh nutrients and removal of waste products.

A modified Robbins device (Nickel et al., 1985) has been used in the study of bacterial infections of soft tissue and medical implants. This device consists of a rectangular-section, Perspex (Plexiglas) tube into which retractable pistons are inserted along the tube length, flush with the internal wall. The device is sterilized by ethylene oxide gas or by flushing with hypochlorite solution. Disks of sample material are glued to the end of the pistons so that the sample material makes up part of the internal wall of the tube. The tube is inoculated with cells by passing a mid-log culture through the length of the tube for several hours. Subsequently, suitable sterile media is passed through the device at a constant rate. Antimicrobial susceptibility of the attached cells can be assessed by incorporation of the agent in the media. Typical studies (Costerton et al., 1987; Evans and Holmes, 1987; Gristina et al., 1987; Nickel et al., 1985) using such tubular models compare the properties of adherent cells with those of the equivalent planktonic population of cells passing through the device. While this technique has provided many valuable demonstrations of recalcitrance in bacterial biofilms, interpretation of such data cannot distinguish between the effects of growth rate and adherence (Brown et al., 1988).

Annular Reactor

The annular reactor, or RotoTorque®, (Figure 3, Chapter 3.8; Characklis, 1988) operates as an open system and consists of two concentric cylinders. Sample materials, in the form of removable slides, are attached to the internal walls of the outer, static cylinder. The inner cylinder rotates as media is fed into the intercylinder space through recessed tubes in the inner cylinder. Following inoculation of the reactor, biofilm will develop on the cylinder walls and sample pieces. Complete mixing of fresh media occurs, so there are no problems of nutrient gradients developing within the liquid phase. As the biofilm develops, however, gradients will be created within the biofilm, as described for the Robbins device. This system is not growth-rate controlled.

The annular reactor exerts well-controlled, reproducible shear forces on the developing biofilm. This property is exploited in the observation of biofilm formation and anti-biofouling strategies by the continuous measurement of fluid frictional resistance calculated from torque measurements, taken as the drag force on the surface of the inner cylinder, and rotational speed. Although too complex for routine use, this system is useful in some engineering applications where increased resistance to fluid flow occurs as a result of biofilm accumulation.

Open Growth Models (Steady State)

Systems such as the Robbins device and submerged tiles lack effective growth rate control and are not at steady state. When used in these ways, they do not differentiate between those properties attributable to growth rate and those associated with adherence, neither can antimicrobial, nutrient, pH, and oxygen gradients generated within the biofilms be effectively reproduced. In order to facilitate such studies, steady state models of biofilms must be employed. Two such models have been described.

Constant Thickness Biofilms

This device consists of a rotating PTFE disk holding fifteen sample pans, each of which contains five recessed plugs of sample material (Figure 5, Chapter 3.7; Peters and Wimpenny, 1988). Media flow is directed at the center of the disk, where it is distributed over the surface by a scraper blade which sweeps the surface. Following inoculation, cells are scraped off the surface by the blade, but biofilm formation can progress in the recessed plugs up to a known depth. The device allows a large number of identical biofilms to be obtained of any thickness within a range up to 300 μm. Sample plugs may be removed aseptically for subsequent analysis. As biofilm thickness is controlled, it has been suggested that this technique allows the determination of the net growth rate of the sessile cells, although the device itself does not allow direct control of the biofilm growth rate. Electron microscopy, following cryosectioning of biofilms generated using this device, and microelectrode probing has generated important information on spatial gradients within biofilms (Kinniment and Wimpenny, 1992).

Perfused Biofilms

A device (Gilbert et al., 1989), modified from that of Helmstetter and Cummings (1963) for the continuous collection of daughter cells, establishes a biofilm on the underside of a bacteria-proof cellulose membrane. The membrane is perfused with fresh medium from the sterile side and cells eluted from the biofilm collected. A steady state is developed where the size of the biofilm population remains constant, and dispersed cells collect in the spent medium. At steady state, the rate of perfusion with fresh medium controls the overall growth rate of the culture. The dispersed cells coincidentally correspond to newly divided daughter cells and divide synchronously when transferred to fresh medium. Sample biofilms, prepared by low-temperature stage-freeze techniques, for scanning electron microscopy showed dispersed cells attached to the filter matrix yet embedded within an extracellular polymer matrix which closely resemble *in vivo* samples (Gilbert et al., 1989; Duguid et al., 1992a). Thus, although the initial method of cellular attachment by pressure filtration is artificial, the biofilms so produced at steady state resemble those isolated *in vivo.*

It is important to note that in these systems the biofilm populations are perfused with nutrients from their underside. While this is atypical of *in situ* biofilms on inanimate surfaces, it is representative of bacterial surface infections of soft tissues. This method enables control of the rate of growth of adherent populations of microorganisms and has proved to be a useful model of *in vivo* infection. The technique has now been sucessfully employed for several bacterial species, including clinical, mucoid isolates of *P. aeruginosa* (Evans et al., 1991a) and gram-positive organisms such as *S. epidermidis* (Evans et al., 1991b, DuGuid et al., 1992a). Biocide (Evans et al., 1990a) and antibiotic (Evans et al., 1990b; DuGuid et al., 1992a, b; Evans et al., 1991a, b) susceptibility of bacterial biofilms has been investigated with such models.

CONCLUSIONS

While there have been relatively few direct studies of the antibiotic susceptibility of biofilms, it has recently been demonstrated that sensitivity may be profoundly affected when growth occurs as an adherent biofilm, rather than as planktonic cells, and that such resistance might contribute towards the recalcitrance of particular infections. There is now direct evidence that growth rate, nutrient limitation, and exclusion by glycocalyx, together with expression of adherence phenotypes each play a role in this recalcitrance. Workers in the area should be aware of these contributory factors and select appropriate *in vitro* models and control populations.

REFERENCES

Abbanat, D.R., Godchaux, W. and Leadbetter, E.R. 1988. Surface-induced synthesis of new sulphonolipids in the gliding bacterium Cytophaga johnsonae. *Archives of Microbiology* **149,** 358-364.

Al-Hiti, M.M.A. and Gilbert, P. 1983. A note on inoculum reproducibility: solid versus liquid culture. *Journal of Applied Bacteriology* **55**, 173-176.

Allison, C. and Hughes, C. 1992. Bacterial swarming: an example of prokaryotic differentiation and multicellular behaviour. *Science Progress* **75,** 403-421.

Anwar, H., van Biesen, T., Dasgupta, M.K., Lam, K. and Costerton, J.W. 1989a. Interaction of biofilm bacteria with antibiotics in a novel *in vitro* chemostat system. *Antimicrobial Agents and Chemotherapy* **33**, 1824-1826.

Anwar, H., Dasgupta, M.K., Lam, K. and Costerton, J.W. 1989b. Tobramycin resistance of mucoid *Pseudomonas aeruginosa* biofilm grown under iron-limitation. *Journal of Antimicrobial Chemotherapy* **24**, 647-655.

Anwar, H., Dasgupta, M.K. and Costerton, J.W. 1990. Testing the susceptibility of bacteria in biofilms to antibacterial agents. *Antibacterial Agents and Chemotherapy* **34,** 2043-2046.

Anwar, H., Strap, J.L. and Costerton, J.W. 1992. Establishment of aging biofilms: possible mechanism of bacterial resistance to antimicrobial chemotherapy. *Antimicrobial Agents and Chemotherapy* **36**, 1347-1351.

Bale, M.J., Fry, J.C. and Day, M.J. 1988. Transfer and occurrence of large mercury resistance plasmids in river epilithon. *Applied and Environmental Microbiology* **54**, 972-978.

Belas, R., Mileham, A., Simon, M. and Silverman, M. 1984. Transposon mutagenesis of marine *Vibrio* spp. *Journal of Bacteriology* **158,** 890-896.

Belas, R., Simon, M. and Silverman, M. 1986. Regulation of lateral flagella gene transcription in *Vibrio parahaemolyticus. Journal of Bacteriology* **167**, 210-218.

Bergeron, M.G., Simard, P. and Provencher, P. 1987. Influence of growth medium and supplement on the growth of *Haemophilus influenzae* and on antibacterial activity of several antibiotics. *Journal of Clinical Microbiology* **25,** 650-655.

Brown, M.R.W. and Williams, P. 1985. The influence of environment on envelope properties affecting survival of bacteria in infections. *Annual Reviews of Microbiology* **39**, 527-556.

Brown, M.R.W., Allison, D.G. and Gilbert, P. 1988. Resistance of bacterial biofilms to antibiotics: a growth rate related effect? *Journal of Antimicrobial Chemotherapy* **22**, 777-789.

Brown, M.R.W., Costerton, J.W. and Gilbert, P. 1991. Extrapolating to bacterial life outside the test tube. *Journal of Antimicrobial Chemotherapy* **27**, 565-567.

Brown, M.R.W. and Gilbert, P. 1993. Sensitivity of biofilms to antimicrobial agents. *Journal of Applied Bacteriology, Symposium Supplement* **74,** 87s-97s.

Chan, R.J., Lam, S., Lam, K. and Costerton, J.W. 1984. Influence of culture conditions on expression of the mucoid mode of growth of *Pseudomonas aeruginosa. Journal of Clinical Microbiology* **19**, 8-10.

Characklis, W.G. 1988. Model biofilm reactors. In *Handbook of Laboratory Model Systems for Microbial Ecosystems*, vol 1, ed. Wimpenny, J.W.T. pp. 155-174. CRC Press Inc., Boca Raton, Florida.

Cheng, K.J., Irvin, R.T. and Costerron, J.W. 1981. Autochthonous and pathogenic colonisation of animal tissues by bacteria. *Canadian Journal of Microbiology* **47,** 461-490.

Cheung, A.L. and Fischetti, V.A. 1988. Variation in the expression of cell wall proteins of *Staphylococcus aureus* grown on solid and liquid media. *Infection and Immunity* **56,** 1061-1065.

Costerton, J.W., Irvin, R.T. and Cheng, K.J. 1981. The bacterial glycocalyx in nature and disease. 1981. *Annual Reviews of Microbiology* **35**, 399-424.

Costerton, J.W. and Lashen, E.S. 1984. Influence of biofilm on efficacy of biocides on corrosion-causing bacteria. *Materials Performance* **9,** 13-17.

Costerton, J.W., Cheng, K.J., Geesy, G.G., Ladd, T.I., Nickel, J.C., Dasgupta, M. and Marrie, T.J. 1987. Bacterial biofilms in nature and disease. *Annual Reviews in Microbiology* **41**, 435-464.

Critchley, I.A. and Basker, M.J. 1988. Conventional laboratory agar media provide an iron-limited environment for bacterial growth. *FEMS Microbiology Letters* **50,** 35-39.

Dagostino, L., Googman, A.E. and Marshall, K.C. 1991. Physiological responses induced in bacteria adhering to surfaces. *Biofouling* **4,** 113-119.

Dazzo, F.B. 1984. Bacterial adhesion to plant root surfaces. In *Microbial Adhesion and Aggregation*, ed. Marshall, K.C. pp. 85-93. Springer-Verlag, Berlin.

DeMatteo, C.S., Hammer, M.C., Baltch, A.L., Smith, R.P., Sutphen, N.T. and Mitchelsen, P.B. 1981. Susceptibility of *Pseudomonas aeruginosa* to serum bactericidal activity. *Journal of Laboratory and Clinical Medicine* **98,** 511-518.

Deretic, V., Dikshit, R., Konyecsni, W.M., Chakrabarty, A.N. and Misra, T.K. 1989. The AlgR gene which regulates mucoidy in *Pseudomonas aeruginosa* belongs to a class of environmentally responsive genes. *Journal of Bacteriology* **171,** 1278-1283.

Dickinson, G.M. and Bisno, A.L. 1989a. Infections associated with indwelling devices: concepts of pathogenesis; infections associated with intravascular devices. *Antimicrobial Agents and Chemotherapy* **33**, 597-601.

Dickinson, G.M. and Bisno, A.L. 1989b. Infections associated with indwelling devices: infections related to extravascular devices. *Antimicrobial Agents and Chemotherapy* **33**, 602-607.

Dix, B.A., Cohen, P.S., Laux, D.C. and Cleeland, R. 1988. Radiochemical method for evaluating the effects of antibiotics on Escherichia coli biofilms. *Antimicrobial Agents and Chemotherapy* **32,** 770-772.

Duguid, I.G., Evans, E., Brown, M.R.W. and Gilbert, P. 1992a. Growth-rate-independent killing by ciprofloxacin of biofilm-derived *Staphylococcus epidermidis*; evidence for cell-cycle dependency. *Journal of Antimicrobial Chemotherapy* **30**, 791-802.

Duguid, I.G., Evans E., Brown, M.R.W. and Gilbert, P. 1992b. Effect of biofilm culture upon the susceptibility of *Staphylococcus epidermidis* to tobramycin. *Journal of Antimicrobial Chemotherapy*, **30**, 803-810.

Dunlap, P.V. and Greenberg, E.P. 1988. Analysis of the mechanisms of *Vibrio fischerii* luminescence gene regulation by cyclic AMP and cyclic AMP receptor protein in *Escherichia coli*. *Journal of Bacteriology* **170**, 4040-4046.

Eberhard, A. 1972. Inhibition and activation of bacterial luciferase synthesis. *Journal of Bacteriology* **109**, 1101-1105.

Eberhard, A., Widrig, C.A., McBath, P. and Schineller, J.B. 1986. Analogs of the autoinducer of bioluminescence in *Vibrio fischerii*. *Archives of Microbiology* **146**, 35-40.

Engebrecht, J. and Silverman, M. 1984. Identification of genes and gene products necessary for bacterial luminescence. *Proceedings of the National Academy of Science* **81**, 4154-4158.

Elliott, T.S.J. 1988. Intravascular-device infections. *Journal of Medical Microbiology* **54**, 75-87.

Evans, R.C. and Holmes, C.J. 1987. Effect of vancomycin hydrochloride on *Staphylococcus epidermidis* biofilm associated wtih silicone elastomer. *Antimicrobial Agents and Chemotherapy* **31**, 889-894.

Evans, D.J. Allison, D.G., Brown, M.R.W. and Gilbert, P. 1990a. Effect of growth rate on resistance of Gram-negative biofilms to cetrimide. *Journal of Antimicrobial Chemotherapy* **26**, 473-478.

Evans, D.J., Brown, M.R.W., Allison, D.G. and Gilbert, P. 1990b. Susceptibility of bacterial biofilms to tobramycin: role of specific growth rate and phase in the division cycle. *Journal of Antimicrobial Chemotherapy* **25**, 585-591.

Evans, E., Duguid, I.G., Brown, M.R.W. and Gilbert, P. 1991a. Surface properties and adhesion of *Staphylococcus epidermidis* in batch and continuous culture. *Abstracts of the 91st. Annual Meeting of the American Society for Microbiology* **D-53.**

Evans, D.J., Allison, D.G., Brown, M.R.W. and Gilbert, P. 1991b. Susceptibility of *Pseudomonas aeruginosa* and *Escherichia coli* biofilms towards ciprofloxacin: effect of specific growth rate. *Journal of Antimicrobial Chemotherapy* **27**, 177-184.

Evans, E., Brown, M.R.W. and Gilbert, P. 1994. Iron chelator, exopolysaccharide and protease production on *Staphylococcus epidermidis*: a comparative study of the effects of specific growth rate in biofilm and planktonic culture. *Microbiology* **140**, 153-157.

Favero, M.S., Bond, W.W., Petersen, N.J. and Cook, E.H. 1983. Scanning electron microscopic observations of bacteria resistant to iodophor solutions, pp 158-166. *Proceedings of the International Symposium on Povidone*, University of Kentucky, Lexington, USA.

Fletcher, M. 1984. Comparative physiology of attached and free-living bacteria. In *Microbial Adhesion and Aggregation*, ed. Marshall, K.C. pp. 223-232. Springer-Verlag, Berlin.

Fletcher, M. 1986. Measurement of glucose utilisation by *Pseudomonas fluorescens* that are free living and that are attached to surfaces. *Applied and Environmental Microbiology* **52**, 672-676.

Gambello, M.J. and Inglewski, B.H. 1991. Cloning and characterisation of the *Pseudomonas aeruginosa LasR* gene, a transcriptional activator of elastase expression. *Journal of Bacteriology* **173,** 3000-3009.

Gambello, M.J., Kaye, S. and Inglewski, B.H. 1993. *LasR* of *Pseudomonas aeruginosa* is a transcriptional activator of the line protease gene *(apr)* and an enhancer of exotoxin A expression. *Infection and Immunity* **61**, 1180-1184.

Gibbons, R.J. and van Houte, J. 1975. Bacterial adherence in oral microbial ecology. *Annual Reviews of Microbiology* **29**, 19-44.

Gilbert, P., Allison, D.G., Evans, D.J., Handley, P.S. and Brown, M.R.W. 1989. Growth rate control of adherent bacterial populations. *Applied and Environmental Microbiology* **55**, 1308-1311.

Gilbert, P., Collier, P.J. and Brown, M.R.W. 1990. Influence of growth rate on susceptibility to antimicrobial agents: biofilms, cell cycle and dormancy. *Antimicrobial Agents and Chemotherapy* **34**, 1865-1868.

Gilbert, P., Evans, D.J. and Brown, M.R.W. 1993. Formation and dispersal of bacterial biofilms in-vivo and in-situ. *Journal of Applied Bacteriology, Symposium Supplement* **74,** 87s-78s.

Giwercman, B., Jensen, E.T., Hoiby, N. Kharazmi, A. and Costerton, J.W. 1991. Induction of β-lactamase production in *Pseudomonas aeruginosa* biofilms. *Antimicrobial Agents and Chemotherapy* **35**, 1008-1010.

Gordon, C.A., Hodges, N.A. and Marriott, C. 1988. Antibiotic interaction and diffusion through alginate and exopolysaccharide of cystic fibrosis derived *Pseudomonas aeruginosa. Journal of Antimicrobial Chemotherapy* **22,** 667-674.

Gristina, A.G., Hobgood, C.D., Webb, L.X. and Myrvik, Q.N. 1987. Adhesive colonisation of biomaterials and antibiotic resistance. *Biomaterials* **8,** 423-426.

Gristina, A.G., Jennings, R.A., Naylor, P.T., Myrvik, Q.N., Barth, E. and Webb, L.X. 1989. Comparative in-vitro antibiotic resistance of surface colonising coagulase negative *Staphylococci. Antimicrobial Agents and Chemotherapy* **33,** 813-824.

Helmstetter, C.E. and Cummings, D.J. 1963. Bacterial synchronisation by selection of cells at division. *Proceedings of the National Academy of Science* **50,** 767-774.

Herbert, D., Ellsworth, R. and Telling, R.C. 1956. The continuous culture of bacteria: a theoretical and experimental study. *Journal of General Microbiology* **14,** 601-622.

Hohl, P. and Felber, A.M. 1988. Effect of method, medium, pH and inoculum on the in-vitro antibacterial activities of Fleroxacin and Norfloxacin. *Journal of Antimicrobial Chemotherapy* **22** (SupplD), 71-80.

Hoyle, B.D., Jass, J. and Costerton, J.W. 1990. The biofilm glycocalyx as a resistance factor. *Journal of Antimicrobial Chemotherapy* **26**, 1-6.

Humphrey, B.A., Dixon, M.R. and Marshall, K.C. 1979. Physiological and in-situ observations on the adhesion of gliding bacteria to surfaces. *Archives of Microbiology* **120,** 231-238.

Jones, S., Yu, B., Bainton, N.J., Birdsall, M., Bycroft, B.W., Chhabra, S.R., Cox, A.J.R., Golby, P., Reeves, P.J., Stephens, S., Winson, M.K., Salmond, G.P.C., Stewart, G.S.A.B. and Williams, P. 1993. The *lux* autoinducer regulates the production of exoenzyme virulence determinants in *Erwinia carotovora* and *Pseudomonas aeruginosa. The EMBO Journal* **12**, 2477-82.

Kaiser, D. and Losick, R. 1993. How and why bacteria talk to each other. *Cell* **163,** 1210-1214.

Keevil, C.W., Bradshaw, D.J., Dowsett, A.B. and Feary, T.W. 1987. Microbial film formation: dental plaque deposition on acrylictiles using continuous culture techniques. *Journal of Applied Bacteriology* **62,** 129-138.

Kinniment, S.L. and Wimpenny, J.W.T. 1992. Measurements of the distribution of adenylate concentrations and adenylate energy charge across *Pseudomonas aeruginosa* biofilms. *Applied and Environmental Microbiology* **58**, 1629-1635.

Kurian, S. and Lorian, V. 1980. Discrepancies between results obtained by agar and broth techniques in testing of drug combinations. *Journal of Clinical Microbiology* **11,** 527-529.

Lappin-Scott, H.M. and Costerton, J.W. 1989. Bacterial biofilms and surface fouling. *Biofouling* **1**, 323-342.

Lee, C.A. and Falkow, S. 1990. The ability of *Salmonella* to enter mammalian cells is affected by bacterial growth state. *Proceedings of the National Academy of Sciences, USA*, **87,** 4304-4308.

Lee, J.V. and West, A.A. 1991. Survival and growth of *Legionella* species in the environment. *Journal of Applied Bacteriology* **70** (Suppl), 121S-130S.

Lorian, V. 1989. In-vitro simulatio of in-vivo conditions: Physical state of the culture medium. *Journal of Clinical Microbiology* **27,** 2403-2406.

Marshall, K.C., Stout, R. and Mitchell, R. 1971. Mechanism of the initial events in the sorption of marine bacteria to surfaces. *Journal of General Microbiology* **68**, 337-348.

McCarter, L., Hilmen, M. and Silverman, M. 1988. Flagellar dynamometer controls swarmer cell differentiation of *Vibrio parahaemolyticus*. *Cell* **54,** 345-351.

McCoy, W.F., Bryers, J.D., Robbins, J. and Costerton, J.W. 1981. Observations in fouling biofilm formation. *Canadian Journal of Microbiology*, **27**, 910-917.

McFeters, G.A., Egil, T., Wilberg, E., Adler, A., Schneider, R., Snozzy, M. and Geiger, W. 1990. Activity and adaptation of nitriloacetate(NTA)-degrading bacteria: Field and laboratory studies. *Water Research* **24,** 875-881.

Meighen, E.A. and Dunlop, P.V. 1993. Physiological, biochemical and genetic control of bacterial luminescence. *Advances in Microbial Physiology* **112,** 1-67.

Millward, T.A. and Wilson, M. 1989. The effect of chlorhexidine on *Streptococcus sanguis* biofilms. *Microbios* **58,** 155-164.

Nealson, K.H. 1977. Autoinduction of bacterial luciferase: occurrence, mechanism and significance. *Archives of Microbiology* **112,** 73-79.

Nichols, W.W., Dorrington, S.M., Slack, M.P.E. and Walmsley, H.L. 1988. Inhibition of tobramycin diffusion by binding to alginate. *Antimicrobial Agents and Chemotherapy* **32,** 518-523.

Nichols, W.M., Evans, M.J., Slack, M.P.E. and Walmsley, H.L. 1989. The penetration of antibiotics into aggregates of mucoid and non-mucoid *Pseudomonas aeruginosa*. *Journal of General Microbiology* **135**, 1291-1303.

Nickel, J.C., Ruseska, I., Wright, J.B. and Costerton, J.W. 1985. Tobramycin resistance of *Pseudomonas aeruginosa* cells growing as a biofilm on urinary catheter material. *Antimicrobial Agents and Chemotherapy* **27**, 619-624.

Paerl, H.W. 1975. Microbial attachment to particles in marine and freshwater ecosystems. *Microbial Ecology* **2**, 73-83.

Peters, A. and Wimpenny, J.W.T. 1988. A constant depth laboratory model film fermenter. In *Handbook of Laboratory Model Systems for Microbial Ecosystems*, vol 1, ed. Wimpenny, J.W.T. pp. 175-195. CRC Press, Boca Raton, Florida.

Piper, K.R., Beck von Bodman, S, and Farrand, S.K. 1993. Conjugation factor of *Agrobacter tumifasciens* regulates Ti plasmid transfer by autoinduction. *Nature (London)* **362,** 448-450.

Power, K. and Marshall, K.C. 1988. Cellular growth and reproduction of marine bacteria on surface-bound substrate. *Biofouling* **1,** 163-174.

Prosser, B.T., Taylor, B., Dix, B.A. and Cleeland, R. 1987. Method of evaluating effects of antibiotics upon bacterial biofilms. *Antimicrobial Agents and Chemotherapy* **31,** 1502-1506.

Robertson, B.D. and Meyer, T.F. 1992. Genetic variation in pathogenic bacteria. *Trends in Genetics* **8**, 422-427.

Ruseska, I., Robbins, J., Lashen, E.S. and Costerton, J.W. 1982. Biocide testing against corrosion causing oilfield bacteria helps control plugging. *Oil and Gas Journal* **80**, 253-264.

Shadel, G.S. and Baldwin, T.O. 1991. The *Vibrio fischerii LuxR* protein is capable of bidirectional stimulation of transcription and both positive and negative regulation of the *LuxR* gene. *Journal of Bacteriology* **173,** 568-574.

Shapiro, J.A. 1987. Organisation of developing *Escherichia coli* colonies viewed by scanning electron microscopy. *Journal of Bacteriology* **168,** 142-158.

Shaw, J.C., Bramhill, B., Wardlaw, N.C. and Costerton, J.W. 1985. Bacterial fouling in a model core system. *Applied and Environmental Microbiology* **49**, 693-701.

Slack, M.P.E. and Nichols, W.W. 1981. The penetration of antibiotics through sodium alginate and through the exopolysaccharide of a mucoid strain of *Pseudomonas aeruginosa. Lancet* **11,** 502-503.

Slack, M.P.E. and Nichols, W.W. 1982. Antibiotic penetration through bacterial capsules and exopolysaccharides. *Journal of Antimicrobial Chemotherapy* **10,** 368-372.

Stickland, D., Dolman, J., Rolfe, S. and Chawla, J. 1989. Activity of antiseptics against *Escherichia coli* growing as biofilms on silicone surfaces. *European Journal of Clincal Microbiology and Infectious Diseases* **8,** 974-978.

Stickler, D. and Hewitt, P. 1991. Activities of antiseptics against biofilms of mixed bacterial species growing on silicone surfaces. *European Journal of Clinical Microbiology and Infectious Diseases* **10,** 157-162.

Sutherland, I.W. 1977. Bacterial polysaccharides — their nature and production. In *Surface Carbohydrates of the Procaryotic Cell*, ed. Sutherland, I.W. pp. 27-96. Academic Press, London.

van Loosdrecht, M.C.M., Lyklema, J., Norde, W. and Zehnder, A.J.B. 1990. Influence of interfaces on microbial activity. *Microbial Reviews* **54**, 75-87.

West, A.A., Araujo, R., Dennis, P.J.L., Lee, J.V. and Keevil, C.W. 1989. Chemostat models of *Legionella pneumophila,* pp 107-116. *Airborne deteriogens and pathogens* (Flannagan, B. Ed.), Biodeterioration Society, London.

West, A.A., Rodgers, J., Lee, J.V. and Keevil, C.W. 1990. *Legionella* survival in domestic hot water systems. *Indoor Air 90* (Walkinshaw, D.D. Ed.) pp 37-42. Indoor Air Quality and Climate Inc., Ottawa, Canada.

Williams, P. 1988. Role of the cell envelope in bacterial adaptation to growth *in vivo* in infections. *Biochimie* **70**, 987-1011.

Williams, P., Bainton, N.J., Swift, S., Chhabra, S.R., Winson, M.K., Stewart, G.S.A.B., Salmond, G.P.C. and Bycroft, B.W. 1992. Small molecule-mediated density-dependent control of gene expression in prokaryotes: bioluminescence and the biosynthesis of carbapenem antibiotics. *FEMS Microbiology Letters* **100**, 161-8.

Wimpenny, J.W.T., Peters, A. and Scourfield, M. 1989. Modeling spatial gradients. In *Structure and Function of Biofilms*, eds. Characklis, W.G. and Wilderer, P.A. pp. 111-127. John Wiley, Chichester.

Zobell, C.E. 1943. The effect of solid surfaces upon bacterial activity. *Journal of Bacteriology* **46**, 39-56.

1.7 Sources of Biological Variation and Lack of Inoculum Reproducibility: A Summary

Michael R.W. Brown, Sally F. Bloomfield, and Peter Gilbert

The purpose of this section is briefly to summarize the sources of biological variation to give a kind of aide-mémoire. In essence, the origin of nonreproducibility is that organisms are *alive*. Thus, unlike an inanimate object, an organism will change and adapt with changing circumstances of growth and handling procedures. It may also be damaged by environmental change and may require time and specific circumstances to make repairs. Environmental change may select or even increase the probability of genetic variants, also capable of adaptation. Preservation and storage conditions can cause genetic and phenotypic changes (Chapter 2.1).

CULTURE ORIGINS

In one sense, the origins of any particular culture lie in biological evolution. A "fresh" culture from, say, a culture collection (grown in one medium) may well have two or three subcultures (in another medium) prior to any particular experiment or procedure. Similar considerations apply to clinical or environmental isolates. With initial inocula of about 10^6 cells/ml this might typically involve about 10 generations per subculture (i.e., to about 10^9/ml). This allows ample opportunity for selection of variants more suited to the "new" growth environment chosen, possibly very different to the original, e.g., *in vivo*. Low maximum specific growth rate (μ_{max}) and a high affinity (low K_S) for the substrate favors selection at low nutrient levels. One approach could be

0-8493-4752-1/95/$0.00+$.50

to grow a potential test strain under different conditions of specific nutrient depletion and determine growth conditions optimizing the desired property (e.g., susceptibility) and also considering relevance where appropriate. Thus, if a slow-growing, iron-limited chemostat culture generates cells with appropriate characteristics, then such a chemostat, cultured for numerous generations (to encourage selection) could be used to generate replicates preserved so as to avoid damage/genetic change (Chapter 2.1). Recovery conditions (for slants or cryopreserved cultures) could then be specified to be the same or similar to the original growth conditions and thus minimize the opportunity for further selection.

The well-known example of loss of virulence after *in vitro* "passage" of a strain obtained from an *in vivo* source illustrates the dangers.

GROWTH CONDITIONS OF THE INOCULUM

Before considering phenotypic changes, it must again always be borne in mind that all environments potentially select variants more suited to the particular growth conditions than is the intended inoculum.

There is a massive, well-recognized literature demonstrating phenotypic change with temperature, pH, Eh, and osmolarity. Specific nutrient depletion is referred to in Chapters 1.1, 1.2, and 1.4. Growth rate is well documented as determining crucial cell properties whether in suspension or as a biofilm. Depending on the causes of stationary phase and cell density at onset, microbes will be changing (e.g., starvation) from the point of onset (Chapter 1.3). Biofilm growth per se brings about physiological changes independent of growth rate (Chapter 1.6). Numerous cell properties change over the course of the division cycle and contribute to heterogeneity during logarithmic growth. The use of synchronous culture at specific stages is an underdeveloped technique for enhancing reproducibility.

Depending on cell density, cultures may (or may not) contain adequate concentrations of intercell signals which induce numerous metabolic changes including antibiotic and virulence factor production (see Chapters 1.2, 1.6, and 3.1). Hence, cell concentrations outside the critical concentration (about 5×10^8/ml) are likely to be less variable.

It is also important to bear in mind that no ingredient, even in a chemically defined medium, is absolutely pure. Furthermore, fresh pure water is commonly, in its final container, contaminated to a varying extent, e.g., by cations.

DORMANCY

The presence (or not) of small, relatively dormant cells (e.g., viable, nonculturable — see Chapter 1.3), often after slow growth or starvation, may especially contribute to lack of reproducibility of a heterogenous culture.

POST-GROWTH HANDLING PROCEDURES

Post-growth handling procedures such as centrifugation, vortexing and, indeed, any change, especially if rapid, such as resuspending in a liquid of different composition, pH, osmolarity, or temperature to the original growth conditions can cause changes to a culture. These can involve damage, altered susceptibility to host defenses or to antimicrobial agents, or even direct and substantial loss of viability (see Chapter 2.2).

PROCEDURE/TEST CONDITIONS

It is well established that the physical and chemical test conditions influence the outcome of susceptibility tests. Care must be taken neither to interfere with the activity of the antimicrobial agent (e.g., cations and aminoglycosides) nor to provide an additional insult to the inoculum, e.g., testing susceptibility of log phase *P. aeruginosa* in cold distilled water when, in the *absence* of any chemical antimicrobial agent typical death could be 50%, with the remainder damaged.

The length of an experiment can contribute to variability, even at room temperature. On the one hand, inocula may change and grow in nutrient media, and on the other hand, inocula in nongrowth media, e.g., phosphate-buffered saline or Ringers solution, may express starvation proteins or otherwise change (see Chapters 1.3 and 3.1).

Post-Growth Conditions

2.1 The Preservation of Inocula

Gerald D.J. Adams

INTRODUCTION

The preservation of stock-cultures of microorganisms is prerequisite to a wide range of industrial and research applications. The essence of any preservation process is to provide a viable, stable preparation using simple, well-validated techniques which will assure a high probability of success and batch reproducibility. Other considerations include the requirements that the process should be safe and cost-effective while providing an inoculum which can be stored, transported, and reconstituted with ease. It is doubtful whether any technique currently in use will satisfy all the criteria outlined above. Consequently a compromise will have to be made when selecting an appropriate technique based on individual user requirements and the response of an organism to preservation. When making this decision, the final application of the inoculum should be clearly identified. In some cases, it may be essential to maintain genetic stability; in other instances colony characteristics may be less important than the production of a specified bioproduct. The criteria used to assess preservation efficacy are often taken as reproducibility, biochemical properties, and genetic stability (Bank and Schmehl, 1989).

REPRODUCIBILITY

The ability of an inoculum to grow and reproduce to form typical colonies on medium, is often used to assess a preservation technique (Heckly, 1978). Postprocessing viabilities can be estimated simply by the presence or absence of growth after adding an ampoule of preserved culture to medium. Greater precision for validating the process may be achieved by replicate plating of serially diluted samples before and following preservation in a defined medium (Heckly, 1978). It is oversimplistic to regard the inoculum as containing only living or killed cells, and very important to appreciate that damaged individuals may constitute a high proportion of the total population where injured cells can

0-8493-4752-1/95/$0.00+$.50

progress to killed or repaired status depending on the method of sample reconstitution (Morichi, 1969; Ray et al., 1976; Cox, 1987). Surviving organisms may exhibit a significant increase in the lag phase of growth (Mackey and Derrick, 1982), may be less salt tolerant (Morichi and Irie, 1973) or require resuscitation prior to assay (Morichi and Irie, 1973; de Valdez et al., 1985), and each of these factors can influence the observed level of survival.

Biochemical Properties

Reproduction or colony formation may be inappropriate for assessing the success of a preservation technique, since organisms which exhibit high survival after preservation may lose the ability to produce a required end product (Sharp, 1984), and in such cases biochemical assay may be a more suitable test. Examples where end product may be influenced by preservation include lactic acid production in dairy cultures (Oliver et al., 1988), dough and brewing characteristics of yeast (Westerman and Huige, 1979), antibiotic production (Macdonald, 1972), and the preparation of therapeutic enzymes (Sharp, 1984).

Genetic Stability

Although genetic stability is often regarded as an essential requirement for a preservation process, complete genetic integrity is rarely assessed, and consequently genetic variation may remain undetected. Practically, it is important that the inoculum remains stable with respect to specified characteristics, and the preservation techniques should be validated accordingly. Population variation can occur either as a consequence of selection of a preservation-tolerant variant, by contamination of the culture during processing, by selective damage to a plasmid, or by the induction of mutations by the process used.

Continuous subculturing is often regarded as an unsatisfactory method of preserving a cell line because of the increased probability of population variance, although other techniques, often favored as alternatives to subculturing, including freeze-drying, can also induce mutational changes.

Living organisms are able to tolerate only a narrow range of physical conditions. Metabolic processes, including growth and reproduction, are arrested at extremes of pH, temperature, or desiccation, and advantage may be taken of these extremes to preserve microbial inocula (Franks et al., 1990).

PRESERVATION BY COOLING

When the temperature of the medium is reduced below an optimum, living organisms initially lose the ability to reproduce, and at lower temperatures metabolism ceases. Individual cells may be resuscitated from this state of dormancy with varying degrees of success depending on the organism species and method of treatment.

Except for chill-sensitive species (Morris and Watson, 1984; Strange, 1976; Morris, 1987), in the absence of ice formation, microorganisms can typically withstand chilling, and high survival can be obtained when cells are stored at +4°C. Below 0°C, the microorganism suspension will supercool, and while its composition remains unchanged it becomes susceptible to ice formation (Farrant, 1980; Mackenzie, 1977; Mazur, 1970).

Freezing is a two-step process during which water must first nucleate followed by the proliferation of ice crystals throughout the suspension to produce a mixture of ice and a solute concentrate (Franks, 1985). In practice, ice will invariably nucleate heterogeneously by crystallizing around microscopic impurities within the suspension (Mackenzie, 1977; Franks, 1985). Heterogenous nucleation is a statistical event which is dependent on the nature, size, and number of particulate impurities within the suspension, and is encouraged by reducing temperature and agitating the cooled suspension (Adams, 1991a).

Behavior of Solutes During Freezing

The response of a solute during freezing may be summarized by one of the following patterns of behavior (Mackenzie, 1985):

1. Solute fails to crystallize regardless of cooling conditions but remains associated with unfrozen water as an amorphous concentrate (glass formation).
2. Solute can crystallize but will do so only after the cooled suspension has been warmed to decrease its viscosity sufficiently to induce crystallization (heat annealing) (Pikal, 1991a).
3. Solute is able to crystallize only when the suspension is frozen using a slow rate of cooling.
4. Solute is able to crystallize readily, regardless of cooling rate or freezing regime used.

In typical solutions or suspensions containing several solutes, individual components will interact in a variety of ways, crystallizing or remaining amorphous depending on the freezing protocol used and the nature, concentration, and ratios of each solute within the medium. Regardless of the precise freezing pattern, individual cells will be exposed to an increased solute concentration as ice formation progresses, and cellular injury results from exposure to hypertonic solution concentration effects rather than reduced temperature. Ice can either nucleate within a cell (intracellularly), or throughout the surrounding medium (extracellularly), and because ice propagation is limited by the cell membrane, injury by intracellular freezing will be restricted to individual cells. In contrast, since extracellular ice will proliferate throughout the medium, solute concentration will influence all the cells in the suspension (Farrant, 1980). Cell contents typically supercool by 10 to 15°C even in the presence of extracellular ice. Since supercooled water exhibits a higher vapor pressure than ice, water will diffuse from the cell in an attempt to add to the extracellular ice (Meryman et al., 1977). Water diffusion is influenced by a number of factors, including cooling rate, medium composition, and the permeability characteristics of cell membranes (Farrant, 1980; Franks, 1985; Mazur, 1977; Diller et al., 1972; Toupin et al., 1989).

Rates of Cooling

Cooling rates are defined as "slow" (suboptimal), "rapid" (supraoptimal) or "optimal", as assessed by criteria such as postfreezing survival.

"Slow" Cooling Injury

At slow cooling rates, water can diffuse freely from the cell until all freezable water has been removed. As ice forms outside of the cell, resulting in an increase in solute concentration in the suspending medium, the osmotic gradient generated across the cell membrane will cause the cell to shrink. Cell shrinkage can only be tolerated to a limited extent, and below a critical value (approximately 60% of the original cell volume), further increases in the extracellular solute concentration will result in membrane damage and cell death (Meryman, 1968; Pegg and Diaper, 1989).

There is little evidence that extracellular ice can directly damage cells (Farrant, 1980), although Mazur has suggested that damage under slow-cool conditions could result partly from the crowding of cells into the narrow, solute-rich capillaries within the ice matrix (Mazur and Cole, 1989). In summary, potential causes of slow-cooling injury include:

1. Temperature reduction (in chill-sensitive species).
2. Increase in extracellular solute concentration.
3. Variations in the ionic strength and pH of the medium as solutes concentrate or crystallize (Cammack and Adams, 1985; van den Berg and Rose, 1959; Taylor, 1981).
4. Abnormal imbalances in transmembrane solute transport caused by extracellular medium changes and biochemical changes induced by reduced metabolism and cell dehydration at low temperatures (Lee and Chapman, 1987).

It is important to appreciate that increases in solute concentrations may be appreciable during freezing (for example a 0.9% saline solution will concentrate to 30% prior to eutectic freezing; Cammack and Adams, 1985), and damage will be exacerbated by prolonged exposure to concentrate conditions resulting from slow cooling (Grout et al., 1990).

Optimal Cooling Rate

Resistance to water diffusion will increase as the cooling rate increases and cell contents may supercool before all the freezable water has been removed (Meryman et al., 1977). Under these conditions, cell shrinkage and membrane damage will be minimized.

Rapid Cooling Injury

At cooling rates faster than an optimum, cell shrinkage is not observed and resistance to dehydration increases to the point where sufficient water remains within the cell to initiate intracellular ice formation (Farrant, 1980). Rapid cooling induces the formation of numerous, small ice crystals which can reform into larger crystals upon warming, and it is probable that rapid cooling injury is a consequence of ice recrystallization during thawing rather than ice formation during freezing (Farrant, 1980; Grout et al., 1990; Boutron, 1987).

In summary, cells may be injured during freezing by solution effects predominating during slow cooling, and by intracellular ice formation at rapid cooling rates (Mazur et al., 1972).

Rate of Sample Thawing

In a similar manner to definitions of "fast" and "slow" cooling, thawing rates are also loosely defined as "fast" or "slow". Cells will be exposed to osmotic shock as the suspension is thawed, resulting in a progressive decrease in the extracellular solute concentration as the ice melts. Rapid thawing combined with an optimal cooling rate will usually provide conditions for maximal survival although slow thawing may provide optimal conditions for preserving certain cell types, outside the scope of this article (for example, embryos, (Farrant, 1980). Small-volume inocula (0.5 to 2.0 ml) can be thawed rapidly by immersing and agitating containers of frozen samples in a water bath at 37°C.

In order to prevent toxic injury when dimethyl sulfoxide has been included as a cryopreservative, care should be taken not to hold samples at high temperatures for prolonged periods after thawing.

FACTORS AFFECTING SURVIVAL

Organism Species

Individual bacterial species vary markedly in their tolerance to freezing and thawing (Yamasato et al., 1973). Resistant organisms include bacterial spores (Fry, 1966), while sensitive species include gram-negative bacteria such as *Pseudomonas* spp., (Ashwood-Smith and Warby, 1971), psychrophilic bacteria (Yamasato et al., 1973) being particularly sensitive.

Influence of Stages in Growth Cycle on Cryosensitivity

In practice, cells from young, logarithmic-phase cultures do not generally survive freezing and thawing as well as those cells harvested at high concentration from the stationary phase of growth (Heckly, 1978). However, the method of cell culture, inoculum concentration, efficiency of washing cells prior to preservation, etc., will all influence survival.

Cell Concentration and Cryotolerance

Survival often improves as the cell concentration increases. This self-protection may result from the release of soluble, cryoprotective compounds from lysed, killed cells (Bretz and Ambrosini, 1966).

Cryoprotection

Survival of vegetative microorganisms frozen in water is often low, and damage is often exacerbated by suspending the cells in simple salt solutions prior to freezing (Adams and Warnes, 1994). Polge et al. (1949), first demonstrated the cryoprotective properties of glycerol. Since then a wide range of compounds have been identified as cryoprotectants (Farrant, 1980; Meryman et al., 1977; Albrecht et al., 1973; Grout and Morris, 1987). Although varying in their chemical properties, cryoprotectants are characteristically highly soluble in water, remain as a concentrate after freezing, are capable of hydrogen bonding to promote water structuring and display low toxicity (Farrant, 1980).

Cryoprotectants function colligatively by reducing ice formation, function kinetically by delaying water diffusion from the cell, thereby producing conditions similar to those experienced when cells are frozen at an optimal rate (Meryman et al., 1977), or act directly on cell components (for example, by stabilizing cell membranes by substituting for water between polar residues (Anchordoguy et al., 1987).

Cryoprotectants are classed as penetrating protectants which protect extra- and intracellularly, or as nonpermeant when unable to pass through the outer membrane and therefore protect cells only extracellularly (McGann, 1978).

Penetrating protectants are low molecular weight compounds and include additives such as glycerol, dimethyl sulfoxide, alcohols, etc. (Ashwood-Smith, 1980a). Cryoprotectants such as these reduce cell injury by altering the physical conditions of ice and solute in the medium (Farrant, 1980). Shepard et al. (1976) demonstrated that addition of glycerol could substantially reduce salt concentrations at any given subzero temperature. This supported Lovelock's original suggestion that permeants protect by colligative mechanisms (Lovelock, 1954).

Nonpermeants may function partly by impeding cell dehydration through increased viscosity of the extracellular medium. This provides conditions similar to those experienced when cells are cooled at optimal rates (Meryman et al., 1977). Suggested mechanisms based solely on colligative properties (McGann, 1978), are perhaps oversimplistic, since several authors have demonstrated that cryostabilization requires a direct interaction between the protectant and the cell membrane (Meryman et al., 1977; Anchordoguy et al., 1987). Using purified proteins, Carpenter et al. (1992) related cryoprotection to the ability of solutes to bind with biomolecules. Solutes which readily bind to proteins, causing them to unfold, destabilize, while solutes which are excluded from the protein stabilize (Carpenter et al., 1992). Attempts to improve cell survival by incorporating butane or propane diols into the suspending medium to suppress ice formation and induce a completely amorphous state within cells after cooling have shown promise for improving survival (MacFarlane and Forsyth, 1990).

The temperature of incubation of the cells in the cryoprotective medium, before freezing, may influence the effectiveness of a cryoprotectant. Incubation of eukaryote cells in glycerol, at +4°C, may inhibit the penetration of the additive into the cell because of increased medium viscosity at the lower temperature (Taylor et al., 1974).

Toxic Effects of Cryoprotectants

A prerequisite for cryoprotection is that the additive should be nontoxic, although additives, such as methanol, frequently exhibit toxicity at concentrations above an optimum. In this laboratory (unpublished data), methanol has proved an excellent cryoprotectant for *Escherichia coli* frozen in nutrient broth at 6% w/w methanol, but not above 7.5% w/w, when toxicity is exhibited. Concentration of the solute during freezing is an important factor which may exacerbate the toxicity of an additive (Fahy, 1986). Dimethyl sulfoxide (DMSO) exhibits interesting behavior: the additive is nontoxic at low temperatures (below +4°C) but becomes toxic at higher temperatures (+37°C). It has been suggested that this behavior is because DMSO is excluded from biomolecules at low temperatures but binds at temperatures above 20°C (Arakawa et al., 1990).

FREEZE-PRESERVATION IN PRACTICE

Bulk inocula should not be subjected to repeated cycles of freezing and thawing since each cycle will potentially reduce viability. It is therefore preferable to divide the bulk into manageable aliquots for freeze-preservation.

Small-volume inocula can be conveniently frozen in ampoules, although care should be taken when sealing glass ampoules to prevent leakage during storage (Greiff et al., 1975). This is particularly important when samples are stored in liquid nitrogen, since an imperfectly sealed ampoule may explode upon thawing due to vaporization of nitrogen, which may have entered the container. Rubber-stoppered vials should not be stored in gas- or liquid-phase nitrogen, since stoppers will lose elasticity at these temperatures (Barbaree and Smith, 1981).

Glass containers are prone to fracture by thermal stress when cooled or warmed. In order to prevent accidental contamination by breakage, vials or ampoules should be contained in a shatter-reducing net or cellophane wrap, or thawed within a cabinet. Sterilizable, plastic cryotubes provide a safer alternative to glass containers (Simione et al., 1977). Small volumes of inocula can be frozen on agar plugs in drinking straws (Dietz, 1975), while glass or ceramic beads coated with culture provide a convenient means of storing frozen inocula where individual beads can be removed for subsequent culture (Nagel and Kunz, 1972). Alternatively, culture can be dripped or sprayed into liquid nitrogen or a cryogenic liquid to produce individually frozen culture pellets (Cox, 1968). Grivell and Jackson (1969) have obtained high survival by freezing microorganism cultures on dry beads of silica gel.

Purpose built aluminium or stainless steel containers can be used for preserving large volumes of inocula. Sharp (1984) has described the alternative use of pliable, plastic blood-transfusion bags for long-term freeze-preservation of culture volumes of up to 400 mls.

Methods of Cooling

Cultures are often frozen to terminal temperatures simply by placing in a deep-freeze or by immersion in a cryogenic liquid. In such instances resistant organisms may survive at a high level, after rapid thawing, in the presence of a suitable cryoprotectant. When preserving more sensitive organisms (for example, algae or protozoa), a two-step freezing cycle may be more appropriate. In the first stage the inoculum is cooled slowly (1 to 2°C/min) to –20 or –30°C, and held at this temperature to permit dehydration and cell shrinkage while avoiding intracellular ice formation. The second stage involves rapid freezing to final storage temperatures of –70 or –136°C (James, 1984; Lesson et al., 1984).

Storage Temperatures

Domestic Deep Freezers

Storage of liquid cultures in nutrient media at domestic deep-freezer temperatures (–13 to –20°C), generally results in poor survival. Such temperatures are above the eutectic at which an aqueous solution of sodium chloride (a common cell diluent, eutectic temperature –21.7°C) completely crystallizes, exposing the inocula to salt concentrations approaching 30% w/w during storage. Other disadvantages of storing cultures in a domestic deep-freeze have been noted by several authors (Cammack and Adams, 1985).

Storage at –60°C to –80°C

Reliable, relatively inexpensive laboratory freezers are available which operate at approximately –70°C. For practical purposes, metabolic processes may be regarded as arrested at this temperature (Rightsel and Greiff, 1967), and inocula can be stored for extended periods without reduction in viability. Frozen inocula can be transported in solid carbon dioxide (temperature –70°C), to maintain stability.

Liquid/Gas-Phase Nitrogen Storage

Storage in gas-phase (–136 to –178°C) or liquid-phase (–180 to 190°C) nitrogen provides the lowest practical storage temperatures, but is not without risk. Apart from the risks associated with container breakage (see above), personnel operating deep-freeze storage vessels should be protected against freezer burn. Additional hazards associated with liquid nitrogen storage include impairment of operator vision by the condensate clouds formed when the liquid nitrogen cabinet is opened, the risk of asphyxiation by nitrogen gas, and increased risks of freezer burn caused by splashing with liquid nitrogen or permeation of the liquid refrigerant through porous protective clothing. (See also Gareis et al., 1969.)

Cold Storage in the Absence of Ice Formation

Because loss of inoculum viability during freezing results from ice formation and associated solute-concentration effects, high survival may be achieved if freezing is prevented. The most obvious way to avoid ice-freezing is to store the inoculum at 4°C. Although certain cell types may be injured by chilling, bacterial species which are sensitive to freezing damage, for example *Pseudomonas* spp., can be successfully preserved at 4°C. One disadvantage of such storage is that molds and psychrophilic contaminants are able to grow at this temperature, resulting in spoilage and enzymatic degradation. Supercooling inocula below 0°C, to prevent ice formation, can provide inocula with increased viability. Unpublished data from this laboratory has compared the survival of *E. coli* WP2 stored at –18°C in the supercooled (100% survival) and frozen state (77% survival).

Attempts to supercool an inoculum under laboratory conditions are frequently unpredictable since it may be difficult to maintain supercooled conditions. More controllable supercooling can be induced by prefiltering the suspending medium to remove ice-promoting impurities. The suspension should then be emulsified in oil to produce aqueous droplets of culture which are relatively free of nucleating particles. This will suppress heterogenous ice nucleation and produce a stable, supercooled suspension (Franks, 1985). Hatley et al. (1987) have developed this technique for storing cryosensitive biomaterials at subzero temperatures.

The disadvantages of storing inocula in the cooled or frozen state include the expense of maintaining inocula at low temperatures, the inconvenience of transporting frozen or chilled cultures, and the potential for total loss of the culture collection following refrigerator failure. For these reasons, strategically important culture repositories should be duplicated.

FREEZE-DRYING

Freeze-drying/lyophilization is an alternative to freeze-preservation and in principle will produce a dried, shelf-stable inoculum which can be stored, distributed, and reconstituted with ease. Suspensions are prefrozen to form a mixture of ice and a solute-rich phase, and dried under vacuum to sublime the ice from the frozen mass (Rowe, 1970; Mellor, 1978; Adams, 1991a). Operationally the process is considerably more complex than this. In practice, freeze-drying is a multistep process in which the individual steps should be regarded as a discrete, though interrelated, stresses imposed on an organism, each of which might lead to damage unless controlled. The stages of the freeze-drying drying process may be summarized as, (1) product preparation, (2) sample prefreezing, (3) primary drying (sublimation), (4) secondary drying (desorption), (5) stoppering, (6) removal, (7) storage, and (8) reconstitution.

FREEZE-DRYING EQUIPMENT

Laboratory-scale freeze-driers typically comprise a robust chamber which can be evacuated to approximately 10^{-1} mBar (the operating pressure of the drier). Water, subliming from the sample, is trapped either by a chemical desiccant (i.e., phosphorus pentoxide) or on a refrigerated surface (termed a condenser), maintained at a lower temperature than the sample. The drier may be fitted with shelves upon which inocula can either be prefrozen or heated to encourage sample dehydration within a convenient time. Other desirable design features may include stoppering devices for sealing the vials prior to removal, automatic cycle control and an autoclave system for decontaminating the drier at the end of the run (Rowe, 1971).

RATE OF FREEZING PRIOR TO FREEZE-DRYING

Cooling rate is critical in minimizing cell injury during freezing (see above). While any of the freezing regimes noted above may be used to prefreeze cells for drying, samples are often prefrozen in vials or ampoules on refrigerated shelves within the freeze-drier. Engineering considerations restrict the cooling capabilities of a freeze-drier to freezing rates of between 0.25 and 2°C/min. Slow cooling rates such as these (resulting in freezing rates of 0.5 – 1.0 mm cake depth per minute) are generally preferred in freeze-drying since they produce a large, open ice crystal structure which offers low impedance to the migration of water vapor from the drying sample. It is not possible to maintain this preferred cooling rate throughout a sample whose depth exceeds 1 cm. Consequently 1 cm should be regarded as the maximum fill depth (Rowe and Snowman, 1978). A batch of vials frozen on the freeze-drier shelf will display a diversity of ice crystallization patterns, as the contents of some vials freeze slowly from the base while others supercool extensively before suddenly crystallizing. While there has been debate concerning the optimal degree of supercooling (Jennings, 1980; Pikal, 1991c), the different freezing patterns obtained will significantly alter drying rates from vial to vial. Pikal (1991c) demonstrated that introduction of a warming step in the freezing cycle, followed by a second cooling, would induce small ice crystals to recrystallize into larger structures, enabling samples to dry more rapidly and more uniformly. Such processes are termed heat-annealing (Pikal, 1991a).

Heat and Mass Transfer

The sublimation interface (freeze-drying front) may be observed, macroscopically, as a sharp, discrete boundary which progresses through the frozen sample. Heat energy must be provided to sustain sublimation, and, in the main, reaches the freeze-drying front by conduction through the frozen mass from the heated shelf. Additional heat is radiated into the sample from the upper shelf and from the chamber walls. As the thickness of the dry layer increases, impedance to vapor flow will also increase so that it becomes progressively more difficult to remove heat from the sample as water vapor. The temperature at the sublimation interface will consequently increase and may rise above the sample collapse temperature such that the drying product softens and distorts. In extreme cases, for example, when an impervious skin forms on the surface of the sample, impedance to vapor flow may be so high that the sample melts.

Shelf Temperatures during Primary Drying

In the past it has been suggested that samples should be dried below their "eutectic" temperatures, since exposure to higher temperatures would reduce the survival of inocula during primary drying (Meryman, 1966). It is more probable that the critical temperature determining safe processing during freeze-drying is the collapse temperature (Mackenzie, 1975; Franks, 1990), which is often incorrectly defined as the "eutectic" temperature (below).

Secondary Drying (Desorption Drying)

Completion of primary drying (sublimation) can be judged by several criteria, including noting the point at which the product and shelf temperatures coincide (Adams, 1991a). After primary drying, the residual moisture content of the sample is usually too high (7 to 15% w/w) for optimal stability. The drying cycle is therefore often extended by secondary drying, where water is removed by desorption, to give residual moisture contents below 2% w/w. Secondary drying is frequently facilitated by raising the shelf temperature, but survival may be reduced by using high shelf temperatures during secondary drying (Takano and Terui, 1969), and high desorption temperatures may aggregate purified proteins (Pikal, 1991a).

Sample Prefreezing — Collapse

Solutes can behave in a number of ways when suspensions are frozen. They may crystallize or remain in an amorphous state depending on the nature of the solute, freezing conditions, etc. Persistence of an amorphous state may reduce cellular injury (Orndorf and Mackenzie, 1973). During freeze-drying, however, the persistence of noncrystallizing solutes as an amorphous concentrate can present problems during sublimation (Bellows and King, 1972). Below the glass transition temperature (T_g'), an amorphous concentrate (glass) will exhibit high viscosity and behave as a brittle solid. Upon warming, during primary drying, the viscosity of the glass may decrease sufficiently for the concentrate to soften, deform, and in extreme cases lose structure completely as the drying progresses. This will form a structureless, sticky residue within the vial. This behavior is termed "collapse" (Bellows and King, 1972; Mackenzie, 1975) and can occur whenever the temperature/water content ratio exceeds a critical value (Franks et al., 1991). During freeze-drying, collapse is often observed at the sublimation interface (Mackenzie, 1977) where the temperature rises as sublimation cooling decreases and where the drying boundary exhibits a high moisture content. Collapse temperatures can be determined by measuring the temperature at which the brittle amorphous concentrate begins to soften (T_g'), (Levine and Slade, 1988) or the temperature at which the drying sample is microscopically observed to distort, (collapse temperature, T_c), (Mackenzie, 1977; Adams and Irons, 1993). Collapse temperatures for excipients commonly included in freeze-drying formulations are listed in Table 1. To avoid collapse, the shelf temperature should be controlled to maintain the sublimation front below T_c. This will prevent the concentrate from softening during drying but will necessarily prolong the cycle time. Alternatively, the suspension may be reformulated with additives which display higher T_c values (Orndorf and Mackenzie, 1973; Levine and Slade, 1988; Franks, 1989). Collapsed products may suffer reduced viability, be difficult to reconstitute and exhibit an unacceptably high moisture content leading to poor shelf-stability (Adams and Irons, 1993).

TABLE 1
Collapse Temperatures (T_c) for Common Freeze-Drying Excipients Determined using the Freeze-Drying Microscope

Excipient	Collapse Temperature (°C)
Sorbitol	–46.0
1% Lactose plus 0.3% sodium chloride	–45.0
Glucose	–41.0
Sucrose	–31.0
Lactose	–30.5
Trehalose	–28.5
Polyvinylpyrrolidone (MW 70,000)	–24.0
Dextran (MW 70,000)	–11.0
Mannitol (Eutectic Temperature)	–1.4

Containers and Sample Removal

Any of the containers described for freeze-preservation could be used for freeze-drying although inocula are usually processed in rubber stoppered vials or glass ampoules. The vial is undoubtedly the most convenient format for freeze-drying since it enables samples to be sealed, within the drier at the end of cycle, with predetermined moisture contents. Additionally, after secondary drying, vials can remain under vacuum or be back-filled with an inert gas prior to stoppering and removal. Dried inocula can then be stored under appropriate conditions before distribution, reconstitution, and use (Rowe, 1970; Adams, 1994). Rubber-stoppered vials are, however, susceptible to leakage during storage, particularly at subzero temperatures where the stopper might lose elasticity (Cammack and Adams, 1985). All-glass sealed ampoules overcome these problems but must be removed from the drier prior to sealing. Additional procedures are therefore required to prevent sample exposure to atmosphere (Rowe, 1971).

FACTORS AFFECTING THE SURVIVAL OF FREEZE-DRIED MICROORGANISMS

Freeze-dried inocula are shelf-stable and retain viable cells for tens of years if processed correctly (Kupletskaia, 1987; Smith, 1984). Dried cultures will, however, invariably lose some viability during processing, storage, and reconstitution (below).

Microorganism Species

As a general rule, organisms become more sensitive to freeze-drying as they become larger and more complex (Burns, 1964; Heckly, 1978). Animal cells do not generally survive freeze-drying (Jeyendran et al., 1981) and are best preserved by freezing in a suitable cryoprotective medium. Despite early reports to the contrary (Burns, 1964), marine and freshwater algae can be freeze-dried (Tsuru, 1973) but survival is poor compared to freeze-preservation in DMSO or glycerol. Yeasts vary in their sensitivity to freeze-drying (Smith, 1984; Kirsop, 1984) with *Saccharomyces cerevisiae* being moderately resistant (Heckly, 1978).

Although structurally the simplest living organisms, viruses vary markedly in their sensitivity to freeze-drying. These variations reflect differences in suspending media, freeze-drying methodology, the physico-chemical nature of individual species, and the inability of viruses to repair freeze-injury and dehydration damage. Rightsel and Greiff (1967) classified viruses into eight groups depending on their physico-chemical characteristics and sensitivity to freezing or drying. Only three groups — picornaviruses, poxviruses, and polyoma viruses — were stable to freeze-drying, although the authors remarked on the variability of responses within each group depending on the suspending medium and freeze-drying protocol adopted.

Bacteria can generally be successfully freeze-dried, although there is often considerable species variation. Since freezing is a prerequisite to freeze-drying, it follows that freeze-sensitive organisms (particularly obligate psychrophils) are likely to be sensitive to freeze-drying (Yamasato et al., 1973). Gram-positive organisms are generally more resistant to freeze-drying damage than are gram-negative ones, possibly reflecting the more complex nature of the gram-negative cell envelope. Spore-forming bacteria are among the most resistant organisms to freeze-drying damage (Heckly, 1978), but their vegetative cells are often particularly sensitive to injury just prior to sporulation. Early reports, suggesting anaerobic species to be particularly sensitive to freeze-drying (Haynes et al., 1955), have now been refuted by reports of high freeze-drying survival for several strict anaerobes (Phillips et al., 1975), although postdrying survival may be markedly influenced by suspending medium composition (Staab and Ely, 1987).

Of the vegetative organisms, *Streptococcus pyogenes* and *Staphylococcus aureus* are particularly resistant to freeze-drying injury, while *E. coli* and species of *Salmonella, Shigella, Brucella, Mycobacterium,* and *Lactobacillus* are moderately or variably resistant (Fry, 1966; Heckly, 1978). Similar trends were described by Peterz and Steneryd (1993) for freeze-dried bacterial mixtures in foods. These authors suggested that coliforms were the most sensitive to freeze-drying. In a comprehensive review of the freeze-drying characteristics of over 70 strains of clinically important bacteria, de Mello and Snell (1985) reported freeze-drying resistance of most strains, including species of *Clostridium, Bacillus, Corynebacterium, Pasteurella, Yersinia,* and *Pseudomonas aeruginosa, P. fluorescens,* and *P. maltophilia*. The authors noted that *Haemophilus influenza, Neisseria gonorrhea, Bacteroides melaninogenicus, Campylobacter coli,* and *Campylobacter jejuni* were the most sensitive to freeze-drying. The Rickettsiaceae appear poorly stable to freeze-drying (Heckly, 1961).

Cell Concentration and Culture Age

Increasing the cell concentration invariably results in improved survival after lyophilization (Heckly, 1978). This is possibly a consequence of the release of cryoprotective lysates from damaged cells (Bretz and Ambrosini, 1966).

Typically cells harvested from the stationary phase of growth are more resistant to freeze-drying than logarithmic-phase cells (Heckly, 1978). Data from this laboratory using *E. coli* confirms this pattern, with batch cultures harvested at 4 h being less resistant to freeze-drying than those harvested at 16 h. There are reports that some species may be more resistant in logarithmic than in stationary phase (Amarger and Jaquemetton, 1972). *S. aureus* appears equally resistant to freeze-drying at all stages of growth (Fry, 1966).

Suspending Medium Composition

Attempts to freeze-dry cells in water or simple salt solutions often result in poor and variable survival. Freeze-drying media should therefore be formulated to protect the organisms from freezing and drying damage, should produce a shelf-stable inoculum, must be compatible with the freeze-drying process, and should dry to form a cosmetically elegant cake. Skimmed milk has been widely used for drying lactic acid and other bacteria. While the inclusion of solutes, such as ascorbic acid, may improve its effectiveness, skim milk is unsatisfactory for preserving the more sensitive organisms. Alternative formulations, based on serum, particularly in combination with nutrient media should be employed. In 1951 Fry and Greaves developed "mist desiccans" (Greaves, 1960), a mixture of serum (75% v/v) and nutrient broth (25% v/v) containing added glucose (5 to 10% w/v), as a universal freeze-drying medium.

Carbohydrates and their derivatives are undoubtedly most effective freeze-drying protectants, particularly when combined with nutrient media or serum. They are useful for preserving a wide range of microorganisms, including foot and mouth disease virus (Dextran and skim milk + 5% glucose; Ferris et al., 1990); luminescent bacteria (2% soluble starch + 13% lactose; Janda and Opekarova, 1989); anaerobes (chopped meat ~ carbohydrate medium; Staab and Ely, 1987) and algae (Tsuru, 1973; Takano et al., 1973).

At the National Collection of Type Cultures (England), Redway and Lapage (1974) compared the effectiveness of over 20 carbohydrates and related compounds as freeze-drying protectants for a variety of bacterial species including sensitive strains such as *Haemophilus suis* and *Neisseria gonorrhea*. Solutes were categorized as protective in the order: meso-inositol (most effective) > carbohydrate alcohols (e.g., mannitol) > nonreducing disaccharides (e.g., trehalose) > reducing disaccharides (e.g., lactose) > monosaccharides (e.g., glucose) > glycerol > pentoses (poorest protectants). This is to be contrasted with the data of Orndorf and Mackenzie (1973) and unpublished data from this laboratory, where aqueous inositol and mannitol were poor protectants. Recently

Adams and Warnes (1994) have described very poor survival of genetically manipulated strains of *E. coli* when freeze-dried in aqueous mannitol (1% w/v) compared with aqueous lactose (1% w/v). Pikal (1991b) has suggested that mannitol is a poor protectant because it crystallizes during freezing and is consequently removed from the frozen suspension. The behavior of mannitol contrasts markedly with that of a stabilizing excipient, such as trehalose, which remains amorphous in the frozen mass. Redway and Lapage (1974) used a serum-rich base in their studies, and it is possible that the serum provided an amorphous (protective) matrix during prefreezing when inocula were formulated with inositol or mannitol. A serum-inositol mixture, augmented with 25% glucose, has proved effective for preserving sensitive *Campylobacter*, where viability losses could be attributed to primary drying rather than to prefreezing or secondary drying (Owen et al., 1989). At the molecular level, Crowe et al. (1988), demonstrated that while a wide range of additives — including amino acids, sugars, and polymers — protect against freeze-damage, only the disaccharides were effective as drying protectants. These act by substituting for water molecules in the cell membrane during desiccation.

Ablation

Media should be formulated to provide a cohesive, structure within the vial which will prevent ablation losses (Adams, 1991b). Ablation is the entrainment of friable particles in the vapor escaping from a sample, and results in loss of vial contents and contamination of the freeze-drier interior. This can result in cross-contamination of different cultures processed as a single batch, and contamination of the filling area. Because of the high vapor flow velocities generated during sublimation, attempts to reduce ablation contamination by plugging containers with cotton (cotton wool) may prove ineffective (Rowe, 1970). Ablation is exacerbated by poor product formulation and is most marked in samples which collapse during drying (Adams, 1991b).

Reconstitution

It is important to appreciate that a freeze-dried culture comprises a mix of dead, living and damaged cells, where the damaged subpopulation can progress to killed or living depending on the method of rehydration or reconstitution (de Valdez et al., 1985). Consequently, microorganisms will often exhibit poor growth on selective media after freeze-drying (Morichi, 1969). Morichi and Irie (1973) reported that reconsituted freeze-dried cultures of *Streptococcus faecalis* comprise two classes of survivors; (1) those which do not require the synthesis of RNA prior to growth and therefore grow on minimal agar, and (2) those which require the addition of a peptide to promote RNA synthesis. These classes of survivors also respond differently to the addition of sodium chloride (6% w/v) to the resuscitation medium, since this concentration of salt inhibits RNA synthesis. Protection from osmotic shock during reconstitution may also influence the observed viability. Choate and Alexander (1967) noted that reconstitution of *Spirillum altanticum* in hypertonic sucrose rather than water improved survival 10,000-fold. These authors also reported higher survival for rehydration completed at low temperature. Vapor-phase rehydration remains controversial (Monk and McCaffrey, 1967). While Cox (1987) reports vapor-phase rehydration as providing higher survival than reconstitution with water, the author notes that by altering the experimental conditions the converse can apply.

LYOPHILIZATION DAMAGE AND STORAGE CONDITIONS

Freeze-drying should produce a stable inoculum where the shelf life can be assessed by periodic bioassay or accelerated storage tests (Greiff and Rightsel, 1965a; Griffin et al., 1981). Cammack and Adams (1985), however, have cautioned that such tests may result in inaccurate predictions of stability.

A detailed description of lyophilization damage is beyond the scope of this article. As freeze-drying progresses, the structure of the sample is constantly changed from from a frozen to a dried state. This makes it difficult to differentiate between freezing, drying, and storage damage. Membrane damage is the major injury during sublimation (Israeli et al., 1974) and freezing (Calcott and Calcott, 1983). However, drying may also cause damage to individual cell components (Gomez et al., 1973), such as proteins, enzymes (Cox, 1987), lipids, and nucleic acids (Novick et al., 1972). Viability determinations, after freeze-drying or storage, must take into account the protracted lag phases which often follow reconstitution.

FACTORS INFLUENCING STABILITY DURING AND AFTER DRYING

Residual Moisture Content

Nei (1978) has suggested that removal of ice, by sublimation, from the extracellular medium does not cause cell damage; rather, injury results as water is withdrawn from the organism. Soper and Davies (1976) have demonstrated that when spores are freeze-dried, water is removed from two different sites in the organism. This dictates whether the dehydration damage is reversible or not. There remains debate over the extent by which cells should be dehydrated. We have been unable to detect any variation in viability immediately after removal from the drier for *E. coli* freeze-dried in lactose (1% w/v), over a moisture range of 0.5 to 7.5% w/v (unpublished data). Long-term stability was, however, affected by the water content, with inocula being less stable at higher moisture content. This supports earlier data for *Brucella abortus* (Fry, 1966) and confirms a generally held view that underdrying should be avoided by secondary drying.

Influenza virus shows optimal survival when freeze-dried in an albumin/ growth medium to 1.6% residual moisture content; underdrying (to 4% moisture) and overdrying (to 0.6% moisture) give reduced stability (Greiff and Rightsel, 1968; Cammack and Adams, 1985). Other workers have not observed such optima for different organisms and therefore generally recommended prolonged and secondary drying (Suzuki, 1973; May et al., 1992).

Suggestions that overdrying may remove "bound" or "structured" water (Phillips et al., 1981) should be treated with caution since the concept of "water structure" is questionable (Beale, 1983). It is possible, however, that overdrying may expose sites on the organism which are sensitive to reactive gases (Greiff and Rightsel, 1969). In this context dried baker's yeast can become more sensitive to oxygen damage when the residual moisture content is reduced from 8% to 4%. Antioxidants such as butyrated hydroxyanisole (BHA), may be included into these formulations to reduce oxygen damage (Heckly, 1978). It is worth noting that water contents in a vial upon removal from the drier represent an averaging of different water contents throughout the dry structure and that these underdried or overdried areas will equilibrate by moisture redistribution during sample storage.

Sealing Gas

Vacuum provides a favorable sealing environment for long-term stability of dried inocula and enables seal integrity to be tested using glow-discharge (Rudge, 1984). Sealing under a reactive gas such as oxygen (air) will markedly reduce the storage life of a freeze-dried culture. Greiff and Rightsel (1969) demonstrated that sealing freeze-dried influenza virus under helium, argon, or vacuum markedly improved stability compared with nitrogen, although the possibility that the nitrogen had been contaminated by traces of gaseous biocides similar to the Open Air Factor (Cox, 1987) were not examined.

Free-Radical Damage

Free-radical activity has been suggested as a cause of injury and death in dried bacteria (Heckly and Dimmick, 1968; Heckly and Quay, 1983). Cox and Heckly (1973) have suggested that free-radical induction may follow rather than cause cell death, since viability losses preceded detection of free radicals. It is possible, however, that the authors failed to detect early, short-lived radicals (Halliwell and Gutteridge, 1987).

Temperature of Sample Storage

Although microorganisms are much more thermostable in the freeze-dried form than in suspension, dried cultures are not immune to thermal inactivation, and storage at low temperatures will improve stability. In a well-formulated

TABLE 2
Solid-State Collapse Temperatures for Common Freeze-Drying Excipients

Excipient	Water Content (% w/w)	Solid-State Collapse Temperature (T_c)
Trehalose	0.5%	+115°C
	1.0%	+110°C
	2.0%	+100°C
	5.0%	+64°C
Lactose	0.5%	+110°C
	1.0%	+105°C
	2.0%	+94°C
	5.0%	+58°C
Sucrose	0.5%	+74°C
	1.0%	+66°C
	2.0%	+55°C
	5.0%	+21°C
Glucose	2.5%	+34°C
	3.5%	+20°C

Note: All samples freeze-dried from 1% solutions.

sample, storage at 4°C should be adequate for long-term stability, and short-term excursions to higher temperatures (for example, during distribution) should not affect the texture of the dried plug or biostability.

Collapse in the Solid State

While the persistence of an amorphous concentrate throughout the freeze-drying process may be a prerequisite for preserving organisms, the cake resulting from drying these concentrates may begin to soften as the storage temperature is raised and/or the sample water content is increased. At temperatures above the solid-state collapse temperature (Tsourouflis et al., 1976) or glass transition temperature (Franks et al., 1991), samples may begin to distort, collapse and, more importantly, will decay at a faster rate than predicted by Arrhenius kinetics (Levine and Slade, 1988). Solid-state collapse temperatures are dependent on the nature of the excipients within the suspension formulation and moisture content of the dried inocula (Table 2). Greiff et al. (1975) have demonstrated that tip-sealed glass ampoules are prone to leakage, while Barbaree and Smith (1981) have demonstrated that rubber stoppers may lose elasticity below −20°C and allow leakage. Container leakage should not be confused with a common problem where moisture can diffuse into dried samples from stoppers which have been inadequately dried following steam sterilization (Adams, 1989, 1990; Pikal and Shah, 1992). Data in Table 2 indicates that water ingress into the sample during storage can markedly reduce the solid-state collapse temperature.

Maillard Reactions

Inocula may exhibit thermal inactivation in the absence of structural changes to the dried cake (collapse). Inactivation of dried cells by the Maillard reaction (Scott, 1960) of reducing sugars with free amino groups results in protein and cell damage. To avoid Maillard denaturation, Greaves (1960) recommends both the avoidance of reducing sugars and the addition of Maillard suppressors, i.e., sodium glutamate, to suspension media for freeze-drying (Greaves, 1960). While Maillard reactions are often associated only with high temperatures, Cox has recently reported that Maillard or Maillard-type interactions may be a more universal mechanism of cell injury at ambient temperatures than may be supposed (Cox, 1991).

CHANGES IN CHARACTERISTICS OF FREEZE-DRIED INOCULA

While it has been advocated that test organisms used for disinfectant evaluation should be freeze-dried to prevent variation in susceptibility (Crowshaw, 1981), there is little doubt that freeze-drying will induce changes in the characteristics of an inoculum. The more important mechanisms are outlined below:

1. Selection of a freeze-drying tolerant contaminant or variant.
2. Loss of plasmid or plasmid function (Sharp, 1984).
3. Contamination during dispensing.
4. Cross-contamination, by ablation, from contaminants resident within the freeze-drier or from simultaneous processing of different organisms.
5. Mutation.

MUTATION CAUSED BY FREEZE-DRYING OR DRYING

Of particular concern is the potential for freeze-drying to induce genotypic changes in the preserved cells (Ashwood-Smith and Grant, 1976). While Calcott and Gargett (1981) report that freezing alone can induce genetic change, most authors suggest that mutation is restricted to the drying stages of freeze-drying (Ashwood-Smith, 1980b). Tanaka et al. (1979) have suggested that such mutation is caused by error-prone enzymatic repair of the DNA, damaged by drying. It is possible that free-radical activity may also play a role, since free-radical scavengers incorporated into freeze-drying media reduce mutation rates (unpublished data). Mutation appears to be associated with all drying techniques (Asado et al., 1980), including aerosolization (Webb, 1967) and L-drying (Banno et al., 1978), and the mutation rate may increase in dehydrated cultures during storage (unpublished data).

Damage by Electromagnetic Radiation

Visible Light

Since visible light has been implicated as a cause of inactivation, then storage of bacteria in lightproof containers may be advocated. Few publications, however, have addressed this issue in detail (de Rizzo et al., 1990). Such evidence as exists supports the view that while dried inocula are sensitive to prolonged exposure to light (Fry, 1966), intermittent exposure will not significantly reduce viability, particularly when dried cultures are boxed and stored in a darkened refrigerator.

Gamma, X-Irradiation, or Background Radiation

While dried inocula are regarded as stable to physical stress, desiccated organisms can be inactivated by exposure to ionizing radiations. The secondary effects of gamma or X-irradiation, such as the induction of free radicals (Auerbach, 1976) and ozone, may be particularly injurious.

Glow Discharge Testing

Container vacuum integrity may be tested using a high-voltage discharge method (Rudge, 1984) which induces a characteristic glow within an evacuated container. We have been unable to detect any damaging effects to dried inocula induced by glow discharge testing (unpublished data), provided that containers are not subjected to frequent or prolonged testing.

The Influence of Cell Dormancy on Postpreservation Survival

While genus type is probably the single most important factor associated with the survival of a preserved inoculum, it is clear from the discussion above that the physiological age of the organism, culture conditions, and medium formulation will significantly influence survival. In this context, cells harvested from the logarithmic phase are generally more sensitive to freezing or drying injury than organisms harvested from the stationary phase.

Stationary-phase cells from asynchronous cultures, typically used as starting inocula for preservation, will contain organisms at all stages of the cell cycle. Such differences in subpopulation physiology will be reflected in wide variations of individual cell susceptibility to stress (Gilbert et al., 1990), including freezing or freeze-drying. Laboratory conditions of culture are atypical of natural growth conditions, where vegetative cells are exposed for prolonged periods to extremes of temperature, salinity, desiccation, or nutrient deprivation. Under such adverse conditions, vegetative cells will be encouraged into a state of dormancy. Perhaps the most widely studied examples of cell dormancy are the spore-forming organisms, which, in response to stress, undergo a number of biochemical and physiological changes resulting in sporulation (Losick and Youngman, 1984). Until recently, dormancy in nonsporulating organisms has not been extensively studied, although the presence of a dormant alternative to the vegetative state has been postulated. Dormancy (or the anabiotic state) is characterized by a marked decrease in metabolism and cessation of cell division, accompanied by the absence of cell wall septum formation, cell elongation and other related processes essential for nucleoid replication. Kaprelyants et al. (1993) describe the transition of "old" cells in stationary phase from the vegetative to dormant state as involving 40 to 80 genes (including "survival" genes). These are responsible for the expression of starvation proteins which inhibit cell division and which may be excreted into the medium to decrease the apparent viability of young cells in the culture. Dormant cells are markedly reduced in size due to the condensation of cytoplasm and have reduced RNA content. Dormant cells may fail to revert to the vegetative form when added to media. This pattern of behavior is very similar to the response of cells made dormant by freeze-drying, where reversibly damaged, nondividing cells may constitute a high proportion of the rehydrated population.

Under natural, starvation conditions, the transition from the vegetative to the dormant state requires a relatively slow metabolic adaptation to stress. In contrast, when organisms are frozen or freeze-dried, they have insufficient time in which to biochemically adapt to the stress. Since endospores are particularly resistant to freezing or freeze-drying, it is possible that dormant vegetative cells may also be similarly resistant to these stresses. By inducing vegetative stationary-phase cells into a dormant state, using controlled growth conditions prior to preservation, it may be possible to achieve high survival in preserved cultures (M.R.W. Brown, personal communication).

In his treatise on Maillard reactions, Cox (1991) suggested that the conversion of viable, dividing cells into a dormant state may be induced by a reversible Maillard reaction. Subsequent rearrangements of the Schiffs bases and Amardori compounds, formed by the reaction, then result in a progressive reduction in the viability of the dormant cells. The contents of the review further suggest that Maillard reactions will be slower in dried samples stored below the glass transition temperature (T_g) when sample viscosity is high than when inocula are stored above the glass transition temperature when decreased sample viscosity will facilitate Maillard interactions. These hypotheses fit in well with the pattern of events which occur when dried inocula are stored, where viability is observed to decrease slowly with time, the rate of inactivation being accelerated by storage in the presence of oxygen (which potentiates Maillard reactions) or storage at temperatures above the glass transition temperature.

ALTERNATIVES TO VACUUM FREEZE-DRYING

Atmospheric Freeze-Drying

Freeze-drying can be completed at atmospheric pressure rather than under vacuum, although the process is less controllable under atmospheric conditions. Results which compare the two methods suggest that bacterial survival is improved when vacuum rather than atmospheric conditions are used (Wolff et al., 1990). (See also Burlacu et al., 1989.)

Drying in the Absence of Freezing

Drying techniques which avoid prefreezing may be advantageous for preserving cryosensitive organisms. Traditional drying often compares unfavorably with freeze-drying for preserving inocula, although spray-drying has been used successfully for bulk drying of starter cultures (Peri and Pompei, 1976). Annear (1970) described a technique where freeze-dry-sensitive organisms (i.e., *Leptospira* spp. and Treponemes) could be preserved by "puff-drying" onto peptone-glucose plugs. Similar techniques, including L-drying, have been used to preserve freeze-dry-sensitive microorganisms by desiccating inocula on filter paper. L-drying has proved particularly useful for drying complex cell species (Banno et al., 1978). Franks et al. (1991) describe a drying technique for sensitive biomolecules which avoids freezing while providing a shelf-stable product. Xerophilic fungi have been successfully stored by mixing cultures of the organism with sterile bentonite (Dallyn and Fox, 1980).

A number of techniques used as alternatives to both freezing and freeze-drying have been reviewed by Kirsop and Snell (1984), and storage of suspensions under sterile mineral oil has been successfully used for a number of organisms sensitive to freeze-drying (Arkad'eva et al., 1983).

CULTURE COLLECTION PRACTICE

While this review is not intended to be an exhaustive description of all techniques available for preserving inocula, freezing and freeze-drying are the most widely used general methods for preserving microorganisms and have therefore merited particular attention. While freeze-drying provides the convenience of shelf-stable samples which can be distributed and simply reconstituted for use, it does suffer the disadvantages that the process can significantly reduce viability, requires expensive equipment, and perhaps most importantly may induce genetic changes in cultures. In a recent review of culture-collection practices within the ATCC, Simione (1992) has recommended that master stocks should be maintained by low temperature storage (liquid nitrogen or comparable) while working stocks can be frozen or freeze-dried. Practical requirements also include the need for adequate stock control, the validation of procedures used for preserving inocula and the provision of back-up storage facilities.

REFERENCES

Adams, G.D.J. 1989 Rubber closures used to stopper vials containing freeze-dried product: Observations on the effects of various sterilising treatments on water retention by stoppers, pp 491-507, In *Lyophilisation Technology Handbook 1989*. Institute for Applied Pharmaceutical Science, Center for Professional Advancement, East Brunswick, NJ.

Adams, G.D.J. 1991a Freeze-drying of biological materials. *Drying Technology* **9,** 891-925.

Adams, G.D.J. 1991b The loss of substrate from a vial during freeze-drying using *Escherichia coli* as a trace organism. *Journal of Chemical Technology and Biotechnology* **52,** 511-518.

Adams, G.D.J. 1994 Freeze-drying of biohazardous products. In *Biosafety in Industrial Biotechnology* (P. Hambleton, J. Melling, and T.T. Salusburg, Eds.) Blackie Academic and Professional, London, pp. 178-212..

Adams, G.D.J. and Irons, L.I. 1993 Some implications of structural collapse during freeze-drying using *Erwinia caratovora* L-asparaginase as a model. *Journal of Chemistry and Biotechnology* **58,** 71-76.

Adams, G.D.J. and Warnes, A. 1994 The sensitivity of genetically modified organisms to freeze-drying: Influence of recombinant protein "A" on the survival of *Escherichia coli* Jm83. *Journal of Chemical Technology and Biotechnology*, submitted.

Albrecht, R.M., Orndorf, G.R. and Mackenzie, A.P. 1973 Survival of certain microorganisms subjected to rapid and very rapid freezing on membrane filters. *Cryobiology* **10,** 223-239.

Amarger, N. and Jaquemetton, M 1972 Influence of the age of the culture on the survival of *Rhizobium meliloti* after freeze drying and storage. *Archive Mikrobiology* **81**, 361-366.

Anchordoguy, T.J., Rudolph, A.S., Carpenter, J.F. and Crowe, J.H. 1987 Modes of interaction of cryoprotectants with membrane phopholipids during freezing. *Cryobiology* **24,** 324-331.

Annear, D.I. 1970 Preservation of microorganisms from the liquid state. In *Proceedings of the First International Conference on Culture Collections*, (H. Iizuka and I. Hasegawa, Eds). University of Tokyo Press.

Arakawa, T., Carpenter, J.F., Kita., Y.A. and Crowe, J.H. 1990 The basis for toxicity of certain cryoprotectants: A hypothesis. *Cryobiology* **27,** 401-415.

Arkad'eva, Z.A., Karryeva, D.A. and Baranova, N.A. 1983 Preservation of lactic acid bacteria *Biologicheskie Nauki (Moskow)* **2,** 101-104.

Asado, S., Takano, M. and Shibasaki, I. 1980 Mutation induction by drying of *Escherichia coli* on a hydrophobic filter membrane. *Applied and Environmental Microbiology* **40**, 274-281.

Ashwood-Smith, M.J. 1980a Low temperature preservation of cells, tissues and organs, pp. 19-44, in *Low Temperature Preservation in Medicine and Biology* (M.J. Ashwood-Smith and J. Farrant, Eds). Pittman Medical, Tonbridge Wells, UK.

Ashwood-Smith, M.J. 1980b Preservation of microorganisms by freezing, freeze-drying and desiccation, pp 219-252, in *Low Temperature Preservation in Medicine and Biology* (M.J. Ashwood-Smith and J. Farrant, Eds). Pitman Medical, Tonbridge Wells, UK.

Ashwood-Smith, M.J. and Warby C. 1971 A species of *Pseudomonas*, a most useful bacterium for cryobiological studies. *Cryobiology* **8,** 208-210.

Ashwood-Smith, M.J. and Grant, E. 1976 Mutation induction in bacteria by freeze-drying. *Cryobiology* **13**, 206-213.

Auerbach, C. 1976 *Mutation Research*. Chapman and Hall, London.

Bank, H.L. and Schmehl, M.K. 1989 Parameters for evaluation of viability assays: Accuracy, precision, sensitivity and standardization. *Cryobiology* **26**, 203-211.

Banno, L., Sakane, T. and Iijima, T. 1978 Mutation problems in preservation of bacteria *(Escherichia coli)* by L-Drying. *Cryobiology* **15,** 692-693.

Barbaree, J.M. and Smith, S.J. 1981 Loss of vacuum in rubber stoppered vials stored in a liquid nitrogen vapour phase freezer. *Cryobiology* **18,** 528-531.

Beale, P.T. 1983 Water in biological systems. *Cryobiology* **20,** 324-334.

Bellows, R.J. and King, C.J. 1972 Freeze-drying aqueous solutions: Maximum allowable operating temperatures. *Cryobiology* **9,** 559-561.

Boutron, P. 1987 Non-equilibrium formation of ice in aqueous solutions: Efficiency of polyalcohol solutions for vitrification, pp 201-236, in *The Biophysics of Organ Cryopreservation* (D.E. Pegg and A.M. Karow, Eds). Plenum Press, New York.

Bretz, H.W. and Ambrosini, R.A. 1966 Survival of *Escherichia coli* frozen in cell extracts. *Cryobiology* **3**, 40-46.

Burlacu, E., David, C. and Vasilescu, T. 1989 Comparative evaluation of three *in vacuo* desiccation procedures on bacterial cultures. *Archives of Roumanian Pathology and Experimental Microbiology* **48**, 65-78.

Burns, M.E. 1964 Cryobiology viewed by the microbiologist. *Cryobiology* **1**, 18-39.

Calcott, P.H. and Gargett, A.M. 1981 Mutagenicity of freezing and thawing. *FEMS Microbiology Letters* **10,** 151-155.

Calcott, P.H. and Calcott, K.N. 1983 Involvement of outer membrane proteins in freeze-thaw resistance of *Escherichia coli. Canadian Journal of Microbiology* **30,** 339-344.

Cammack, K.A. and Adams, G.D.J. 1985 Formulation and storage, pp 251-288, in *Animal Cell Biotechnology* Vol II, (R.E. Spier and J.B. Griffiths, Eds), Academic Press, London.

Carpenter, J.F., Arakawa, T., and Crowe, J.H. 1992 Interactions of stabilizing additives with proteins during freeze-thawing and freeze-drying. *Developmental Biological Standards* 74, 225-238.

Choate, R.V. and Alexander, M.T. 1967 The effect of the rehydration temperature and rehydration medium on the viability of freeze-dried *Spirillum atlanticum. Cryobiology* **3,** 419-422.

Cox, C.S. 1968 Method for routine preservation of microorganisms. *Nature (London)* **220,** 1139.

Cox, C.S. 1987 *The Aerobiological Pathway of Microorganisms*. John Wiley & Sons, Chichester, UK.

Cox, C.S. 1991 *Roles of Maillard Reactions in Disease.* HMSO Publications, London.

Cox, C.S. and Heckly, R.J. 1973 Effects of oxygen upon freeze-dried and freeze-thawed bacteria: Viability and free-radical studies. *Canadian Journal of Microbiology* **19,** 189-194.

Crowe, J.H., Crowe, L.M., Carpenter, J.F., Rudolph, A.S., Aurell Wistrom, C., Spargo, B.J. and Anchordoguy, T.J. 1988 Interactions of sugars with membranes. *Biochimica et Biophysica Acta* **947,** 367-384.

Crowshaw, B. 1981 Disinfectant testing — with particular reference to the Rideal-Walker and Kelsey-Sykes tests, pp 1-15, in *Disinfectants: Their Use and Evaluation of Effectiveness* (C.H. Collins, M.C. Allwood, S.F. Bloomfield and A. Fox, Eds), Academic Press, London.

Dallyn, H. and Fox, A. 1980 Spoilage of materials of reduced water activity by xerophilic fungi, pp 129-157, in *Microbial Growth and Survival in Extremes of Environment* (G.W. Gould and J. E. L. Corry, Eds). Academic Press, London.

De Mello, J.V. and Snell, J.J.S. 1985 Preparation of simulated clinical material for bacteriological examination. *Journal of Applied Bacteriology* **59,** 421-436.

de Rizzo, E., Pereira, C.A., Fang, F.L., Takata, C.S., Tenorio, E.C., Pral, M.M., Mendes, I.F. and Gallina, N.M. 1990 Photosensitivity and stability of freeze-dried and/or reconstituted measles vaccines. *Reviews Saude Publications* **24**, 51-59.

de Valdez, G.F., de Giori, G.S., de Ruiz-Holgado, A.P. and Oliver, G. 1985 Effect of the rehydration medium on the recovery of freeze-dried lactic acid bacteria. *Applied and Environmental Microbiology* **50**, 1339-1341.

Dietz, A. 1975 In *Round Table Discussion on Cryogenic Preservation of Cell Cultures* (A.P. Rinfert and A.B. LaSalle, Eds). National Academy of Science, Washington, DC, pp 22-36.

Diller, K.R., Cravalho E.G., and Huggins, C.E. 1972 Intracellular freezing in biomaterials. *Cryobiology* **9,** 429-440.

Fahy, G.M. 1986 The relevance of cryoprotectant "toxicity" to cryobiology" *Cryobiology* **23,** 1-13.

Farrant, J. 1980 General observations on cell preservation, pp 1-18, in *Low Temperature Preservation in Medicine and Biology* (M.J. Ashwood-Smith and J. Farrant, Eds), Pitman Medical, Tonbridge Wells, UK.

Ferris, N.P., Philpot, R.M., Oxtoby, J.M. and Armstrong, R.M. 1990 Freeze-drying foot and mouth disease antigens. I. Infectivity studies. *Journal of Virological Methods* **29,** 43-52.

Franks, F. 1985 *Biophysics and Biochemistry at Low Temperatures.* Cambridge University Press, Cambridge, UK.

Franks, F. 1989 Improved freeze-drying: An analysis of the basic scientific principles. *Process Biochemistry* **24,** iii-vii.

Franks, F. 1990 Freeze-drying: From empiricism to predictability. *Cryo-letters* **11**, 93-110.

Franks, F., Mathias, S.F. and Hatley, R.H. 1990 Water, temperature and life. *Philosophical Transactions of the Royal Society of London. Biology* **326**, 517-533.

Franks, F., Hatley, R.H.M. and Mathias, S.F. 1991 Materals Science and the production of shelf-stable biologicals. *Pharmaceutical Technology International* **3**, 24-34.

Fry, R.M. 1966 Freezing and drying of bacteria, pp 665-696, in *Cryobiology* (H.T. Meryman, Ed). Academic Press, London.

Gareis, P.J., Cowley, C.W. and Gallisdorfer, H.R. 1969 Operating characteristics of biological storage vessels maintained with liquid nitrogen. *Cryobiology* **6,** 45-56.

Gilbert, P., Collier, P.J. and Brown, M.R.W. 1990 Influence of growth rate on susceptibility to antimicrobial agents: Biofilms, cell cycle, dormancy, and stringent response. *Antimicrobial Agents and Chemotherapy* **34,** 1865-1868.

Gomez, R., Takano, M. and Suskey, A.J. 1973 Characteristics of freeze-dried cells. *Cryobiology* **10,** 368-374.

Greaves, R.I.N. 1960 Some factors which influence the stability of freeze-dried cultures, pp 203-217, in *Recent Research in Freezing and Drying*, (A.S. Parkes and A.U. Smith, Eds). Blackwell Scientific Publications, Oxford.

Greiff, D. and Rightsel, W.A. 1965a An accelerated storage test for predicting the stability of suspensions of measles virus dried by sublimation *in vacuo. Journal of Immunology* **94,** 395-400.

Greiff, D. and Rightsel, W. 1965b Stabilities of suspensions of viruses after vacuum sublimation and storage. *Cryobiology* **3,** 435-443.

Greiff, D. and Rightsel, W.A. 1968 Stability of *influenza* virus dried to different contents of residual moisture by sublimation *in vacuo. Applied Microbiology* **16,** 835-840.

Greiff, D. and Rightsel, W.A. 1969 Stabilities of dried suspensions of influenza virus sealed in vacuum or different gases. *Applied Microbiology* **17**, 830-835.

Greiff, D., Melton, H. and Rowe, T.W.G. 1975 On the sealing of gas-filled ampoules. *Cryobiology* **12**, 1-14.

Griffin, C.W., Cook, E.C. and Mehaffey, M.A. 1981 Predicting the stability of freeze-dried *Fusobacterium montiferum.* Proficiency testing samples by accelerated storage tests. *Cryobiology* **18,** 420-425.

Grivell, A.R. and Jackson, J.F. 1969 Microbial culture preservation with silica gel. *Journal of General Microbiology* **58,** 423-425.

Grout, B.W.W. and Morris G.J. (Eds) 1987 *The Effects of Low Temperatures on Biological Systems*, Edward Arnold, London.

Grout, B., Morris, J. and McLellan, M. 1990 Cryopreservation and the maintenance of cell lines. *Tibtech* **8,** 293-297.

Halliwell, B. and Gutteridge, J.M. 1987 *Free Radicals in Biology and Medicine*. Oxford Scientific Publications, Oxford.

Hatley, R.H.M., Franks, F. and Mathias, S.F. 1987 The stabilization of labile biochemicals by undercooling. *Process Biochemistry* Dec. 1987, 169-172.

Haynes, W.C., Wickerham, L.J. and Heseltine, C.W. 1955 Maintenance of industrially important microorganisms. *Applied Microbiology* **3,** 361-368.

Heckly, R.J. 1961 Preservation of bacteria by lyophilization. *Advances in Applied Microbiology* **3,** 1-76.

Heckly, R.J. 1978 Preservation of microorganisms. *Advances in Applied Microbiology* **24,** 1-53.

Heckly, R.J. and Dimmick, R.L. 1968 Correlations between free radical production and viability of lyophilized bacteria. *Applied Microbiology* **16,** 1081-1085.

Heckly, R.J. and Quay, J. 1983 Adventitious chemistry at reduced water activities: Free radicals and polyhydroxy agents. *Cryobiology* **20,** 613-624.

Israeli, E., Giberman, E and Kohn, A. 1974 Membrane malfunctions in freeze-dried *Escherichia coli. Cryobiology* **11,** 473-477.

James, E. R. 1984 Maintenance of parasitic protozoa by cryopreservation, pp 161-175, in *Maintenance of Microorganisms, a Manual of Laboratory Methods* (B.E. Kirsop and J.J.S. Snell, Eds). Academic Press, London.

Janda, I. and Opekarova, M. 1989 Long term preservation of active luminous bacteria by lyophilization. *Journal of Bioluminescence and Chemiluminescence* **3**, 27-29.

Jennings, T.A. 1980 Optimization of the lyophilization schedule. *Drugs and Cosmetics Industry*, Nov. 1980, 43-53.

Jeyendran, R.S., Graham, E.F. and Schmehl, M.K.L. 1981 Fertility of dehydrated bull semen. *Cryobiology* **18,** 292-300.

Kaprelyants, A.S., Gottschal, J.C. and Kell, D.B. 1993 Dormancy in non-sporulating bacteria. *FEMS Microbiology Reviews* **104,** 271-286.

Kirsop, B.E. 1984 Maintenance of yeasts, pp 109-130, in *Maintenance of Microorganisms* (B.E. Kirsop and J.J.S. Snell, Eds). Academic Press, London.

Kirsop, B.E. and J.J.S. Snell 1984 *Maintenance of Microorganisms*. Academic Press, London.

Kupletskaia, M.B. 1987 Results of the storage of freeze-dried microbial cultures for 25 years. *Mikrobiologiia* **56,** 488-491.

Lee, D.C. and Chapman, D. 1987 The effects of temperature on biological membranes and their models, pp 35-52, in *Temperature and Animal Cells* (K. Bowler and B.J. Fuller, Eds). Number XXXXI, Symposia of the Society for Experimental Biology. Company of Biologists Ltd., Cambridge, UK.

Lesson, E.A., Cann, J.P. and Morris, G.J. 1984 Maintenance of algae and protozoa, pp 131-161, in *Maintenance of Microorganisms* (B.E. Kirsop and J.J.S. Snell, Eds). Academic Press.

Levine, H. and Slade, L. 1988 Water as plasticizer: Physico-chemical aspects of low moisture polymeric systems. *Water Science Reviews* **5,** 79-185.

Losick, R. and Youngman, P. 1984 Endospore formation in bacillus, pp 63-88, in *Microbial Development* (R. Losick and P. Youngman, Eds). Cold Spring Harbor Monograph, Cold Spring Harbor, New York.

Lovelock, J.E. 1954 The protective action of neutral solutes against haemolysis by freezing and thawing. *Biochemistry* **56,** 265-270.

Macdonald, K.D. 1972 Storage of conidia of *Penicillium chrysogenum* in liquid nitrogen. *Applied Microbiology* **23,** 990-993.

MacFarlane, D.R. and Forsyth M. 1990 Recent insights on the role of cryoprotective agents in vitrification. *Cryobiology* **27,** 345-358.

Mackenzie, A.P. 1975 Collapse during freeze-drying-qualitative and quantitative aspects, pp 277-307, in *Freeze-Drying and Advanced Food Technology* (S.A. Goldblith, L. Rey and W.W. Rothmeyer, Eds). Academic Press, London.

Mackenzie, A.P. 1977 The physico-chemical basis for the freeze-drying process. *Developments in Biological Standards* **36,** 51-67

Mackenzie, A.P. 1985 A current understanding of the freeze-drying of representative aqueous solutions, pp 21-34, in *Refrigeration Science and Technology: Fundamentals and Applications of Freeze-Drying to Biological Materials, Drugs and Foodstuffs*. International Institute of Refrigeration, Paris.

Mackey, B.M. and Derrick, C.M. 1982 The effect of sublethal injury by heating, freezing, drying and gamma-radiation on the duration of the lag phase of *Salmonella typhimurium*. *Journal of Applied Bacteriology* **53,** 243-251.

May, J.C., Wheeler, R.M., Etz, N. and Del-Grosso, A. 1992 Measurement of final container residual moisture in freeze-dried biological products. *Developments in Biological Standards* **74,** 153-164.

Mazur, P. 1970 Cryobiology: The freezing of biological systems. *Science* **168,** 939-949.

Mazur, P. 1977 Slow-freeze injury in mammalian cells, pp 19-48, in *Ciba Foundation Symposium Number 52,* January 18–20.

Mazur, P., Leibo, S.P. and Chu, C.H.Y. 1972 A two factor hypothesis of freezing injury. Evidence from Chinese Hamster tissue cells. *Experimental Cell Research* **71,** 345-355.

Mazur, P. and Cole, K.W. 1989 Roles of unfrozen fraction, salt concentration and changes in cell volume in the survival of frozen human erythrocytes. *Cryobiology* **26,** 1-29.

McGann, L.E. 1978 Differing actions of penetrating and non-penetrating cryoprotective agents. *Cryobiology* **15,** 382-390.

Mellor, J.D. 1978 *Fundamentals of Freeze-Drying*, Academic Press.

Meryman, H.T. 1966 Freeze-drying, pp 610-660, in *Cryobiology* (H.T. Meryman, Ed). Academic Press, London.

Meryman, H.T. 1968 Modified model for mechanism of freezing injury in erythrocytes. *Nature (London)* **218,** 333-336.

Meryman, H.T., Williams, R.J. and St. J. Douglas, M. 1977 Freezing injury from solution effects and its prevention by natural or artificial cryoprotection. *Cryobiology* **14**, 287-302.

Monk, G.W. and McCaffrey, PA. 1967 Effect of sorbed water on the death rate of *Serratia marcescens. Journal of Bacteriology* **73,** 85-88.

Morichi, T. 1969 Metabolic injury in frozen *Escherichia coli*, pp 53-68, in *Freezing and Drying of Microorganisms* T. Nei (Ed). University of Tokyo Press, Tokyo.

Morichi, T. and Irie, R. 1973 Factors affecting repair of sublethal injury in frozen or freeze-dried bacteria. *Cryobiology* **10,** 393-399.

Morris, G.J. 1987 Chill injury, pp 120-146, in *The Effects of Low Temperatures on Biological Systems.* (B.W.W. Grout and G.J.Morris, Eds). Edward Arnold, London.

Morris, G.J. and Watson, P.F. 1984 Cold shock bibliography. *Cryoletters* **5,** 352-372.

Nagel, J.G. and Kunz, L.J. 1972 Simplified storage and retrieval of stock cultures. *Applied Microbiology* **23,** 837-838.

Nei, T. 1978 Some aspects of freezing and drying of microorganisms on the basis of cellular water. *Cryobiology* **10,** 403-408.

Novick, O., Israeli, E. and Kohn, A 1972 Nucleic acid and protein synthesis in reconstituted *Escherichia coli* exposed to air. *Journal of Applied Bacteriology* **35,** 185-191.

Oliver, G. de Giori, G.S. and de Valdez, G.M. 1988 Cheese industry development and research in Argentina. *Critical Reviews in Food Science and Nutrition* **26,** 225-241.

Orndorf, G.R. and Mackenzie, A.P. 1973 The function of the suspending medium during freeze-drying preservation of *Escherichia coli. Cryobiology* **10,** 475-487.

Owen, R.J., On, S.L. and Costas, M. 1989 The effect of cooling rate, freeze-drying suspending fluid and culture age on the preservation of *Campylobacter pylori. Journal of Applied Bacteriology* **66,** 331-337.

Pegg, D.E. and Diaper, M.P. 1989 The "unfrozen fraction" hypothesis of freezing injury to freezing injury: A critical examination of the evidence. *Cryobiology* **26,** 30-43.

Peri, C. and Pompei, C. 1976 Search of optimum survival conditions for lactic acid bacteria in powders obtained by spray drying of yoghurt. *Rivista di Scienza e Technologia degli Alimenti e di Nutrizione Umana* **6,** 231-236.

Peterz, M. and Steneryd, A.C. 1993 Freeze-dried mixed cultures as reference samples in quantitative and qualitative microbiological examinations of food. *Journal of Applied Bacteriology* **74,** 143-148.

Phillips, B.A., Latham, M.J. and Sharpe, M.E. 1975 A method for freeze drying rumen bacteria and other strict anaerobes. *Journal of Applied Bacteriology* **38,** 319-322.

Phillips, G.O., Harrop, R., Wedlock, D.J., Srbova, H., Celba, V. and Drevo, M. 1981 A study of the water binding in lyophilized viral vaccine systems. *Cryobiology* **18,** 414-419.

Pikal, M.J. 1991a Freeze-drying of proteins I: Process design. *Pharmaceutical Techology International* **3,** 37-43.

Pikal, M.J. 1991b Freeze-drying of proteins II: Formulation selection. *Pharmaceutical Technology International* **3,** 40-43.

Pikal, M.J. 1991c Freeze-drying of pharmaceuticals, pp 111-175, In, *Lyophilization Technology Handbook,* Vol. II. Institute for Applied Pharmaceutical Science, Center for Professional Advancement, East Brunswick, NJ.

Pikal, M.J. and Shah, S. 1992 Moisture transfer from stopper to product and resulting stability implications. *Developments in Biological Standards* **74,** 165-177.

Polge, C., Smith, A.U. and Parkes, A.S. 1949 Revival of spermatozoa after vitrification and dehydration at low temperatures. *Nature (London)* **164,** 666.

Ray, B., Speck, M.L. and Dobrogosz, W.J. 1976 Cell wall lipopolysaccharide damage in *Escherichia coli* due to freezing. *Cryobiology* **13,** 153-160.

Redway, K.F. and Lapage, S.P. 1974 Effect of carbohydrates and related compounds on the long-term preservation of freeze-dried bacteria. *Cryobiology* **11,** 73-79.

Rightsel, W.A. and Greiff, D. 1967 Freezing and freeze-drying of viruses. *Cryobiology* **3,** 423-431.

Rowe, T.W.G. 1970 Freeze-drying of biological materials, pp 61-138, In, *Current Trends in Cryobiology* (A.U. Smith, Ed). Plenum Press.

Rowe, T.W.G. 1971 Machinery and methods in freeze-drying. *Cryobiology* **8,** 153-172.

Rowe, T.W.G. and Snowman, J.W. 1978 *Edwards Freeze-Drying Handbook.* Edwards High Vacuum (publishers), Crawley, West Sussex, UK.

Rudge, R.H. 1984 Maintenance of bacteria by freeze-drying, pp 23-35, in *Maintenance of Microorganisms,* (B.E. Kirsop and J.J.S. Snell, Eds). Academic Press, London.

Scott, W.J. 1960 Mechanisms causing death during storage of dried microorganisms pp 188-202, in *Recent Research in Freezing and Drying* (S. Parkes and A.U. Smith, Eds). Blackwell, Oxford.

Sharp, R.J. 1984 The preservation of genetically unstable microorganisms and cryopreservation of fermentation seed cultures. *Advances in Biotechnological Processing* **3,** 81-109.

Shepard, M.L., Goldston, C.S., and Cocks, F.H. 1976 The H_2O-NaCl-Glycerol phase diagram and its application in cryobiology. *Cryobiology* **13,** 9-23.

Simione, S.P. 1992 Key issue relating to the genetic stability and preservation of cells and cell banks. *Journal of Parenteral Science and Technology* **46,** 226-232.

Simione, F.P., Daggert, P., McGrath, M.S. and Alexander, M.T. 1977 The use of plastic ampoules for freeze-preservation of microorganisms. *Cryobiology* **14,** 500-502.

Smith, D. 1984 Maintenance of fungi, pp 83-107, In, *Maintenance of Microorganisms.* (B.E. Kirsop and J.J.S. Snell, Eds). Academic Press, London.

Soper, C.J., and Davies, D.J.G. 1976 The response of bacterial spores to vacuum treatments. I. Design and characterisation of the vacuum apparatus. *Cryobiology* **13,** 61-70.

Staab, J.A. and Ely, J.K. 1987 Viability of lyophilized anaerobes in two media *Cryobiology* **24,** 174-178.

Strange, R.E. 1976 Microbial response to mild stress. *Patterns in Progress* **6,** 44-61.

Suzuki, M. 1973 Stability and residual moisture content of dried vaccinia virus *Cryobiology* **10,** 432-434.

Takano, M. and Terui, G. 1969 Correlation of dehydration with death of microbial cells in the secondary stage of freeze-drying, pp 131-142, in *Freezing and Drying of Microorganisms*, (T. Nei, Ed). University of Tokyo Press.

Takano, M., Sado, J-I., Ogaa, T. and Terui, G. 1973 Freezing and freeze-drying of *Spirulina platensis. Cryobiology* **10,** 440-444.

Tanaka, Y., Yoh, M., Takeda, Y. and Miwatani, T. 1979 Induction of mutation in *Escherichia coli* by freeze-drying. *Applied and Environmental Microbiology* **37,** 369-372.

Taylor, M.J. 1981 The meaning of pH at low temperature. *Cryobiology* **2,** 231-239.

Taylor, R., Adams, G.D.J., Boardman, C.F.B. and Wallis R.G. 1974 Cryoprotection — permeant vs non-permeant additives. *Cryobiology* **11,** 430-438.

Toupin, C.J., LeMaguer, M. and McGann, L.E. 1989 Permeability of human granulocytes to water: Rectification of osmotic flow. *Cryobiology* **26,** 431-444.

Tsourouflis, S., Flink, J.M. and Karel, M. 1976 Loss of structure in freeze-dried carbohydrate solutions: The effect of temperature, moisture content and composition. *Journal of Science, Food and Agriculture* **27,** 509-519.

Tsuru, S. 1973 Preservation of marine and freshwater algae by means of freezing and freeze-drying. *Cryobiology* **10,** 445-452.

van den Berg, L. and Rose, D. 1959 Effect of freezing on the pH and composition of sodium and potassium phosphate solutions to the reciprocal system KH_2-Na_2HPO_4-H_2O. *Archive Biochimica et Biophysica* **81,** 319-329.

Webb, S.J. 1967 Mutation of bacterial cells by controlled desiccation. *Nature (London)* **213,** 1137-1139

Westerman, D.H. and Huige, N.J. 1979 Beer brewing, in *Microbial Technology*, Vol. II, (H.G. Peppler and D. Perlman, Eds). Academic Press, London.

Wolff, E., Delisle, B. and Corrieu, G. 1990 Freeze-drying of *Streptococcus thermophilus*: A comparison berween the vacuum and atmospheric method. *Cryobiology* **27,** 569-575.

Yamasato, K., Okuno, D. and Ohtomo, T. 1973 Preservation of bacteria by freezing at moderately low temperatures. *Cryobiology* **10,** 453-463.

2.2 Influence of Post-Growth Procedures on the Properties of Microorganisms

(Out of the Test Tube into the Frying Pan)

Peter Gilbert, Phillip J. Collier, Julie Andrews, and Michael R.W. Brown

INTRODUCTION

Many procedures are adopted within standard tests, and within routine laboratory experiments, which prepare inocula for performance tests, after their growth in suitable media (see Growth Conditions Section). These procedures, perceived to increase the reproducibility of the response of the inoculum, are also intended to standardize and possibly concentrate the cell density, to replace the growth medium with a suitable test menstrum, and also to separate the cells from associated "debris" which might influence the outcome of the chosen study (Gilbert and Brown, 1991). Washed cell suspensions, for example, routinely prepared by successive centrifugation and resuspension steps, are often used in evaluations of antimicrobial effectiveness and studies of microbial physiology. Each of these procedures, intended to increase reproducibility, has possible adverse effects on the cells, and might influence outcome when they are challenged in subsequent tests. The results of the tests might be highly reproducible yet irrelevent.

PRETREATMENT OBJECTIVES

It is widely established that the cell density within a test system has a marked effect upon the result, particularly in assessments of antimicrobial

0-8493-4752-1/95/$0.00+$.50

effectiveness. Thus, determinations of minimum growth-inhibitory concentration often stipulate that no more than 10^5 cells are added to each tube/well. This is intended not only to eliminate any quenching of the antimicrobial, by added cell mass, but also to minimize the genotypic pool represented within the inoculum. Similarly, the shapes of survival curves after exposure of test suspensions to bactericidal agents are often changed by the initial cell numbers, making valid comparisons of D-value unlikely. In order to standardize cell density, procedures are generally adopted that relate cell numbers to optical opacity/density and subsequently either dilute the cells, with fresh medium or diluent, or concentrate them. Concentration of the cells might be achieved by centrifugation or filtration processes. These processes have the added advantage that they separate the cells from their growth medium and thereby facilitate change/control of the test suspension medium. Care must be taken, however, that the separation procedures are not themselves damaging to the cells and also that the resuspension/dilution processes are not injurious (below).

The suspension medium in which an inoculum of cells is challenged or tested is often standardized with repect to its pH, tonicity, and temperature in an attempt to mimic the application of the product/biocide. Cells are variously suspended in sterile distilled water, "standard hardness water", normal saline, 1/4 strength Ringer's solution, and various buffers, in attempts to standardize/control the test. Additionally these suspension media often lack the nutrients which are necessary to sustain the viability of the inoculum (Chapter 3.1). The cells which are included within the test suspension have, on the other hand, generally been grown under very different conditions, chosen to maximize and optimize their growth. Cells taken from an actively growing culture might, therefore, in the quest for relevance and reproducibility, have been inadvertantly subjected to sudden starvation by removal of the carbon substrate (Chapter 1.3), osmotic shock through changes in tonicity of their environment, and temperature shock (below). During a lengthy process of inoculum preparation, the organisms might be subjected to any number of such stresses relating to the resuspension and "washing" of the cells. Once prepared, suspensions of cells are, in extreme circumstances, stored in a refrigerator, or on the bench, for many hours, if not days, before use. This further exacerbates problems associated with relevance and reproducibility for the performance of the inoculum (below and Chapter 1.7).

METHODS OF HARVESTING OF CULTURED BACTERIA

CENTRIFUGATION

Washed cell suspensions are routinely prepared, for evaluations of antimicrobial effectiveness and for studies of microbial physiology, by successive centrifugation and resuspension steps. Indeed, centrifugation procedures may even be used to separate intact cells from their associated exopolymers (Allison and Sutherland, 1984). Centrifugal separation of the cells from their suspending medium subjects them to *g*-forces commonly in the range 5 to 15 k*g*, which in centrifuge tubes also translates to high hydrostatic pressures. During the spin, the cells are additionally exposed to shear stresses as they are propelled through the liquid. Successive decompression and resuspension in various diluents could lead to cells with surface characteristics grossly distorted from those typical of the growth system from which they were derived. These changes could, in turn, greatly affect the outcome of subsequent tests.

There are conflicting reports in the literature concerning the sensitivity of cells to centrifugation. Farwell and Brown (1971) comprehensively reviewed the various and significant effects of inoculum history upon the results of tests on antimicrobial susceptibility. Little data existed on centrifugation injury, and they concluded that such damage was slight. Smith and Weiss (1969) and Winslow and Brooke (1927) reported progressive losses in the viabilities of *Azotobacter* spp., and *Bacillus cereus, Bacillus megaterium,* and *Serratia marcescens* respectively, when subjected to centrifugation stress. More recently, Gilbert, Caplan and Brown (1991) have shown that significant reductions in viability are observed for logarithmic phase *Pseudomonas aeruginosa* populations (Figure 1) when subjected to centrifugation (*g*-forces of 5 to 20 k for 5 to 20 min). In contrast, no loss in viability was observed for either stationary phase *P. aeruginosa* or for any cultures of *Escherichia coli* or *Staphylococcus epidermidis*. Failure to observe decreases in viability does not, however, mean that the cells are uninjured by the treatment (Gilbert, 1984).

Sublethal injury of microorganisms implies damage to structures within the cells, the expression of which entails some loss or alteration of cellular function (Gilbert, 1984). Injuries can take many forms and the consequences of them to the organisms can vary, from negligible to severely debilitating, according to their physiological environment. Sublethally injured cells are unable to grow under certain environmental conditions and are usually more susceptible than noninjured cells to secondary stresses, such as selective agents (Roberts and Ingram, 1966; Corry, Kitchell and Roberts, 1969). Sublethal injury is well documented within the survivors of populations subjected to inimical processes such as heat-sterilization. Sublethal injury of cells during inoculum preparation might alter the results of tests subsequently performed upon them.

Injuries can be regarded as metabolic and/or structural (Beauchat, 1978). For enumeration purposes, metabolic injury is often taken as an inability to form colonies on minimal salts media, while retaining colony-forming ability in the presence of complex nutrients. Sublethal structural injury can be taken as an inability to proliferate or survive in media containing selective agents that have no apparent inhibitory action upon unstressed cells. By their nature, media selective for structurally injured cells are heavily biased towards damage to the cytoplasmic membrane and cell wall, and detect failure of the envelope adequately to osmoregulate the cell or to exclude toxic materials. Differential media such as these have been widely used to identify sublethal

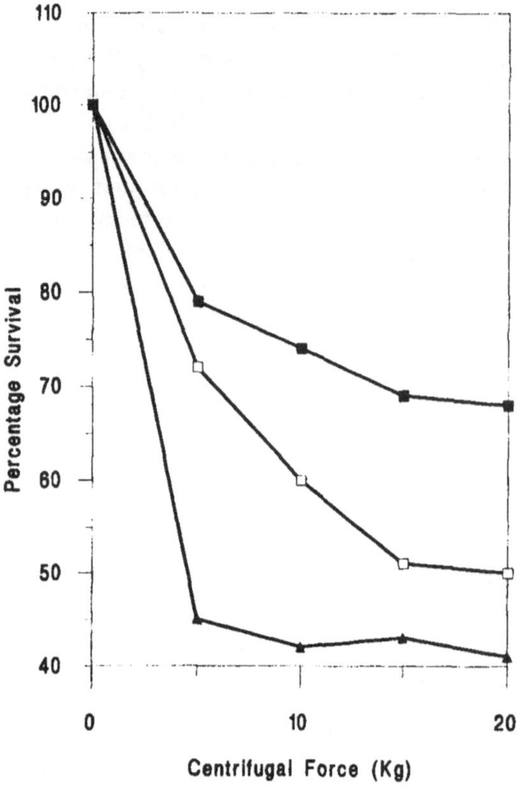

FIGURE 1 The effects of centrifugal force (■, 5 kg; □, 10 kg; ▲, 20 kg) and centrifugation time on the survival of exponential phase *Pseudomonas aeruginosa*, grown in nutrient broth. (Data from Gilbert, P., Caplan, F., and Brown, M.R.W., *Journal of Antimicrobial Chemotherapy*, 27, 859–862.)

injuries induced through treatment of cells with physical agents such as heat (Allwood and Russell, 1970; Adams, 1978), freezing (Ray and Speck, 1973), freeze-drying (Gomez et al., 1973) and irradiation (Bridges, 1976; Krinsky, 1976) as well as by chemicals (Nadir and Gilbert, 1982) and preservatives (Corry, van Doorne and Mossell, 1977). Figure 2 shows the effects, upon a mid-logarithmic phase, nutrient broth culture of *Escherichia coli*, of a 20 min centrifugation at various *g*-forces. Significant sublethal injury was detected at all the *g*-forces tested and was observed to increase in extent, from ca. 30% to 98% as the *g*-force was increased. Ability to form colonies on the control nutrient agar plates and thereby viability of the inoculum as conventionally defined was unaffected by this treatment. Clearly, if the ability to survive a relatively mild salt stress is so affected by centrifugation then the susceptibility towards chemical agents such as biocides and antibiotics will be markedly changed.

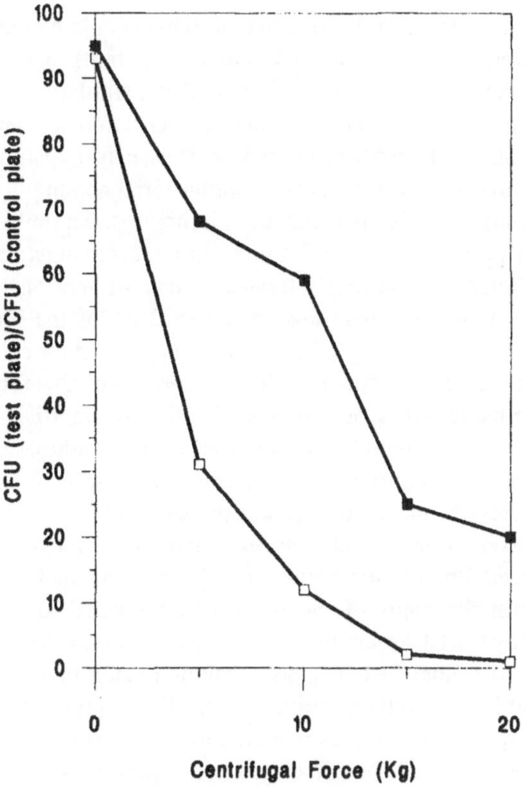

FIGURE 2 Effect of centrifugation (20 min) at various *g* forces upon the ability of *Escherichia coli* to form colonies on nutrient agar plates containing either 2.0% w/v NaCl (■) or benzalkonium chloride, 6×10^{-4}% w/v (□). Colony-forming ability is expressed as a percentage relative to colony formation of similarly treated suspensions on control nutrient agar. (Unpublished data from P. Gilbert.)

In this respect, Gilbert, Pemberton, and Wilkinson (1990) reported that while the viability of log and stationary phase cultures of *Escherichia coli* was unaffected by centrifugation at 10 k*g*, significant amounts of lipopolysaccharide were shed to the medium when cells were harvested in this way. This was also associated with dramatic increases (100-fold) in the susceptibility of the centrifuged cells towards polymeric-biguanide antiseptics. In this instance cells harvested and washed using filter membranes were recalcitrant to concentrations of biocides which gave measurable and significant bactericidal activity towards centrifuged preparations. Similarly, Gilbert, Caplan, and Brown (1991), studying the bactericidal effects of cetrimide and chlorhexidine upon *P. aeruginosa*, noted a significant dependence of such activity upon the centrifugation forces and times employed in inoculum preparation. In studies of the antibacterial potency of some glycolmonophenylethers, Gilbert et al. (1978)

noted that while significant bactericidal activities could be measured for four of these analogs (D-values at 35°C ca. 2 to 5 min), against washed cell suspensions prepared by centrifugation at 15 k*g*, similar concentration failed even to slow the growth in batch culture. Suspension media and cell densities were kept constant between the two studies. Thus, centrifugation clearly affects the results of many *in vitro* studies of antibacterial action.

It is not only salt tolerance and antimicrobial susceptibility that might be influenced by post growth centrifugation. In a recent study, examining transformation efficiency of variously treated *E. coli* suspensions, the number of transformants, following exposure on ice to 2 ng of the plasmid pBR322 (AmpR) and heat shock (42°C, 2 min), were compared for cells harvested by filtration or at various centrifugation force. Numbers of transformants obtained from the filter-harvested suspension were in all cases significantly greater than the centrifuged ones (Figure 3, unpublished data J. Andrews and P. Gilbert).

The nature of the injury induced by centrifugal stress remains unclear. At least three effects of the centrifugation process could be implicated: (1) the shear forces exerted on the cells through their forced movement through the suspending medium, (2) distortion of the cells and their envelopes when pelleted against the walls of the tubes and subjected to increased *g*, and (3) increased hydrostatic pressures caused by *g* and the head of liquid above the cell pellet. In the latter instance, suspensions treated in 15 ml tubes with a 10 cm head of liquid would be subject to ca. 10 bar pressure (100 m head) at 10 k*g*. There have been a number of studies made, particularly in the area of marine microbiology, of the effects of hydrostatic pressure upon microbial physiology (Dring, 1976). In this respect cells incubated under high (300 bar) pressures grow more slowly (Zobell and Johnson, 1949), are often morphologically altered (e.g., filamentation) (Zobell and Cobet, 1964) and are often more susceptible to some antibiotics (Marquis, 1973). Additionally, rapid compression and decompression (up to 500 bar) has been reported to reduce the viability of some microorganisms (Foster, Cowan and Maag, 1962). Pressures required in these studies to affect microbial physiologies were significantly greater than those associated with centrifugation. The injurious stresses reported in the various studies discussed here are, perhaps, more likely to be associated with shear of the cell surface and distortion of the envelope of the pelletted cells than with hydrostatic pressures.

In view of the widespread use of centrifugation in the preparation of microbiological inocula, and indeed biology in general, it is suprising that so few reports of centrifugation-induced injury appear in the microbiological literature.

Filtration

Filtration is often used as an alternative to centrifugation for the separation of bacterial cells from growth medium. Cells are collected onto

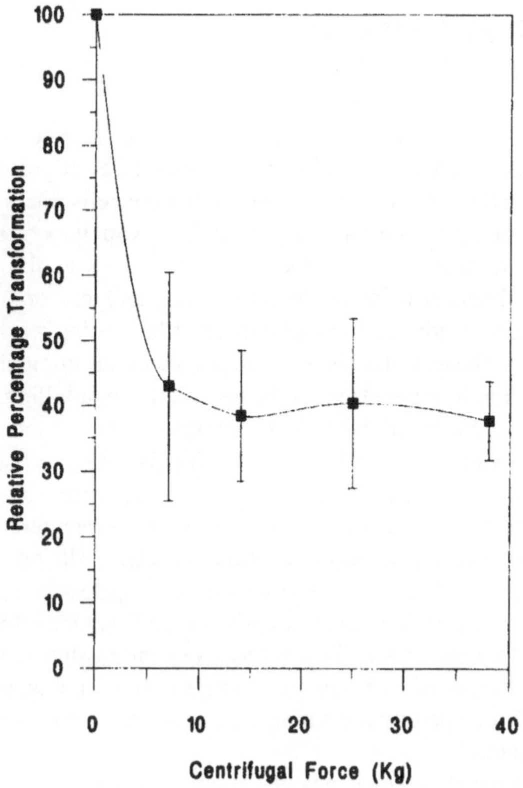

FIGURE 3 Effect of centrifugation (20 min) at various *g* forces upon the transformation efficiency of *Escherichia coli*. The number of transformants is expressed as a percentage relative to the number of transformants of similarly treated competent cells prepared by filtration (Unpublished data from J. Andrews and P. Gilbert.)

bacteria-proof filter membranes (<0.45 μm porosity) by the application of negative pressure to the underside of the filter or positive pressure to the culture. Cells become entrapped within the filter matrix as the spent cell-free medium passes through. The entrapped cells may be washed, to remove residual traces of medium, by passing appropriate volumes of diluents through the filter. Provided that care is taken with the choice of washing and resuspension media such that temperature, nutrient, and osmotic shocks are minimized, then such processes are not expected to elicit any damaging effect. In this respect data presented in Figure 3 shows the transformation efficiency of filtered *E. coli* to be 2.5 times greater than for centrifuged cells. Membrane filters become quickly blocked when challenged with moderate numbers of bacterial cells. Filteration is not, therefore, an appropriate method of cell separation when high cell densities are required. The separated cells may be collected and resuspended from the membrane by scraping and vortexing.

WASHING AND RESUSPENSION OF HARVESTED CELLS

After harvesting it is common practice for the cells to be given one or more washes in order to cleanse them of any residual traces of growth media, waste products, or cell debris. Perceived as cleaning the cells, these processes actually remove many of the naturally occurring exopolymers and appendages associated with them. The degree of injury relates to the severity of the harvesting and resuspension method, the nature and temperature of the washing/resuspension medium, and the number of harvesting/washing steps.

Clumps or clusters of cells are undesirable in bactericidal determinations since they will reduce the extent of the observed rate of killing. Many standard protocols, therefore, require that the pellets resulting from centrifugation or the filter membranes be vigorously vortexed with glass beads or subjected to mild sonication treatment. Curiously, both of these approaches are also used to prepare cell-free extracts, and similar treatments are reported to remove extracellular slime layers and polysaccharide capsules (Allison and Sutherland, 1984). Repeated reflux of suspensions through pipettes is also often used to facilitate the resuspension of centrifuged cell pellets. One effect of such treatment will be to subject the cells to successive compression and decompression analogous to the use of a French press. There have been few, if any, studies on the effects of such potentially damaging treatments on the performance properties of the cells.

Washes are undertaken in a wide variety of diluents, including, distilled or deionized water, saline, buffer solutions, buffered saline solutions, fresh or spent culture media, and various biological fluids (e.g., blood, serum, urine, albumin, etc.). It is unlikely that these diluents will exactly match the physicochemical environment of the cells at the moment they were harvested. The possibility exists, therefore, that the cells might be subjected to a variety of stresses, including cold-shock, heat-shock, osmotic-shock, and stringent response to starvation. Stringent responses will increase expression on extended storage of the suspension. Storage time and conditions must also, therefore, be considered.

INJURY RESULTING FROM COLD-OSMOTIC SHOCK

Injury or death of microorganisms following their subjection to rapid chilling is termed "cold-shock" (MacLeod and Calcott, 1976), and might result from the inadvertant use of diluents taken from a refrigerator for use with suspensions grown and held at 35°C or the transfer of cells to a chilled centrifuge rotor. Sudden changes in the osmolarity of the suspending medium are termed "osmotic-shock", and will cause plasmolysis when cells are transferred to more hypertonic environments, or increased intracellular turgor pressure when cells are transferred to more hypotonic environments. While in both instances damage to the cell envelope results, causing an increased susceptibility of the cells towards other stresses and antimicrobial agents, transfer to hypotonic media induces an additional loss of many periplasmic enzymes. Gram-negative bacteria are regarded as being particularly susceptible to osmotic-shock, cold-shock, and cold-osmotic-shock (Brown, 1971; Gorrill and McNeil, 1960). The extent of cellular injury resulting from changes in temperature and osmolarity of the suspension medium can be minimized by reducing the rate of change (Farwell and Brown, 1971).

The primary events leading to cold-shock-mediated death are thought to involve disorganization of the outer (Brown, 1975) and cytoplasmic membranes (Leder, 1972; Meynell, 1958; Strange and Dark, 1962) through phase changes in the membrane lipids (Farrell and Rose, 1968). Since this leads to a decreased osmoregulatory capability of the cells (Adikari, 1975; Gilbert, Dickinson and Brown, 1979), associations have often been made between cold- and osmotic-shock. Autodegradation of nucleic acid polymer (Smeaton and Elliot, 1967) and inactivation of DNA ligase (Sato and Takahashi, 1970) have also been implicated as mechanisms of cold-shock-mediated killing.

Phenotypic variation in the extent of cold-shock-mediated killing is widely reported. In batch culture, mid-log phase cultures are most sensitive (Adikari, 1975; Hegarty and Weeks, 1940), with the extent of killing being dependent upon the complexity of the media (Farrell and Rose, 1968; Meynell, 1958) and growth temperature (Farrell and Rose, 1968; Ring, 1965). In continuous cultures, direct relationships between the extent of cold-shock-mediated cell death and specific growth rate have been observed (Green, 1978; Kenward and Brown, 1978). In such instances, time–survivor curves were biphasic (Green, 1978) with the extent of the biphasicity increasing with increasing growth rate. This has led to the suggestion that sensitization to cold-shock is related to a cell-cycle event such as DNA replication or cell-wall constriction (Gilbert, Dickinson and Brown, 1979).

Sensitization to cold-shock has been related to inactivation of magnesium-dependent DNA ligase (Sato and Takahashi, 1970), which in turn activates endonucleases.

Starvation Storage of Cell Suspensions

Once cells have been removed from their growth environment then they will become rapidly deprived of nutrients, not only for further growth, but also for endogenous metabolism and survival. The responses of bacterial populations to sudden starvation are currently extensively studied and various aspects of it have been reviewed elsewhere in this volume (Chapters 1.2 and 1.3). In terms of the use of challenge suspensions for performance tests, the lesson to be learned is that any procedure which extends the time between the harvesting of a culture and its use in a test will tend to alter the response demonstrated. Even time intervals relatively short by human standards (e.g., lunch) are often considerable in the bacterial time scale. Messenger RNA is turned over many times faster in prokaryotic cells than it is in eukaryotic ones, stringent response genes can be transcriptionally activated in minutes, and, as a consequence, the properties of the cells may become changed very rapidly.

REFERENCES

Adams, D.M. (1978) Heat injury of bacterial spores. *Advances in Applied Microbiology* **23,** 245-261.

Adikari, P.C. (1975) Sensitivity of cholera and El Tor vibrios to cold shock. *Journal of General Microbiology* **87,** 163-166.

Allison, D.G. and Sutherland, I.W. (1984) The role of exopolysaccharides in adhesion of freshwater bacteria. *Journal of General Microbiology* **133,** 1319-1327.

Allwood, M.C. and Russell, A.D. (1970) Mechanisms of thermal injury in non-sporulating bacteria. *Advances in Applied Microbiology* **12,** 89-119.

Beauchat, L.R. (1978) Injury and repair of Gram-negative bacteria, with special consideration of the cytoplasmic membrane. *Advances in Applied Microbiology* **23,** 219-241.

Bridges, B.A. (1976) Survival of bacteria following exposure to ultraviolet and ionising radiations, pp. 183-208, in *The Survival of Vegetative Microbes* (Eds. Gray, T.R.G. and Postgate, J.R.). Cambridge, UK, Cambridge University Press.

Brown, M.R.W. (1971) Inhibition and Destruction of *Pseudomonas aeruginosa.* p 358, in *Inhibition and Destruction of the Microbial Cell* (Ed. W.B. Hugo) London, Academic Press.

Brown, M.R.W. (1975) The role of the cell envelope in resistance, p. 97, in *Resistance of* Pseudomonas aeruginosa. (Ed. Brown, M.R.W.) John Wiley, Chichester.

Corry, J.E.L., Kitchell, A.G. and Roberts, T.A. (1969) Interactions in the recovery of *Salmonella typhimurium* damaged by heat or gamma radiation. *Journal of Applied Bacteriology* **32,** 415-428.

Corry, J.E.L., van Doorne, H. and Mossell, D.A.A. (1977) Recovery and revival of microbial cells, especially those from environments containing antibiotics, pp 174-196, in *Antibiotics in Agriculture* (Ed. Woodbine, M.). London, Butterworths.

Dring, G.J. (1976) Some aspects of the effects of hydrostatic pressure on microorganisms, pp 257-277, in *Inhibition and Inactivation of Vegetative Microbes* (Eds. Skinner, F.A. and Hugo, W.B.). London, Academic Press.

Farwell, J.A. and Brown, M.R.W. (1971) The influence of inoculum history on the response of microorganisms to inhibitory and destructive agents, pp 703-752, in *Inhibition and Destruction of the Microbial Cell* (Ed. Hugo, W.B.). London, Academic Press.

Farrell, J. and Rose, A.H. (1968) Cold-shock in a mesophilic and psychrophilic pseudomonad. *Journal of General Microbiology* **50,** 429-439.

Foster, J.W., Cowan, R.M. and Maag, T.A. (1962) Rupture of bacteria by explosive decompression. *Journal of Bacteriology* **83,** 330-334.

Gilbert, P. (1984) The revival of microorganisms sub-lethally injured by chemical inhibitors, pp 175-197, in *The Revival of Injured Microbes* (Eds. Andrew, M.H.E. and Russell, A.D.). London, Academic Press.

Gilbert, P. and Brown, M.R.W. (1991) Out of the test-tube into the frying pan: post-growth, pre-test variables. *Journal of Antimicrobial Chemotherapy* **27,** 859-862.

Gilbert, P., Beveridge, E.G., Byron, P., Boyd, I. and Crone, P.B. (1978) Correlations between the physico-chemical properties and antimicrobial activity for some glycolmonophenyl ethers. *Microbios* **22,** 203-216.

Gilbert, P., Caplan, F. and Brown, M.R.W. (1991) Centrifugation injury of microorganisms. *Journal of Antimicrobial Chemotherapy* **27,** 550-551.

Gilbert, P., Dickinson, N.A. and Brown, M.R.W. (1979) Interrelation of DNA replication, specific growth rate and growth temperature in the sensitivity of *Escherichia coli* to cold-shock. *Journal of General Microbiology* **115,** 89-94.

Gilbert, P., Pemberton, D. and Wilkinson, D.E. (1990) Synergism within polyhexamethylene biguanide biocide formulations. *Journal of Applied Bacteriology* **69,** 593-598.

Gomez, R.F., Sinskey, A.J., Davies, R. and Labuza, T.P. (1973) Minimal medium recovery of heated *Salmonella typhimurium* LT2. *Journal of General Microbiology* **74,** 267-274.

Gorrill, R.H. and McNeil, E.M. (1960) The effect of cold diluent on the viable count of *Pseudomonas pyocyanea. Journal of General Microbiology* **22,** 437-442.

Green, J.A. (1978) The effects of nutrient limitation and growth rate in the chemostat on the sensitivity of *Vibrio cholera* to cold shock. *FEMS Microbiology Letters* **4,** 217-219.

Hegarty, C.P. and Weeks, O.B. (1940) Sensitivity of *Escherichia coli* to cold shock during logarithmic growth phase. *Journal of Bacteriology* **39,** 475-484.

Kenward, M.A. and Brown, M.R.W. (1978) Relation between growth rate and sensitivity to cold shock of *Pseudomonas aeruginosa. FEMS Microbiology Letters* **3,** 17-19.

Krinsky, N.I. (1976) Cellular damage initiated by visible light, pp 209-240, in *The Survival of Vegetative Microbes* (Eds. Gray, T.R.G. and Postgate, J.R.). Cambridge, UK, Cambridge University Press.

Leder, I.G. (1972) Interrelated effects of cold shock and osmotic pressure on the permeability of the *Escherichia coli* membrane to permease accumulated substrates. *Journal of Bacteriology* **111,** 211-219.

Macleod, R.A. and Calcott, P.H. (1976) Cold shock and freezing damage to microbes. *Society for General Microbiology Symposium* **26,** 81-109.

Marquis, R.E. (1973) The physiological bases for microbial barotolerance. Office of Naval Research, Contract N00014-67-A-0398-0013. Work Unit No. NR 136-924. Technical Report Number 1.

Meynell, G.G. (1958) The effect of sudden chilling on *Escherichia coli. Journal of General Microbiology* **19,** 380-389.

Nadir, M.T. and Gilbert, P. (1982) Injury and recovery of *Bacillus megaterium* from mild chlorhexidine treatment. *Journal of Applied Bacteriology* **52,** 111-115.

Ray, B. and Speck, M.L. (1973) Freeze-injury in bacteria. *Critical Reviews in Clinical Laboratory Sciences* **4,** 161-213.

Ring, K. (1965) The effect of low temperatures on permeability in *Streptomyces hydrogenans. Biochemica et Biophysica Research Communications* **19,** 579-581.

Roberts, T.A. and Ingram, M. (1966) The effect of sodium chloride, potassium nitrate and sodium nitrate on the recovery of heated bacterial spores. *Journal of Food Technology* **1,** 147-163.

Sato, M. and Takahashi, H. (1970) Cold-shock of bacteria IV. Involvement of DNA ligase reaction in recovery of *Escherichia coli* from cold shock. *Journal of General and Applied Microbiology* **16,** 279-290.

Smeaton, J.R. and Elliot, W.H. (1967) Selective release of ribonuclease-inhibitor from *Bacillus subtilis* cells by cold-shock treatment. *Biochemica et Biophysica Research Communications* **26,** 75-81.

Smith, D.D. and Weiss, O. (1969) The rapid loss of viability of *Azotobacter* in aqueous solutions. *Antonie van Leewenhoek* **35,** 84-96.

Strange, R.E. and Dark, F.A. (1962) Effect of chilling on bacteria in aqueous suspension. *Journal of General Microbiology* 29, 719-730.

Winslow, C.E.A. and Brooke, O.R. (1927) The viability of various species of bacteria in aqueous suspensions. *Journal of Bacteriology* **234,** 705-709.

Zobell, C.E. and Cobet, A.B. (1964) Growth, reproduction and death rates of *Escherichia coli* at increased hydrostatic pressures. *Journal of Bacteriology* **84,** 1228-1236.

Zobell, C.E. and Johnson, F.H. (1949) The influence of hydrostatic pressure on the growth and viability of terrestrial and marine bacteria. *Journal of Bacteriology* **57,** 179-189.

Applications

3.1 Factors Affecting the Reproducibility and Predictivity of Performance Tests

Peter Gilbert and Michael R.W. Brown

INTRODUCTION

In the preceding chapters consideration has been given to those factors, relating to the growth of microrganisms, which influence their physiological status. Accordingly, detailed analyses have been presented of the growth of microorganisms in batch and continuous culture (Chapters 1.2 and 1.4), starvation responses (Chapter 1.3), dormancy (Chapter 1.5), and the effects of adhesion to surfaces (Chapter 1.6). Each of these has been considered in terms of the phenotypic and genotypic change that they might facilitate. The objective of the present chapter is to put this information into a context of inocula preparation for performance tests. These must not only be reproducible in their response, but they must also respond to the tests in a meaningful way. While much of what will be described can be inferred from the literature, few direct studies have been made, or accounts given, relating to the reproducibility and predictiveness of microbial challenge inocula. This chapter is intended to redress the balance, others address the key issues of post-growth, pretest treatment of the inoculum (Chapter 2.2), and storage of test strains (Chapter 2.1).

REPRODUCIBILITY AND PREDICTIVITY OF INOCULA GROWN IN BATCH CULTURE

Nutritionally rich, complex growth media have evolved and become commonplace in our laboratories, not only because they will support the growth of a wide range of species (i.e., multipurpose media), but also because they maximize the rates of growth. This is often for the convenience of the laboratory

0-8493-4752-1/95/$0.00+$.50

and its staff, rather than part of a considered approach to definition of the inoculum (Gilbert, Brown and Costerton, 1987). The growth of microorganisms in batch culture and the consequences of closed nutrient environments have been described extensively in other contributions to this volume (Chapters 1.2 and 1.3). In essence, growth commences in the presence of nutrients which, in terms of their nature and availability, meet and probably exceed the requirements of the organisms. The initial, adaptive phase of growth (lag), customizes the cells for the available sources of critical nutrients. The tendency, after this, to define cultures as early, mid- or late-logarithmic phase is a recognition that, in complex media, the physiology of the organisms changes as the growth of the culture progresses. Typically inocula for test procedures are cultured in either the logarithmic or the stationary phase of growth.

Reproducibility of Logarithmic Phase Inocula

Once growth and multiplication of the cells has started then the concentrations of each nutrient will steadily decrease, and secondary metabolites and waste products of growth will accumulate. So long as the concentration of each critical nutrient remains well in excess of its Ks, then exponential increases in cell number/biomass will be observed, within the population. Under such circumstances the rate of growth will be at μ_{max} for the nutrient which is present and utilized most slowly (i.e., the growth-rate-limiting step involves some aspect of the uptake and metabolism of that substrate — Monod, 1950; Herbert, Ellsworth and Telling, 1956). While μ_{max} is, by definition, the specific growth rate of a population when all nutrients are in excess (see Chapters 1.2 and 1.4), it will vary according to the origins of each essential nutrient. Thus, growth rates will change if glucose is supplemented by glycerol or succinate, or if nitrogen is provided as a mixture of amino acids rather than as inorganic ammonium salts. Thus, it is common for cells taken from their log phase, within different complex media, to have substantially different growth rates, and consequently different physiologies.

The rate of consumption of each nutrient from the medium will depend upon the growth rate of the cells and the conversion efficiency of that nutrient into cellular mass (molar yield, Y). Thus, carbon and nitrogen sources will be consumed more rapidly than trace metals, which will disappear only slowly from the growth medium. Concentrations of nutrients which restrict the rate of growth of individual cells are independent of their molar yield. Thus, during the initial phase of growth in batch culture, rates of growth might be limited by the available concentration of a nutrient which possesses a high Ks and high molar yield. As growth proceeds, then nutrients with lower molar yields will be consumed more rapidly. When the available concentration of one of these reduces to a level where it becomes rate limiting for growth, then conditions in the culture will change. In such instances the growth curve would appear to be bi- or multiphasic (Morita, 1988). If inocula are harvested close to or during such a transition period, then day to day, batch to batch reproducibility of the inoculum will suffer adversely. Multiphasicity such as this should not be confused with diauxy, where a number of alternative carbon or nitrogen sources are used succesively by the cells with interjoining secondary lag periods.

If inocula are to be prepared from batch cultures, then care must be exercised that the cells are consistently harvested when they are in an unambiguous physiological state. Clearly the medium that is employed must be fully understood.

Provided that definition of the medium is not mistakenly confused with a complete description of the medium, then chemically defined growth conditions can offer such a possibility. In this respect there are numerous defined media described in the literature which simply list the ingredients together with their starting concentrations. Through detailed analysis, it would be possible fully to define Brain Heart Infusion Broth, but such definition would not make the cells, harvested in mid-log phase, any more reproducible from batch to batch.

Definition of a medium (see also Chapter 1.2) should ensure that a full characterization of the organism's requirements for growth has been made. Particularly such a characterization should ensure that

1. Oxygen demand by the culture does not, at any time during the growth, exceed the rate of dissolution of oxygen into the medium.
2. A single nutrient (i.e., carbon or nitrogen), derived from a single source, should determine the rate of growth of the culture throughout the log phase.
3. Each element required for growth is gained from a single source, or all of the available sources are present to an excess of the overall culture requirements (i.e., still in excess in stationary phase).
4. The medium is sufficiently buffered that pH does not change.

Oxygen restriction will not cause stationary phase; rather, growth will proceed at a rate determined by the rate of dissolution of oxygen to the culture. This is in turn affected by, for example, flask shake rates, size, shape, and contents. As growth slows and cell density increases, then the steady-state oxygen concentration will fall. This will give rise to a protracted onset of a pseudostationary phase of growth, with cells becoming progressively more oxygen deprived. Oxygen-limited growth is difficult to define in batch culture and should therefore be discouraged if reproducibility of inocula is sought. Particularly, facultative anaerobes, such as *Escherichia coli*, will adapt their physiology from oxidative to fermentative as oxygen becomes scarce. This change in physiology will affect the performance of the inocula in challenge tests. In nutritionally rich, complex media, oxygen is often the cause of stationary phase. It is noteworthy that for many of these media, organisms may be removed from stationary-phase cultures, by filtration or centrifugation, permitting regrowth in the medium up to similar cell densities as were obtained in the primary culture.

Provided that the conditions listed above (1–4) are met, then cells taken in logarithmic phase will have a reproducible physiology for use in most performance tests. This is provided that the number of divisions from inoculation is defined, and that growth at the time of harvesting is sufficiently removed from the onset of stationary phase that nutrient limitation/depletion is not imminent. It has been suggested (Domingue, Schwarzinger and Brown, 1989) that at least three generations should have occurred in logarithmic phase, and that at least three generations remain to onset of stationary phase, in order for the phenotype characteristic of the medium to be expressed.

Such conditions are easier to establish in minimal media (Brown and Williams, 1985a, b). Some of these are media described, for a range of organisms, in Chapters 1.2 and 1.6. In these, all nutrients with the exception of one (the growth-limiting nutrient), are present to a 5 to 10 times excess of that required to take growth to the stationary phase. The nutrient causing onset of stationary phase is present to a concentration which allows growth to an optical density (E_{470}) of 1 or less. This cell density (ca. 5×10^9) is critical, in that, for shake-flask cultures, growth of strongly aerobic organisms such as *Pseudomonas* to cell densities above this will become limited by the availability of oxygen.

Harvesting cells in their logarithmic phase of growth is convenient and can lead to reproducible performance of the inocula. It should be remembered, however, that the harvesting procedures often separate cells from the growth medium (Chapter 2.2). This will impose nutrient stresses far in excess of those usually experienced by the cells as they enter stationary phase. While many of the properties of the suspension will reflect those of the original culture (growth rate, etc.), cells will undergo stringent responses to sudden imposition of starvation conditions (Chapter 1.3; Kolter, Siegele and Tormo, 1993). Expression of the stringent response will increase with the length of time the suspension is stored prior to use and will be affected by the composition of the suspension medium. In order to increase reproducibility of the performance characteristics of the inoculum, harvesting by removal from a growth environment should be avoided. It would be preferable, wherever possible, to use the untouched, mid-log-phase culture directly as the inoculum for testing.

Predictivity of Logarithmic Phase Inocula

Unfortunately, most complex media have been designed for convenience rather than relevance. Media suit the easy life-style of inoculations being made at the end of one working day and being used at the beginning of the next (i.e., the ubiquitous 16 h overnight stationary phase culture). This is justified by a belief that an organism grown at or near to its maximum rate has grown optimally. In truth such cultures have been so pampered that they are unrepresentative of the same organism in the real world. Under such circumstances the organism can grow rapidly only because it does not require the expression of the many biochemical systems required for growth under less "optimal" conditions. The produced cells are, in reality, deficient in many critical physiologies. If these physiologies influence the outcome of the screeening process then misleading results will be obtained.

The physiology and susceptibility of actively dividing bacteria towards antimicrobials is substantially affected by their rate of growth. Ideally the growth rate of any challenge inocula should replicate the growth rate *in vivo/ in situ* (Brown, 1977). The best way of achieving this is to culture the organisms in continuous, chemostat cultures (below and Chapter 1.4). Where this is not possible then the batch culture media should be chosen for which the cell division rate in mid-log phase most closely matches that *in situ*.

Growth rate is not the only factor that will influence the ability of an inoculum to match the physiologies found in the real world. Complex culture media are generally employed as a means of quickly and conveniently generating high cell densities. Relevence of the cells as an inoculum for performance tests will clearly be increased as the culture media used for their preparation are changed to match more closely that of the *in situ/in vivo* environment. An extreme example of this is *in vitro* culture using materials from the environment as the medium. Thus, spoilage organisms will maintain their spoilage capacity if cultured in product (Beveridge, 1975). Such media might be unpreserved or underpreserved formulations of cosmetics. In a similar fashion water borne communities will reproduce to some extent their *in situ* characteristics if cultured in natural water taken from the site and filtered (Lee and West, 1991). Combining continuous culture techniques with the use of such media is often highly predictive of outcome in the field. In such instances factors such as biomass, speed, and convenience are sacrified for relevant inocula that will perform in a manner that is predictive of the application.

Provided that some thought is given to the physiological status of the organisms *in situ*, then the relevance and predictive value of inocula can be increased without going quite to these extremes. The following general points should be borne in mind for inocula derived from the logarithmic phase of growth.

Temperature

The growth temperature should match that found *in situ/in vivo*. While it is right that studies into the effectiveness of antibiotics should culture the inocula at 37°C, it would be wrong to culture test organisms for environmental hygiene tests at the same temperature. Growth temperature affects the phospholipid composition of the cell envelopes and will markedly influence the activity of membrane-active biocides (Gilbert, 1988; Beveridge, 1975). Growth of the test inoculum at room temperature could be far more predictive of outcome in the field.

Carbon Substrate

Facultative anaerobes will often ferment particular carbon substrates rather than oxidize them, even if oxygen concentrations are relatively high (hence fermentation tests for species identification). Thus, growth of such organisms on media containing glucose will produce a fermentative phenotype, rendering the organism less susceptible towards agents acting against oxidative metabolism (i.e., uncouplers), whereas growth on succinate will result in oxidative growth. Glycerol as a carbon source is metabolized oxidatively, but it is transported into the cells by passive diffusion rather than active transport. Thus, if cells are growing in glycerol and challenged with an agent that inhibits group translocation then the cells will appear less sensitive (Al-Hiti and Gilbert, 1980).

Complex media often lack a carbohydate source. In such media cells must de-aminate the amino acids/peptones in order to provide carbon skeletons for anabolism and energy. This will have a profound influence on their physiology but can be used to match *in vivo* physiologies. Thus, *in vivo,* cells growing in plasma/tissues will find excess carbon in the form of glucose, and growth is likely to be affected by the availability of iron (Chapter 1.2; Brown and Williams, 1985b; Griffiths, 1983). In the bladder glucose is virtually absent, and carbon and energy must be obtained following deamination.

Nitrogen Source

Different species and genera of microorganism are capable of deriving nitrogen from a range of sources. At one extreme are those species which can fix atmospheric nitrogen; at the other are those which must have most, if not all, of their nitrogen requirements met in the form of complex organic molecules. The *in situ* source of such nutrients must be borne in mind when selecting ingredients for inclusion within laboratory media. While it might be necessary to include particular amino acids within a growth medium for the growth of auxotrophic strains, it must be remembered that their use solely for anabolism cannot be guaranteed and that such materials will also be available for deamination and catabolism.

Other Nutrients

In a fashion similar to the consideration of carbon and nitrogen sources for inclusion in complex and defined media, care must be exercised in choosing appropriate supplies of the other major elements, notably phosphorus, potassium, sulfur, calcium, magnesium, and growth factors (Kenward, Brown and Fryer, 1979; Cozens, Klemperer and Brown, 1983). With respect to trace metals it should be borne in mind that while they often represent elements which are abundant in nature, bio-availablity is generally severely restricted. Thus, in freshwater and marine environments the availability of iron is even less than it is in infections (Reid et al., 1993). This low availability is caused through the formation of highly insoluble salts (often phosphates and hydroxides). Cation availability will have profound effects on growth rate and envelope composition (Kenward, Brown and Fryer, 1979; Brown and Williams, 1985a, b; Chapter 1.2) and will affect the properties of dormant forms (i.e., endospores; Chapter 1.5).

Reproducibility of Stationary Phase Inocula

The term stationary phase is a misnomer. It implies that nothing is happening during this period of the growth cycle in batch culture (Kolter, Siegele and Tormo, 1993). This, in turn, induces complacency in the minds of researchers with respect to the timing of harvest and the need for any further definition of physiological status after logarithmic growth has ceased. The description "stationary-phase cultures" is most often used to describe those cultures that have, some time previously, ceased to show an increase in biomass. Generally this cessation of growth has neither been observed nor recorded. Since, in batch culture, the duration of the logarithmic phase of growth will depend upon the size of the initial inoculum (affected by heaviness of the loop-hand) and the physiological status of that inoculum, then cells observed to be in "stationary phase" after "overnight growth" could have stopped logarithmic growth anything from 1 to 8 h previously. Changes effected by cells and populations of cells upon entering stationary phase have been described previously in this volume (Chapter 1.3) and will not be elaborated further. Suffice to say that while overall cell numbers cease to increase in stationary phase, many physiological changes are going on in the culture which will affect the performance of an inoculum taken from it.

In order to increase reproducibility of stationary phase inocula, it is therefore imperative that the passing of cells from logarithmic phase be observed and harvest of the culture timed appropriately. This can be easily achieved by inoculating late at night, by employing a relatively simple medium for which μ_{max} is low (generation time 1 to 1.5 h), or by keeping the initial inoculum very small (ca. 10^3 cells). The culture should then be monitored the following morning and progression through stationary phase followed.

Changes that occur in populations after entering stationary phase vary according to the nature of the nutrient deprivation that has caused them. Thus, cultures deprived of a carbon source will often die rapidly due to their inability to maintain endogenous metabolism, whereas those which have ceased growth due to a deficiency in niitrogen will survive longer through endogenous metabolism of the excess carbon substrate. Stationary-phase cultures which have been generated under nitrogen depletion might therefore be more reproducible than those generated under carbon depletion. In this light, cells resuspended in buffers and diluents that lack the carbon substrate that was present throughout the growth period, faced with an absence of substrate, will mobilize and turnover cytosolic materials (protein, mRNA, rRNA etc.; Mandelstam, 1960).

The transition from the logarithmic to the stationary phase of growth varies in its speed of onset and progression according to the nature of the depleted nutrient. Depleted nutrients which possess high Y and high Ks (characteristically metal cations) will give rise to cultures where the progression into stationary phase is slow and extended. In such instances, since the available concentrations of the depleted nutrient decrease only slowly, the growth rate of the culture is subject to continual slowing over several hours. During this period the depleted nutrient may often be redistributed throughout the population by a process of reductive division. Not only is such a gradual onset into stationary phase difficult to determine experimentally until it is too late, but the physiological status of such stationary-phase cells as are generated is difficult to reproduce from culture to culture.

Depleted nutrients, on the other hand, which possess low Y and low Ks (characteristically nitrogen and carbon sources) give rise to cultures where the progression into stationary phase is sudden. These abrupt stationary phases are relatively easy to detect and the timing of harvest can be made accurately. Taking stationary-phase, nitrogen-limited cultures into diluents containing the carbon substrate but no other nutrient is likely, therefore, to maximize reproducibility for such inocula.

Predictivity of Stationary Phase Inocula

Adaptation of the expressed phenotype, during and following onset of stationary phase, varies according to the nature of the depleted nutrient (Chapters 1.2 and 1.3). Susceptibility to biocides and antibiotics differs substantially between these phenotypes (Brown, Collier and Gilbert, 1990; Gilbert, Collier and Brown, 1990). In order for the expressed physiology to be predictive of outcome in applications, then the relevant nutrient deficiences must be duplicated. While it is easy to speculate about the nature of the least available nutrient for a wide range of environmental niches, there will be no single answer to apply to all situations (Brown, 1977). The depleted nutrient might be determined, however, for a particular application (i.e., preservative system) by adding candidate substances, singly and in combination, to that environment and monitoring the effects upon cell growth and biomass. This empirical approach (Chapter 1.2) to the test situation will then point to the most relevant *in vitro* medium.

REPRODUCIBILITY AND PREDICTIVITY OF INOCULA GROWN IN CONTINUOUS CULTURE

Continuous cultures operate under steady-state conditions where, if a steady state is developed, the growth rate of the culture can be controlled by the rate of addition of fresh medium. Since the cultures are at steady state, then they may be maintained for many months or years. They may either be used to provide a continuous source of inoculum, or they might be used directly as the test system (Chapter 3.6). The operation and performance characteristics of continuous cultures have been described previously (Chapter 1.4; Brown, Collier and Gilbert, 1990), and the range of phenotypes generated and the effects of specific growth rate upon susceptibility to various antimicrobial treatments described. The following discussion relates to the use of continuous culture and chemostats to increase the reproducibility and relevance of inocula.

Reproducibility of Inocula Generated in Continuous Culture

Continuous cultures represent extremely selective/competitive growth environments. Very slight differences in the efficiency of utilization of the rate-limiting nutrient, or in the efficiency with which the nutrient is sequestered from the medium by the cells, will result in the dominance of the advantaged phenotype/genotype. Thus, when spore-forming organims are grown in continuous culture, at slow growth rates where sporogenesis is probable, then selection will be in favor of asporogenous mutants. Since the cultures will often operate continuously over long time periods then even very minor differences in phenotype/genotype that would not cause significant enrichment of one strain over another in batch culture will be selected for. Rather than maintain an initial inoculum, a chemostat will select the most appropriate phenotype/genotype from it.

If the primary purpose of an inoculum is to generate a reproducible response to a given stress over an extended time period, rather than to replicate the reponse of a specified type-strain (i.e., for reference or comparative purposes), then the tendency of chemostats to select for the most competitive phenotype/genotype can be used to advantage. Thus, if a continuous culture has been established for many generations, then much of the selection will be complete. The culture will then provide highly reproducible inocula on a day-to-day basis. If these cells are taken and stored, by freeze-drying or frozen in glycerol (Chapter 2.1), then they may be used to reinoculate the vessel should it have to be dismantled. They may also be used to set up continuous cultures elsewhere and, provided that the media and conditions of the culture are identical, regenerate identical inocula.

The selective pressures in a continuous culture will be greatest when the cells are growing at the extremes of possible growth rate. Thus, in a turbidostat, where growth is maintained at the maximum permissible rate in order to optimize the biomass produced, then those cells which can grow fastest will be selected (Dean et al., 1976). If rates of growth of individual cells are measured in batch culture then it can be seen that division times within the population, like cell size, are not uniform; rather, a range of division times is observed

(Kubitschek and Woldridgh, 1983). The coefficient of variation for the division time is variously reported in the literature but rarely exceeds 20% (Collins and Richmond, 1962). If a continuous culture is maintained with $\mu > 80\% \mu_{max}$, then a proportion of the cells will not be able to divide as rapidly as the culture is diluted and they will therefore disappear. Similarly, in the "realm of slow growth rate" those cells best able to reduce endogenous metabolism will be favored. For long-term stability, therefore, continuous cultures should be operated at between 25 and 75% of μ_{max}.

For some growth-rate-limiting nutrients (i.e., magnesium, oxygen etc.) then utilization by the cells (and as a result Y, the molar conversion to biomass) varies with specific growth rate (see Chapter 1.4). A major use of magnesium within a cell is stabilization of ribosomes. Since fast-growing cells require more ribosomes than slow-growing cells, fast-growing cells have a greater requirement (μg/μg biomass) for magnesium. A direct consequence of this is that the steady-state biomass in continuous culture, for such nutrients, decreases continuously with dilution rate up to washout. This places an additional selective advantage on those cells which can most economize on their utilization of the limiting nutrient. Cultures limited through carbon- or nitrogen-source availability might therefore provide less competitive environments than those limited by cations.

In some circumstances continuous cultures are challenged directly with an antimicrobial and the effect of the agent is determined from the perturbations of steady state (Chapter 3.6). When used in this way it is usual to employ a two-stage fermenter where the primary fermenter maintains an inoculum which is fed to the second stage. The second stage receives fresh supplies of nutrient and may then be the subject of experimentation without compromising the stability of the original inoculum (Keevil, Dowsett and Rogers, 1993).

Since growth rate in a chemostat is controlled ultimately through maintenance in the culture of a residual, limiting concentration of one substrate, it follows that the moment culture is removed from the fermenter or collected from the overflow, then steady-state conditions are lost and the cells progress to stationary phase. In order to increase reproducibility of the response of the inocula it is important to minimize, or at least standardize, the time interval between collection and use of the inocula. As for batch cultures, harvested in stationary phase, media and suspending menstrua which contain carbon substrate are likely to minimize or slow those changes occurring within the cells during storage. If spent medium is collected from a fermenter, rather than making collections of culture from the vessel, then samples may be taken throughout a working day. This is preferable to preparing suspensions and storing them for extended periods. If it is necessary to remove culture from the fermentation vessel, then this will disturb the steady state. Immediately after removal of culture the volume of medium in the vessel (V) will be reduced, and since the speed of the pump delivering fresh medium will have remained unchanged, the dilution rate will increase. It is conventional to allow a minimum of five volume changes to occur before further samples are taken, and to restrict the quantity of culture removed on each occasion to <20% of the fermenter volume. Such procedures ensure that the transitional change in D is no greater than 20%, and that steady state is reestablished between samples.

Predictivity of Inocula Generated in Continuous Culture

If a continuous culture is subjected to some inimical treatment (e.g., biocide/preservative or degraded substrate is added at sub-MIC levels), then the selective pressure will favor the resistant/aggressive phenotype and the retention of degradative and drug resistance plasmids. In this fashion continuous cultures may be used to maintain an appropriate phenotype for use in subsequent testing programs. Similarly, if the culture is grown at dilution rates close to, or in excess of, D_{MAX} then planktonic cells will wash out. Running a fermenter in this manner will select and maintain adherent phenotypes which will remain in the culture through attachment to the vessel wall or to tiles and test-pieces suspended within it (Keevil, Dowsett and Rogers, 1993; and Chapter 3.6).

The major advantage of continuous culture is that it enables growth rate to be controlled while maintaining a constant, but controllable, physico-chemical environment. Since growth rates *in vivo* and *in situ* are often significantly reduced from those in batch culture, then continuous culture facilitates the testing of slow-growing phenotypes. The influence of growth rate upon phenotype has been discussed previously in this volume and will not be reiterated here (Chapter 1.4), but growth rates should be selected which are representative of those *in vivo*. In infections of soft tissues and chronic infections associated with lung and gut, studies indicate division times to be in the order of 20 to 40 h (Brown, Allison and Gilbert, 1988; Eudy and Burroughs, 1973). During the establishment and initial phase of an infection, and also in septicemias and bacteremias, growth rates are probably much faster. Actual growth rates in the general environment, and in biofouling situations, may be estimated from the biological oxygen demand of the system and the generated biomass.

As with stationary phase batch-grown inocula, it is possible, through appropriate design of the growth medium, to select an appropriate nutrient ultimately to restrict growth. The rationale behind the choice of an appropriate nutrient limitation is as before (Brown, 1977). Similarly, the choice of substrate for the provision of nonlimiting nutrients (i.e., carbon source under nitrogen limitation) will enable further selection of phenotypes (i.e., fermentative or oxidative modes of catabolism) and dual provision of carbon and nitrogen source. The nature of the substrate slowest in its utilization by the cells, the pH of the medium, and the incubation temperature will determine μ_{max}/D_{MAX} for the culture. The steady-state concentration of the nutrient least available in the medium reservoir, on the other hand, will control growth rate below μ_{max}/D_{MAX} (Herbert, Ellsworth and Telling, 1956; and Chapter 1.4).

SOLID VERSUS LIQUID CULTURE

The preceding account has been restricted to inocula derived from liquid cultures of one type or another. In practice, most, if not all, laboratories routinely prepare inocula on solidified media. This is necessary in order to make viability assessments and to assess the purity, or otherwise, of the cultures. Agar-based cultures are also useful in that thick suspensions of cells may be prepared from them without recourse to the use of a centrifuge. The resulting suspension will be relatively free of residual medium and as such has been deemed suitable for inclusion in many standard tests. Indeed, studies (Orth, 1979) showing that residues of complex liquid media contaminating preservative challenge tests will influence the outcome have made some regulatory authorities stipulate that inocula be prepared in this manner (Orth, 1979, 1989). Other workers have suggested that colonies of microorganims grown on agar will duplicate some of the properties of adherent populations of bacteria associated with biofilms. Nevertheless, colonies grown on solid surfaces are highly heterogenous in terms of phenotype. Al-Hiti and Gilbert (1983) pointed out that colony size varied markedly across streaked plates, according to the closeness of neighboring colonies. Since nutrient and oxygen gradients would exist within each colony, cells on the periphery would experience a different nutrient environment than those in the center, and those associated with large colonies would be distinct from those cells associated with smaller ones. These workers prepared, in quintuplicate, suspensions of cells from plates seeded to give different colony densities, subjected them to challenge with several preservatives, and determined the reproducibility of response. Not only did the level of reproducibility of the suspensions decrease as the plates were seeded with less cfu, i.e., from confluent lawns to large separate colonies (20 per plate), but susceptibility was increased by an order of magnitude. They concluded that if agar derived challenge inocula were to be used, then these should be obtained from lawn cultures. Suspensions prepared from liquid cultures were, however, the most reproducible from day to day in this study.

BIOFILMS OR PLANKTONIC INOCULA

Methods have been described (Chapters 1.5 and 3.7) which enable the growth of bacteria as biofilms. The various methods enable differing degrees of control over the growth rate and nutrient phenotype expressed. While there can be no doubt that in order to screen predictively for antibiofilm agents then *in vitro* tests which replicate such physiology must be employed, the level of day-to-day reproducibility of the majority of the approaches leaves much to be desired. Of the methods described, the Constant Thickness Biofilm Fermenter (Peters and Wimpenny, 1988) and the Perfused Biofilm Fermenter (Gilbert et al., 1989) are the only techniques which generate steady-state growth conditions and thereby offer the possibility of growth-rate control. While these two methods reproducibly generate biofilm-specific phenotypes, at more or less defined growth rates, they are technically complex.

REFERENCES

Al-Hiti, M.M.A. and Gilbert, P. 1980 Changes in the preservative sensitivity of the "USP Antimicrobial Agents Effectiveness Test" microorganisms. *Journal of Applied Bacteriology* **49:** 119-126

Al-Hiti, M.M.A. and Gilbert, P. 1983 A note on inoculum reproducibility: solid versus liquid culture. *Journal of Applied Bacteriology* **55:** 173-176.

Beveridge, E.G., 1975 The microbial spoilage of pharmaceutical products, pp 213-236 in *Microbial Aspects of the Biodeterioration of Materials* (Lovelock, D.W. and Gilbert, R.J. Eds.) Academic Press, London.

Brown, M.R.W., Collier, P.J. and Gilbert, P. 1990 Influence of growth rate upon the susceptibility to antimicrobial agents: modification of cell envelope, and batch and continuous culture. *Antimicrobial Agents and Chemotherapy* **34:** 1623-1628.

Brown, M.R.W., Allison, D.G. and Gilbert, P. 1988 Resistance of bacterial biofilms to antibiotics: a growth rate related effect? *Journal of Antimicrobial Chemotherapy* **20:** 147-154.

Brown, M.R.W. and Williams, P. 1985a Influence of substrate limitation and growth phase on the sensitivity to antimicrobial agents. *Journal of Antimicrobial Chemotherapy* **15:** Suppl. A., 7-14.

Brown, M.R.W. and Williams, P. 1985b The influence of the environment on envelope properties affecting survival of bacteria in infections. *Annual Reviews of Microbiology* **39:** 527-556.

Brown, M.R.W. 1977 Nutrient depletion and antibiotic susceptibility. *Journal of Antimicrobial Chemotherapy* **3:** 198-201.

Collins, J.F. and Richmond, M.H. 1962 Rate of growth of *Bacillus cereus* between divisions. *Journal of General Microbiology* **28:** 15-33.

Cozens, R.M., Klemperer, R.M.M. and Brown, M.R.W. 1983 The influence of cell envelope on antibiotic activity, pp 61-71 in *Antibiotics: Assessment of Antimicrobial Activity* (Russell, A.D. and Quesnel, L.B. Eds.) Academic Press, London.

Dean, A.C.R., Ellwood, D.C., Evans, C.G.T. and Melling, J. (Eds.) 1976 *Continuous Culture: Applications and New Fields,* Ellis Horwood, Chichester.

Domingue, P.A.G., Schwarzinger, E. and Brown, M.R.W. 1989 Growth rate, iron depletion and a sub-minimal inhibitory concentration of penicillin G affects the surface hydrophobicity of *Staphylococcus aureus,* pp 50-62, in *The Influence of Antibiotics on Host-Parasite Relationships,* Vol. III (Gillissen, G., Operfkuch, W., Peters, G. and Pulverer, G. Eds.) Springer Verlag, Berlin.

Eudy, W.W. and Burroughs, S.E. 1973 Generation times of *Proteus mirabilis* and *Escherichia coli* in experimental infections. *Chemotherapy* **19:** 161-170.

Gilbert, P., Collier, P.J. and Brown, M.R.W. 1990 Influence of growth rate on the susceptibility to antimicrobial agents: biofilms, cell cycle, dormancy and stringent response. *Antimicrobial Agents and Chemotherapy* **34:** 1865-1868.

Gilbert, P., Allison, D.G., Evans, D.J., Handley, P.S. and Brown, M.R.W. 1989 Growth rate control of adherent microbial populations. *Applied and Environmental Microbiology* **55:** 1308-1311.

Gilbert, P. 1988 Microbial Resistance to Preservative Systems, pp 171-194, in *Microbial Quality Assurance in Pharmaceuticals, Cosmetics and Toiletries* (Bloomfield, S.F., Baird, R., Leak, R.E. and Leech, R. Eds.) Ellis Horwood, London.

Gilbert, P., Brown, M.R.W. and Costerton, J.W. 1987 Inocula for antimicrobial sensitivity testing: a critical review. *Journal of Antimicrobial Chemotherapy* **20:** 147-154.

Griffiths, E. 1983 Availability of iron and the survival of bacteria in infection. In *Medical Microbiology*, Vol. 3 (Easmon, C.S.F., Jeljasewicz, J., Brown, M.R.W. and Lambert, P.A. Eds.) Academic Press, London.

Herbert, D., Ellsworth, R. and Telling, R.C. 1956 The continuous culture of bacteria: a theoretical and experimental study. *Journal of General Microbiology* **14:** 601-622.

Keevil, C.W., Dowsett, A.B. and Rogers, J. 1993 *Legionella* biofilms and their control, pp 201-216 in *Microbial Biofilms: Formation and Control* (Denyer, S.P., Gorman, S.P. and Sussman, M. Eds.) Blackwell Scientific Press, London.

Kenward, M.A., Brown, M.R.W. and Fryer, J.J. 1979 The influence of calcium and manganese on the resistance to EDTA, polymyxin B or cold shock, and the composition of *Pseudomonas aeruginosa* grown in glucose or magnesium depleted batch cultures. *Journal of Applied Bacteriology* **47:** 489-503.

Kolter, R., Siegele, D.A. and Tormo, A. 1993 The stationary phase of the bacterial life cycle. *Annual Reviews of Microbiology* **47:** 855-874.

Kubitschek, H.E. and Woldridgh, C.L. 1983 Cell elongation and division probability during the *Escherichia coli* growth cycle. *Journal of Bacteriology* **153:** 1379-1387.

Lee, J.V. and West, A.A. 1991 Survival and growth of *Legionella* species in the environment. *Journal of Applied Bacteriology Suppl. 20* **70:** 121s-130s.

Mandelstam, J. 1960 The intracellular turnover of protein and nucleic acids and its role in biochemical differentiation. *Bacteriology Reviews* **24:** 289-308.

Monod, J. 1950 La technique de culture continue: théorie et applications. *Annals of the Institute of Pasteur, Paris* **79:** 390-410.

Morita, R.Y. 1988 Bioavailability of energy and its relationship to growth and starvation in nature. *Canadian Journal of Microbiology* **34:** 436-441.

Orth, D.S. 1989 Evaluation of preservatives in cosmetic products, pp 403-421 in *Cosmetic and Drug Preservation. Principles and Practice* (Kabarra, J.J. Ed.) Marcel Dekker, New York.

Orth, D.S. 1979 Linear regression method for rapid determination of cosmetic preservative efficacy. *Journal of the Society of Cosmetic Chemists* **30:** 320-332.

Peters, A. and Wimpenny, J.W.T. 1988 A constant depth laboratory model film fermenter, pp 175-195 in *Handbook of Laboratory Model Systems for Microbial Ecosystems*. Vol. 1 (Wimpenny, J.W.T. Ed.) CRC Press, Boca Raton, Florida.

Reid, R.T., Live, D.H., Faulkner, D.J. and Butler, A. 1993 A siderophore from a marine bacterium with an exceptional ferric ion affinity constant. *Nature* **366:** 455-458.

3.2 Preservative Efficacy Testing in the Pharmaceutical Industries

Rosamund M. Baird

DEVELOPMENT OF TESTING METHODS

Methods of testing preservative efficacy in pharmaceutical formulations have evolved over the past two decades. Publication of the United States Pharmacopoeia (USP) XVIII first preservative efficacy test in 1970 and later that of the British Pharmacopoeia (BP) in 1980, as well as the Italian (1985) and German (1986) Pharmacopoeiae provided a framework against which the biological availability of an antimicrobial compound could be measured during the development program of a given product. At the same time the test provided an assessment of a product's ability to withstand contamination during storage and consumer use. In essence, the pharmacopoeial tests have involved the inoculation of test product with reasonably large populations (10^5 to 10^6 colony forming units/ml or g) of four or five individual challenge test organisms and the subsequent withdrawal and examination of samples for surviving organisms at defined intervals over a 28-day test period. Products are deemed to have passed the test if the results fall within the required criteria for preservative efficacy. In all instances the test is designed for use in a product's development program; it is not intended as a final product release test.

Since their inception, these first-generation preservative efficacy tests have been refined in detail; more recently, minor differences between test methodologies have been rationalized in an effort to harmonize pharmacopoeial testing requirements on an international scale. Careful comparison of the texts of the 1990 USP XXII, the 1993 European Pharmacopoeia (EP) Fasicule 16, and the 1988 BP indicates few variations in test organism, incubation conditions, and inoculation procedures, as shown in Table 1. Until recently, the principal differences between the various tests have been in the test criteria for preservative

0-8493-4752-1/95/$0.00+$.50

TABLE 1
Preservative Efficacy Tests of the USP XXII, EP 1993, and BP 1988

Criteria	USP	EP	BP
Test organisms	*Aspergillus niger* (ATCC 16404) *Candida albicans* (ATCC 10231) *Pseudomonas aeruginosa* (ATCC 9027) *Staphylococcus aureus* (ATCC 6538) *Escherichia coli* (ATCC 8739)[a]		
Medium	SCD	SCD, SAB	SCD,SAB
	Inoculum preparation:		
Bacteria 18-24 h	30-35°C	30-35°C	30-35°C
C. albicans 48 h	20-25°C	20-25°C	20-25°C
A. niger 1 week	20-25°C	20-25°C	20-25°C
Harvesting[b]	NaCl	NaCl + P	PW
Test inoculum	$10^{5\text{-}6}$ ml^{-1}	$10^{5\text{-}6}$ ml^{-1}	$10^{5\text{-}6}$ ml^{-1}

[a]EP and BP require only oral products to be challenged with *E. coli*.
[b]When harvesting *A. niger*, add 0.05% Polysorbate 80.

Note: NaCl—sterile saline; P—0.1% peptone; PW—0.1% peptone water.

efficacy. Table 2 summarizes the various pharmacopoeial test requirements for multidose parenteral products, ranging from the stringent requirements of the BP to the much more relaxed criteria of the USP. Recent publication of the 1993 EP test, which now takes precedence over any European national pharmacopoeial test, including the newly published 1993 BP, has subsequently harmonized the differing test criteria within Europe; moreover, in putting forward both target and minimum test criteria, the EP has formally recognized the difficulties which may be encountered in preserving certain product types.

TEST METHOD LIMITATIONS

Simultaneously with their development, these pharmacopoeial tests have been subject to a certain amount of criticism, and their limitations have been well documented over the years.

First, the use of laboratory strains grown in chemically rich media, designed to provide optimal conditions for growth, is not considered to be representative of in-use conditions nor to provide a realistic challenge to preservative efficacy. Such cultures may or may not have the ability to adapt to that particular product niche (Yablonski, 1972).

TABLE 2
Test Criteria for Parenteral Products

Pharmacopoeia	Organism	6 h	24 h	48 h	7 d	14 d	28 d
USP XXII	Bacteria					3	NI
	Fungi					NI	NI
EP 1993	Bacteria:						
	Target	2	3				NR
	Minimum		1		3		NI
	Fungi:						
	Target			2			NI
	Minimum					1	NI
BP 1988	Bacteria	3	NR				NR
	Fungi				2		NI

Note: NI—No increase in number of organisms; NR—No organisms recovered.

All figures are the log reduction in the number of viable microorganisms.

Second, the use of pure cultures for each individual challenge does not reflect the spectrum of contaminant strains likely to be encountered during product use. The latter are more likely to effect a cooperative breakdown of antimicrobial systems by virtue of the sequence of events initiated. On the other hand, mixed culture challenges have reportedly been less effective in indicating inherent weaknesses of preserved formulations (Tagliapietra, 1978), and particularly those which are susceptible to fungal contamination (Yablonski, 1972). Microbial competition and overgrowth may also be encouraged.

Third, a single challenge method fails to simulate the practical conditions of use when a product is repeatedly challenged throughout its shelf life. Repeated challenges have therefore been advocated in some quarters.

Fourth, cell numbers used in pharmacopoeial preservative efficacy tests (10^{5-6} cfu/ml or g in USP XXII and EP 1993, 10^6 cfu/ml or /g in the BP 1988) have been criticized as unrealistic of those cell numbers likely to be introduced into a product made under good manufacturing conditions, or indeed under normal in-use conditions. Such numbers, however, have been justified on the basis that the presence of occasional mutants in large populations may show regrowth in marginally preserved products, but might be otherwise overlooked in small populations of challenge test organisms. Furthermore such numbers have been required to demonstrate the logarithmic reduction values (LRV) demanded by pharmacopoeial test acceptance criteria. For example the 1988 BP Test for parenteral and ophthalmic products required a threefold log reduction in numbers over a 6 h period and no organism to be recovered from a 1 ml sample at 24 h (see Table 2). It should be noted, however, that a recent USP publication has proposed that an inoculation level of 10^3 to 10^4 cfu/ml or /g should be used for antacid products (Anon., 1992).

Finally, the age and uniformity of test inocula can profoundly affect test performance, and again the challenge test inocula of the preservative efficacy test is considered unrepresentative of in-use conditions. Young cells in the stationary phase are known to be more susceptible to antimicrobial effects compared with older, more heterogenous populations. Moore (1978) reported that 18 h cultures of *P. aeruginosa* were more susceptible to the effects of sodium chromoglycate than similar cells cultured for a further 48 h at 20°C.

TEST METHOD MODIFICATIONS

Clearly such limitations are not insuperable, and scientists have over the years sought and found various solutions to these problems. Subsequently many of these modifications have been successfully incorporated into tests used both in-house and in commonly accepted industrial practice. Wild strains, considered to be representative of the microbial flora of raw materials, including water, and the product manufacturing environment have been added to the collection of traditional challenge test organisms (*P. aeruginosa* ATCC 9027, *Escherichia coli* ATCC 8739, *Staphylococcus aureus* ATCC 6538, *Candida albicans* ATCC 10231, *Aspergillus niger* ATCC 16404), as well as wild strains originating from spoiled or contaminated products (Anon., 1990; BP, 1988; Leak and Leech, 1988). Such tests in the form of single-culture challenges incur, however, additional resource costs (staff-time and consumables); mixed culture challenges have therefore been advocated by others as an alternative, convenient approach, and indeed may be considered more representative of in-use conditions. Thus, pools or cocktails of Gram-positive and Gram-negative bacteria, yeasts, and molds have been successfully used. Spooner and Croshaw (1981) reported the use of 41 organisms in four standardized pools. Leak and Leech (1988) found that cocktails of inocula, one for neutral or low pH formulations and one for high pH formulations, provided a more discriminatory test in representing variations in microbial behavior.

Various repeat-inoculation tests have been described, which differ in the frequency and number of inoculations (Lindstrom, 1986; Yablonski, 1972), and include those that repeatedly challenge the presevative system until it fails. The so-called capacity test of Barnes and Denton (1969) described a reinoculation of samples with 10^7 cfu of test organism/g or/ml every 48 h for up to 15 inoculations. The end point was determined by the finding of survivors in three successive inoculations. Repeat inoculation tests are not, however, without their drawbacks; the product itself is progressively diluted and the inevitable introduction of a high organic load can in turn interact with preservative, thereby decreasing its activity (Spooner and Croshaw, 1981).

Reinhardt (1984) recommended the use of different levels of challenge inocula to represent in-use conditions, including a high- (10^{7-8} cfu/ml), an intermediate- (10^{4-5} cfu/ml), and a low-density inoculum (10^{2-3} cfu/ml). The rate of inactivation of test organisms is known, however, to be independent of the concentration of organisms added to the sample. Orth (1979) showed that differing concentrations of *S. aureus* (from 10^3 to 10^6 cfu/ml) gave similar rates of inactivation in preservative efficacy tests of a paraben-preserved lotion.

Finally, cells produced under laboratory culture display a uniformity of age which is not reflected in wild populations (Leitz, 1972), as well as showing an increased susceptibility to preservative activity. Moore (1978) recommended the use of both old and young cultures grown under a variety of test conditions. Pharmacopoeial methods have traditionally adopted an arbitrary test period of 28 days, although it is known that the USP is currently considering an extension of the test period to 56 days (Anon., 1992). While regrowth of challenge test organisms outside this period is always possible, it is also arguable that valid results can be obtained in much shorter times (Leak and Leech, 1988; Orth, 1979). The linear regression method, using the kinetics of cell death, has been succesfully developed and used to predict preservative behavior in formulated products over a 48 h period (Orth, 1979, 1980, 1990). Different organisms have different metabolic and physiological characteristics and therefore may exhibit different death rates when exposed to a given lethal treatment. By determining the decimal reduction time (D-value), i.e., the time required for a log (90%) reduction in the population, a quantitative expression of the death rate for a given population of test organism in the test product is obtained. Moreover, comparisons can also be made between the effectiveness of different lethal treatments. Quantitative acceptance criteria based on D-values have been discussed in detail. Orth (1979) suggested that for pathogenic organisms multiple-use topical products should be self-sterilizing within 24 h, i.e., an inoculum of 10^6 cfu/g should be eliminated within this time and hence a D-value of <4 h was required.

FACTORS CAUSING TEST VARIABILITY

Although test conditions for the various pharmacopoeial preservative efficacy tests are defined to some extent, there is nevertheless considerable scope for variation in detail. Such variations can have a profound effect on test results, in some instances producing poor reproducibility between laboratories; indeed, even within a laboratory conflicting results may be obtained. Such differing results from laboratories, purporting to use the same method and test organism, may well be due to the use of nonstandardized procedures, such as differences in the inoculum preparation, or to the type of inocula used. These are discussed below.

Cultivation of Wild Strains

The behavior of wild strains in preservative efficacy tests is known to be affected by their subculture onto laboratory media. Strains isolated from contaminated water, products, or the environment are known to exhibit a marked or irreversible loss of aggressive behavior on subculture onto standard laboratory media, even for example within two subcultures onto nutrient agar or potato-dextrose agar (Leak and Leech, 1988). Clearly such strains should be maintained in media designed to retain their spoilage potential. Leak and Leech (1988) suggested the use of test strains, isolated from and maintained in unpreserved product, for the detection of marginally preserved systems. While it is recognized that the use of product isolates may be informative, the use of adapted organisms is not universally recommended, particularly in view of the likely problems of standardization (Orth, 1984).

Influence of Nutritional Status of Challenge Inocula

Organisms respond in different ways to changes in growth media and culture conditions, according to their various physiological states and capabilities. Microorganisms in the logarithmic phase of growth are known to be more sensitive to most types of stress compared with those in a resting stage (Mossel et al., 1979), while slowly growing cells can be particularly recalcitrant to chemical inactivation (Brown, 1975; Gilbert and Brown, 1980). Recommended challenge test inocula and wild test strains may both exhibit differing growth rates and duration times in the maximum stationary phase. In the case of the latter, considerable time variations may be observed, according to the growth conditions (pH, temperature, growth rate) and the ability of the organism to withstand metabolic stress created by an unfavorable environment (low pH, accumulation of toxic residues and nutrient depletion).

Al-Hiti and Gilbert (1980) reported on the ability of USP test organisms, grown under carbon, nitrogen, or phosphate depletion, to survive and grow in the presence of varying concentrations of preservatives. Although no universal pattern of nutrient limitation and preservative sensitivity emerged between species, their results suggested that the use of different media within different laboratories could be a primary cause of interlaboratory variation, and indeed influence the outcome of a preservative efficacy test. The performance of culture media, both dried and ready-to-use media of the same description and formula, may well show brand to brand variation or even batch to batch variation (Meynell, 1958). Quality control of culture media and the assessment of its performance has been extensively discussed, particularly with reference to media used in food microbiology (Baird, Corry and Curtis, 1987; Baird et al., 1985, 1989); few parallels have been drawn, however, in the case of media used in pharmaceutical microbiology.

Stability of Test Strains

The condition, handling, and maintenance of test strains, prior to their use in the test, are known to affect their subsequent performance. In a comparative trial, conducted in three commercial laboratories, preservative performance was evaluated for a single batch of product; widely differing results were initially reported (Booth, personal communication). Following a detailed examination and standardization of procedures, however, including the rehydration of freeze-dried cultures, the preparation of master and working slopes and the test inocula itself, good agreement and reproducibility were reported between the laboratories concerned. Rapid rehydration of freeze-dried organisms in water or diluent can be lethal in some cases; recovery is improved if water is added very slowly to the freeze-dried material (Mackey, 1984; Chapter 2.1).

The maintenance of culture strains is poorly defined in the current editions of the EP (1993) and USP (1990) being only mentioned in a passing reference to the use of "freshly" or "recently" grown stock cultures. The BP (1993), on the other hand, notes that several subcultures may be necessary before the organism is in an optimal state for use. Additionally, the effect of continued subculture of test strains on resistance characteristics has yet to be addressed. The USP XXII sterility test states that, in performing the fertility test on culture media, strains of not more than five passages away from the ATCC strain should be used. As yet, there are no similar requirements for test strains used in the preservative efficacy test, although the USP has recently put forward one such proposal for consideration (Anon., 1992). It has also proposed the inclusion of a test to determine the index of resistivity of each test organism, according to its resistance to phenol at 20°C using the AOAC method.

Furthermore, the recovery of test cultures following their exposure in the preservative efficacy tests can also be affected by subsequent methods of handling. Ideally cells should be cultured immediately, but occasionally it might be necessary to refrigerate samples; in such cases the effects on cell numbers should be ascertained; injury and death due to cold shock have been widely reported (Chapter 2.2; Oliver, 1981). Additionally, loss of viability may occur when cells are exposed to distilled water and peptone buffer diluents (Chapter 2.2; Ray and Speck, 1973). Such effects will be compounded for cells already sublethally injured by chemical agents (Gilbert, 1984).

Nature of the Product

Preserved pharmaceutical products comprise a motley collection of formulation types. Although a discussion of these is outside the scope of this chapter, the influence of formulation components upon preservative action and bioavailability is considerable. In particular, it is well recognized that the inherent antimicrobial properties of a formulation in the unpreserved product should be established at an early stage of formulation development, including the influence of individual ingredients, pH, and water activity (Aw). Such testing is not referred to in the pharmacopoeial methods, although it is included in the Cosmetic, Toiletry, and Perfumery Association's *CTPA Limits and Guidelines* (Anon., 1990). A recent USP publication has, however, proposed the inclusion of uninoculated product controls into the test method, but these proposals have yet to be formally accepted (Anon., 1992). For challenge test inocula exposed not only to the injurious effects of an antimicrobial compound, but additionally to the superimposed secondary shock of an inherently hostile environment of the pharmaceutical product, the combined effect may indeed be lethal.

In nonaqueous products the partition coefficient of the individual preservative will determine its distribution between the oil and water phases. The activity of the preservative will, in turn, depend almost entirely upon the availability of preservative in the aqueous phase. The performance of a preservative efficacy test therefore becomes a more complex undertaking compared with that in aqueous products.

Nonaqueous products present additional problems in terms of the recovery of microbial cells, which may well be embedded within the matrix of the product itself. Common recovery methods include the warming of the product to not more than 40°C, often combined with the use of sterile surface-active agents, such as Tween 80®, Tween 20®, or Lubrol W®. Such treatments can, in turn, alter the physico-chemical characteristics of the system and lead to preservative emulgent interactions, altered partitioning behavior and wide variations in cell recovery (Allwood and Hambleton, 1972, 1973; Brown et al., 1986). Alternatively, filter-sterilized solvents, such as isopropyl myristate, n-hexane, and light liquid paraffin, have been used with some success (Allwood and Hambleton, 1973; White, Bowman and Kirschbaum, 1968); isopropyl myristate can, however, become toxic if heat sterilized (Bowman et al., 1972; Tsuji et al., 1970). With regard to the pharmacopoeial methods, little guidance is provided on the recovery of challenge test organisms from nonaqueous products although, interestingly, the above methods (warming and use of Tween 80) are described for the examination of oily products in both the 1993 BP sterility test and tests for microbial contamination.

Addition of Organic Matter

The addition of extraneous matter, to a product or test system, in the form of broth inocula or sterile broth media, can influence the rates of inactivation of test microorganisms during use and in preservative efficacy testing. Orth (1979) and Orth et al. (1989) reported that the addition of Brain Heart Infusion Broth and TSB with or without glucose resulted in significant decreases in inactivation rates, as determined by the use of D-values. Whether such reductions were due to the protective effect of nutrients in the culture medium or to inactivation of a part of the preservative system was, in a sense, immaterial. The overall test outcome provided erroneous results. A recommendation was subsequently made that preservative efficacy tests should be performed using saline suspensions of bacteria grown aerobically on solid agar media (see Chapters 1.1, 1.2, and 3.1).

TABLE 3
Methods of Inactivation of Antimicrobial Compounds

Antimicrobial Compound	Method/Addition of
Alcohols	Dilution
Bronopol	Cysteine hydrochloride
Chlorhexidine	Lecithin + Tween 80/Lubrol W
Halogens	Sodium thiosulfate
Mercurials	Sodium thioglycollate/cysteine
Parabens	Tween 80/dilution
Phenolic	Dilution (+ Tween 80)
Quaternary ammonium compounds	Lecithin + Tween 80/Lubrol W
Sorbic acid	Tween 80/dilution

From Baird, R.M., in *Guide to Microbiological Control in Pharmaceuticals.* Ellis Horwood, Chichester, U.K., 1990. With permission.

Neutralization of Antimicrobial Activity

Removal of preservative antimicrobial activity is clearly an important step when recovering test organisms from contaminated products, yet surprisingly little has been published which compares the effectiveness of the different neutralization methods. Inactivation methods can involve the use of a specific neutralizer or simply the physical removal of an agent, by dilution to a subinhibitory level or by washing using a membrane filtration technique, as shown in Table 3. Most procedures rely upon the removal of residual-free inhibitors. Where inhibitors have already bound firmly onto cells, then these techniques will not be successful in removing antimicrobial activity.

Nonionic surface-active agents, such as Tween 80 and Lubrol W, have been widely used as preservative neutralizers, sometimes in combination with lecithin. The detoxifying action of lecithin is particularly useful for cationic antimicrobials. These cause cell membrane damage and generally have a high affinity for the acidic phospholipids. Orth (1981) compared the recovery of *S. aureus* from a preserved lotion on three different media, standards method agar, Baird-Parker agar, and TSALT containing 0.07% lecithin and 0.5% Tween 80. Although no difference was seen between the three media immediately after inoculation, the recovery of *S. aureus* after 3 h exposure to the preservative was significantly higher on TSALT than with the other media. In addition certain media may have some nonspecific neutralization effect. Media containing serum or meat have been used sucessfully for detecting reasonably high numbers of phenol-exposed cells. Cook and Steel (1959) reported that the addition of serum, before but not after the addition of culture, would efficiently neutralize mercuric chloride. This is presumably because there is no residual-free inhibitor if the cells are added first.

Neutralizers can themselves be toxic, however, as are some nonionic surfactants (Lamikanra and Allwood, 1976) and thioglycoleate (Mossel and vanNetten, 1984). These may cause further damage to cells already injured. When validating the effectiveness of neutralization procedures, it is therefore necessary to show not only that the antimicrobial compound has been removed or effectively neutralized, but also that the neutralizer itself does not possess antimicrobial properties, alone or in combination with the low residual levels of preservative. Complete neutralization can be checked by the introduction of a known number of sensitive cells, usually approximately 10^2 cfu/ml into (1) the inactivated product, (2) a peptone water control sample containing the inactivator, and (3) a peptone water control. Following plating onto solid media and subsequent incubation, the antimicrobial compound can be considered to be adequately neutralized and the inactivating agent itself can be considered not to possess antimicrobial properties if the mean cell counts of (1) and (2) are found to be within 50% of (3). It should be noted, however, that the response of healthy cells will differ from sublethally damaged cells, already exposed to the injurious effects of antimicrobial compounds. Unlike healthy cells, the suitability of recovery methods cannot therefore be assessed in the counting of damaged cells, since there is no way of counting absolute numbers of injured cells.

Recovery Systems and Stressed Cells

Use of an efficient recovery system, involving diluent and plating medium, should ensure the generation of both accurate and sensitive preservative efficacy data. Stressed cells (i.e., those which have been sublethally injured) have specialized recovery requirements, sometimes irrespective of the actual stress involved. The injurious effects of exposure to differing stresses — for example heat, ionizing radiation, freezing, and drying as well as those from contact with antimicrobial compounds — often can result in similar types of damage. Depending on the nature and extent of injury, recovery may or may not take place. Membrane damage is a particularly common lesion, causing increased permeability and resulting leakage of cell contents and sometimes increased uptake of antimicrobial compounds (Silver and Wendt, 1967). An impaired ability to synthesize DNA, RNA, and protein essential for repair and subsequent growth may also be shown.

In resuscitating stressed cells a balance must therefore be struck to provide optimum recovery conditions to give maximum recovery of injured cells, while at the same time preventing the multiplication of repaired cells and distortion of cell numbers apparently recovered. Recommended isolation methods for damaged cells usually involve resuscitation in liquid nonselective media for 5 to 24 h (Allwood and Hambleton, 1973) depending on the type of organism and the extent and severity of sublethal damage. Injured populations form a heterogeneous group, and by definition repair times differ both within and between populations. It is noteworthy that although the USP, EP, and BP have for some considerable time incorporated resuscitation steps into microbial contamination tests for the absence of *Salmonella, P. aeruginosa, E. coli,* and *S. aureus*, there is no similar provision for recovery of organisms exposed to the injurious effects of antimicrobial compounds in the preservative efficacy test.

An additional feature of sublethally injured cells is their associated sensitivity to secondary unrelated stress. Hence cells exposed to one primary stress, whether an antimicrobial compound or the physical effects of cold, freezing, thawing, heat, ultraviolet light or osmotic stress, may become increasingly susceptible to a secondary, superimposed stress, such as low pH, or that resulting from exposure to sodium or potassium chloride, hydrogen peroxide, osmotic shock, or the components of selective media.

In considering the total stress exposure that microbial cells may be subjected to during the course of a preservative efficacy test, it is therefore necessary not only to account for the primary stress of exposure to the antimicrobial compound itself, but also to be aware of possible secondary stress factors, including the inadvertent use of toxic neutralizers, and the intrinsic toxicity of some diluents. Thus, for example, when using diluents of distilled water, saline, or 1/4 strength Ringer's solution in place of peptone water (Corry, van Doorne and Mossel, 1977) additional losses of viability may occur. Temperature fluctuations can also be lethal to damaged cells. Exposure to hot (i.e., above 47°C) agar during plate pouring, the use of ice-cold diluents, and even relatively small temperature drops (e.g., from 37°C to room temperature) can cause cold shock in susceptible organisms (Mackey, 1984). This can be overcome by cooling the sample slowly to the temperature of the diluent or through the addition of low concentrations (5 mmol) of magnesium to the diluent (Moore, 1978). Likewise cells with damaged cytoplasmic membranes are more freely permeable to hydrogen and hydroxyl ions and hence are more susceptible to acid or alkaline conditions. A reduction in culture medium pH, as a result of microbial growth, may therefore be less well tolerated by stressed than fully viable cells.

A further consideration is the possibility that microbial cells may already have been stressed, before exposure to the secondary effects of an antimicrobial compound in the preservative efficacy test. Cultures exposed to a freeze-drying process, followed perhaps by refrigerated storage of working culture stocks, may already carry sublethal injuries (Chapter 2.2).

In spite of a general awareness of the existence of stressed cells in pharmaceutical products, there is a dearth of information on the use of suitable recovery methods for cells with chemically induced injuries. Most studies on the recovery of stressed cells have been conducted with bacteria damaged by physical treatments, owing partly to their importance and significance as microbial inactivation processes and also to their comparative ease of control. Although some parallels can be drawn between physically and chemically induced injuries, the nature of the injury may be less well defined in the case of the latter, in terms of the molecular basis of action (Gilbert, 1984). There is thus an urgent need for comparative evaluation and subsequent standardization of recovery methods for microbial cells exposed to chemically induced injuries (Silliker, Gabis and May, 1979). The potential benefits of incorporating certain detoxifying constituents into media should be assessed. The addition of lysozyme and peroxide and superoxide scavengers, such as catalase, or pyruvate into media used for the recovery of cells stressed by a variety of means, is known to aid cell recovery (Flowers et al., 1977; Martin, Flowers and Ordal, 1976; Rayman, Aris and El Derea, 1978) and may well prove beneficial for the recovery of chemically injured cells.

OVERVIEW

Differing methods of sample treatment in various laboratories can have a profound effect not only on the sensitivity and reproducibility of the test method but also on the results of the test itself. Where sample treatment has not been properly standardized and when resuscitation procedures do not allow the repair of intentional or unintentional injury, large variations in results between laboratories handling the same sample may be expected .

REFERENCES

Al-Hiti, M.A.A., and Gilbert, P. 1980 Changes in preservative sensitivity for the USP antimicrobial agents effectiveness test micro-organisms. *Journal of Applied Bacteriology* **49**: 119-126.

Allwood, M.C. and Hambleton, R. 1972 The recovery of *Bacillus megaterium* spores from white soft paraffin. *Journal of Pharmacy and Pharmacology* **24**: 671-672.

Allwood, M.C. and Hambleton, R. 1973 The recovery of bacteria from white soft paraffin. *Journal of Pharmacy and Pharmacology* **25**: 559-562.

Anon. 1990 Microbial quality management. CTPA limits and guidelines. The Cosmetic, Toiletry and Perfumery Association, London.

Anon. 1992 Pharmacopoeial previews (51): antimicrobial effectiveness testing. *Pharmacy Forum* **18:** 3048-3052.

Baird, R.M., Barnes, E.M., Corry, J.E.L., Curtis, G.D.W. and Mackey, B.M. (eds.) 1985 Quality assurance and quality control of microbiological culture media. *International Journal of Food Microbiology* **2:** 1-138.

Baird, R.M., Corry, J.E.L. and Curtis, G.D.W. (eds.) 1987 Pharmacopoeia of culture media for food microbiology. *International Journal of Food Microbiology* **5**: 187-300.

Baird, R.M., Corry, J.E.L., Curtis, G.D.W., Mossel, D.A.A. and Skovgaard, N.P. (eds.) 1989 Pharmacopoeia of culture media for food microbiology — Additional monographs. *International Journal of Food Microbiology* **9:** 85-144.

Barnes, M. and Denton, G.W. 1969 Capacity tests for the evaluation of preservatives in formulations. *Soap, Perfumery and Cosmetics* **42**: 729-733.

Bowman, F.W., Knoll, E.W., White, M. and Mislivec, P.D. 1972 Survey of microbial contamination of ophthalmic ointments. *Journal of Pharmaceutical Science* **61**: 532-535.

British Pharmacopoeia 1980 Appendix XVIC, A192-194, HMSO, London.

British Pharmacopoeia 1988 Appendix XVIC, A200-203, HMSO, London.

British Pharmacopoeia 1993 Appendix XVIC, A191-192, HMSO, London.

Brown, M.R.W. 1975 The role of the envelope in resistance, pp 71-107 in M.R.W. Brown (ed.), *Resistance of* Pseudomonas areuginosa. John Wiley, London.

Brown, M.R.W., Evans, C., Ford, J.L. and Pilling, M. 1986 A note on the recovery of micro-organisms from an oil-in-water cream. *Journal of Clinical and Hospital Pharmacy* **11**: 117-123.

Cook, A.M. and Steel, K.J. 1959 The antagonism of the antibacterial action of mercury compounds. *Journal of Pharmacy and Pharmacology* **11**: 666-670.

Corry, J.E.L., van Doorne, H. and Mossel, D.A.A. 1977 Recovery and revival of microbial cells, especially those from environments containing antibiotics, pp 174-196 in M. Woodbine (ed.), *Antibiotics in Agriculture*. Butterworths Press, London.

European Pharmacopoeia 1993 Fasicule 16 VIII 14-1–14-4.

Flowers, R.S., Martin, S.E., Brewer, D.G. and Ordal, Z.J. 1977 Catalase and enumeration of stressed *Staphylococcus aureus* cells. *Applied and Environmental Microbiology* **33**: 1112-1117.

German Pharmacopoeia 1986 Part VIII, no. 1, pp 369-370, and Supplement 1989, pp 71-72. Deutsche Apothekerverlag, Stuttgart, Govi-Verlag, GmbH, Frankfurt.

Gilbert, P. 1984 The revival of micro-organisms sublethally injured by chemical inhibitors, pp 175-197 in M.H.E. Andrew and A.D. Russell (eds.), *The Revival of injured Microbes*. Academic Press, London.

Gilbert, P. and Brown, M.R.W. 1980 Cell-wall mediated changes in sensitivity of *Bacillus megaterium* to chlorhexidine and 2-phenoxyethanol, associated with the growth rate and nutrient limitation. *Journal of Applied Bacteriology* **48**: 223-230.

Italian Pharmacopoeia 1985 9th edition, pp 509-512. Insituto Poligrafico e Zecca Dello Starto — Libreria Dello Starto, Roma.

Laminkanra, A. and Allwood, M.C. 1976 The antibacterial activity of non-ionic surface-active agents. *Microbios Letters* **1**: 97-101.

Leak, R.E. and Leech, R. 1988 Challenge testing and their predictive ability, pp 129–146 in S.F. Bloomfield, R. Baird, R.E. Leak and R. Leech (eds.), *Microbial Quality Assurance in Pharmaceuticals, Cosmetics and Toiletries*. Ellis Horwood, Chichester.

Leitz, M. 1972 Critique of USP microbiological test. *Bulletin of the Parenteral Drug Association* **26**: 212-216.

Lindstrom, L.M. 1986 Consumer use testing: assurance of microbiological product safety. *Cosmetics and Toiletries* **101**: 71-73.

Mackey, B.M. 1984 Lethal and sublethal effects of refrigeration, freezing and freeze-drying on micro-organisms, pp 45-75 in M.H.E. Andrew and A.D. Russell (eds.), *The Revival of Injured Microbes*. Academic Press, London.

Martin, S.E., Flowers, R.S. and Ordal, Z.J. 1976 Catalase: its effect on microbial enumeration. *Applied and Environmental Microbiology* **32**: 731-734.

Meynell, G. 1958 The effect of sudden chilling on *Escherichia coli*. *Journal of General Microbiology* **19**: 380-389.

Moore, K.E. 1978 Evaluating preservative efficacy by challenge testing during the development stage of pharmacy. *Journal of Applied Bacteriology* (supplement) **44**: Sxliii-Slv.

Mossel, D.A.A. and van Netten, P. 1984 Harmful effects of selective media on stressed micro-organisms: nature and remedies, pp 329-369 in M.H.E. Andrew and A.D. Russell (eds.), *The Revival of Injured Microbes*. Academic Press, London.

Mossel, D.A.A., Eelderink, I., Koopmans, M.P. and van Rossem, F. 1979 Influence of carbon source, bile salts and incubation temperature on recovery of *Enterobacteriaceae* from foods, using MacConkey-type agars. *Journal of Food Protection* **42**: 470-475.

Oliver, J.D. 1981 Lethal cold stress of *Vibrio vulnificus* in oysters. *Applied and Environmental Microbiology* **41**: 710-717.

Orth, D.S. 1979 Linear regression method for rapid determination of cosmetic preservative efficacy. *Journal of the Society of Cosmetic Chemists* **30**: 320-332.

Orth, D.S. 1980 Establishing cosmetic preservative efficacy by use of D-values. *Journal of the Society of Cosmetic Chemists* **31**: 165-172.

Orth, D.S. 1981 Principles of preservative efficacy testing. *Cosmetics and Toiletries* **96**: 43-52.

Orth, D.S. 1984 Evaluation of preservatives in cosmetic products, pp 403 -421 in J.J. Kabara (ed.), *Cosmetic and Drug Preservation. Principles and Practice.* Marcel Dekker Inc., New York.

Orth, D.S. 1990 Preservative evaluation and testing: the linear regression method, pp 304-312 in S.P. Denyer and R.M. Baird (eds.). *Guide to Microbiological Control in Pharmaceuticals.* Ellis Horwood, Chichester.

Orth, D.S., Lutes, C.M. and Smith, D.K. 1989 Effect of culture conditions and method of inoculum preparation on the kinetics of bacterial death during preservative efficacy testing. *Journal of the Society of Cosmetic Chemists* **40**: 193-204.

Ray, B. and Speck, M.L. 1973 Freeze-injury in bacteria. *CRC Reviews of Clinical Laboratory Science* **4**: 161-213.

Rayman, M.K., Aris, B. and El Derea, H.B. 1978 The effect of compounds which degrade hydrogen peroxide on the enumeration of heat-stressed cells of *Salmonella senftenberg*. *Canadian Journal of Microbiology* **24**: 883-885.

Reinhardt, D.J. 1984 Microbial challenge and *in-use* studies of periocular and ocular preparations, pp 465-479 in J.J. Kabara (ed.) *Cosmetic and Drug Preservation. Principles and Practice.* Marcel Dekker Inc., New York.

Silliker, J.H., Gabis, D.A. and May, A. 1979 ICMSF Methods studies. XI. Collaborative/comparative studies on determination of coliforms using the most probable number procedure. *Journal of Food Protection* **42**: 638-644.

Silver, S. and Wendt, L. 1967 Mechanism of action of phenylethyl alcohol. Breakdown of the cellular permeability barrier. *Journal of Bacteriology* **93**: 560-566.

Spooner, D.F. and Croshaw, B. 1981 Challenge testing — the laboratory evaluation of the preservation of pharmaceutical preparations. *Journal of Microbiology and Serology* **47**: 168-169.

Tagliapietra, L. 1978 Antimicrobial power of liquid and semi-solid pharmaceutical preparations. *Cosmetics and Toiletries* **93**: 23-26.

Tsuji, K., Stapert, E.M., Robertson, J.H. and Waiyaki, P.M. 1970 Sterility test method for petrolatum-based ophthalmic ointments. *Applied Microbiology* **20**: 798-801.

United States Pharmacopoeia 1970 XVIII pp. 845-846 U.S. Pharmacopoeial Convention: Bethesda, Maryland.

United States Pharmacopoeia 1990 XXII pp. 1478-1479 U.S. Pharmacopoeial Convention: Rockville, Maryland.

White, M, Bowman, F.W. and Kirschbaum, K. 1968 Bacterial contamination of some non-sterile antibiotic drugs. *Journal of Pharmaceutical Science* **57**: 1061-1063.

Yablonski, J.I. 1972 Fundamental concepts of preservation. *Bulletin of the Parenteral Drug Association* **26**: 220-227.

3.3 Preservation Efficacy Testing in the Cosmetics and Toiletries Industries

Brian F. Perry

INTRODUCTION

Cosmetic microbiologists require maintenance and preservation of microbial cultures for research, training, and testing purposes. The conservation of viable, uncontaminated cultures without variation or mutation in their original isolated characteristics is essential for environmental isolates (product, factory, etc.), preservation testing (challenge) of antimicrobial agents (e.g., preservatives), efficacy and stability testing, quality control evaluations (e.g., positive controls in media preparation and identification tests of isolates), etc. Culture maintenance implies both viability and purity, whereas preservation and conservation involve retention of phenotypic and genotypic characteristics over a period of time.

This chapter will consider the various short- and long-term methods available for microorganism preservation and will focus on inocula preparation considerations in preservation (challenge) efficacy testing.

EUROPEAN COMMUNITY COSMETICS DIRECTIVE

For the production of high-quality cosmetics and toiletries, and to maintain their shelf-life stability, all stages of the Microbial Quality Management (MQM) system (CTPA, 1993), must be optimized. At the same time, within the European Community (EC), it is necessary to comply with the regulations of the Cosmetics Directive 76/768EEC and subsequent amendments. The Basic Council Directive 76/768/EEC was signed July 27, 1976, and published in *Official Journal of the European Community* September 27, 1976, number L262. The last major summary (i.e., Update Version) was in March 1989 and

0-8493-4752-1/95/$0.00+$.50

is also embodied in Schedule 4 of the UK Cosmetic Products (Safety) Regulations 1989. Since 1976 there have been six Amendments to the Commission Directive: the Sixth 94/35/EEC, signed June 14, 1993, number L151/32, covers amendments to the Articles. There have also been 16 Adapting Commission Directives: 16th 93/47/EEC, signed June 22, 1993, and published in *Official Journal of the European Community*, August 13, 1993, number L203/24, also covers amendments to the Articles but specifically adapts the technical progress of the Directives, i.e., changes in ingredients status.

Amendments to the EC Cosmetics Directive deal with the basic articles of the Directive and the structure of Annexes covering prohibited, restricted, and controlled ingredients. On June 14, 1993, the text of the Sixth Amendment (93/35/EEC) was published. This introduced some fundamental changes to the Directive, and notably included Good Manufacturing Practices (GMP). Microbiological quality standards must be applied due to legal demands. In the context of this chapter some Articles merit mention:

Article 1(1) "A 'cosmetic product' shall mean any substance or preparation intended to be placed in contact with the various external parts of the human body (epidermis, hair systems, nails, lips, and external genital organs) or with the teeth and the mucous membranes of the oral cavity with the view exclusively or mainly to cleaning them, perfuming them, changing their appearance and/or correcting body odours and/or protecting them or keeping them in good condition.

Article 2 "A cosmetic product put on the market within the Community must not cause damage to human health when applied under normal or reasonably foreseeable conditions of use.

Article 6(1c) "The date of minimum durability of a cosmetic product shall be the date until which this product, stored under appropriate conditions, continues to fulfill its initial function and, in particular, remains in conformity with Article 2. Indication of the date of durability shall not be mandatory for cosmetic products, the minimum durability of which exceeds 30 months."

The Directive defines general requirements, such as safety and minimum stability, which necessarily involve working to certain standards of microbial quality. In comparison with the clear qualitative and quantitative guidelines for the pharmaceutical industry, little equivalent information is given with respect to cosmetics (CTPA, 1990b). The MQM Limits and Guidelines of the CTPA and the Microbiology Guidelines of the CTFA are useful references. (CTFA, 1993; CTPA, 1990a).

For microbial quality, it is therefore clear that while manufacture of cosmetics under aseptic conditions is not required, the manufacturer must ensure that cosmetics reaching the market place are of satisfactory microbial quality. He must also ensure as far as possible the continued resistance of the cosmetic to microbial insult and dynamic contamination by microorganisms over its shelf life, and then literally, in the hands of the consumer. Since quality can only be produced in the cosmetic and cannot be introduced into it by means of tests, all procedures that influence quality must be adapted by cosmetic manufacturers in order to comply with current and projected quality requirements.

PRESERVATION

Preservation means including protective mechansms in the cosmetic in order to avoid microbial growth (Bloomfield et al., 1988; Corbett, 1992; CTFA, 1993; CTPA, 1994; Curry, 1987; Denyer and Baird, 1990; Diehl, 1992; Durant and Higdon, 1987; Geis, 1988; Kabara, 1984; Muscatiello, 1993; Orth, 1993; Spooner, 1989; Van Doorne, 1992; Warwick, 1993). If the MQM strategy has been optimally applied, nothing else should stand in the way of effective preservation. The precondition is that during the development of a cosmetic, the correct preservation system has been determined. When formulating preserved systems (Anon., 1993a; Cowen and Steiger, 1977) a general principle is to include as little antimicrobial agent as is technically necessary to achieve effective and adequate preservation. In order to determine whether or not this optimal concentration is bioavailable, a wide variety of different methods, all of which have been described previously (Anon., 1990, 1993b,c,d,e; Brannan, Dille and Kaufman, 1987; Bryan, Fizer and Farrington, 1980; Cowen and Steiger, 1976; CTFA, 1986; Gay, 1983; Kabara, 1984; Levy, 1987; McLaughlin et al., 1984; Sabourin, 1990; Tran and Collier, 1992; Tran, Hitchins and Collier, 1990) may be carried out.

Microbial challenge of formulations form a useful basis for evaluating the bioavailability of preservatives in cosmetics. The preservative efficacy (preservation challenge) test (PET) is based upon the insult of a cosmetic with known microorganisms and monitoring of the levels of viable microorganisms over a specified time period (Figure 1; CTPA, 1994). These tests must be designed to be as reproducible as possible since they will form part of a stability program designed to demonstrate shelf-life expiration dating (IFSCC, 1992). It must be clearly understood that a microbial preservation efficacy test is a laboratory test to demonstrate that raw materials, intermediates, and finished products have the ability to withstand known microbial insult. It cannot be assumed that cosmetics which satisfy specific preservation efficacy criteria may never become contaminated. The best protection is to adhere to an effective MQM strategy. The ability to withstand a known microbial insult simply demonstrates the bioavailability of the preservation system in the formulation.

Official test methods include standardized procedures: British and United States Pharmacopoeial [BP (Anon., 1993b) and USP (Anon., 1990)] methods, the Cosmetic, Toiletry and Frangrance Association (CTFA) M3 (Anon., 1993c) and M4 (Anon., 1993d) methods, and the draft Association of Official Analytical

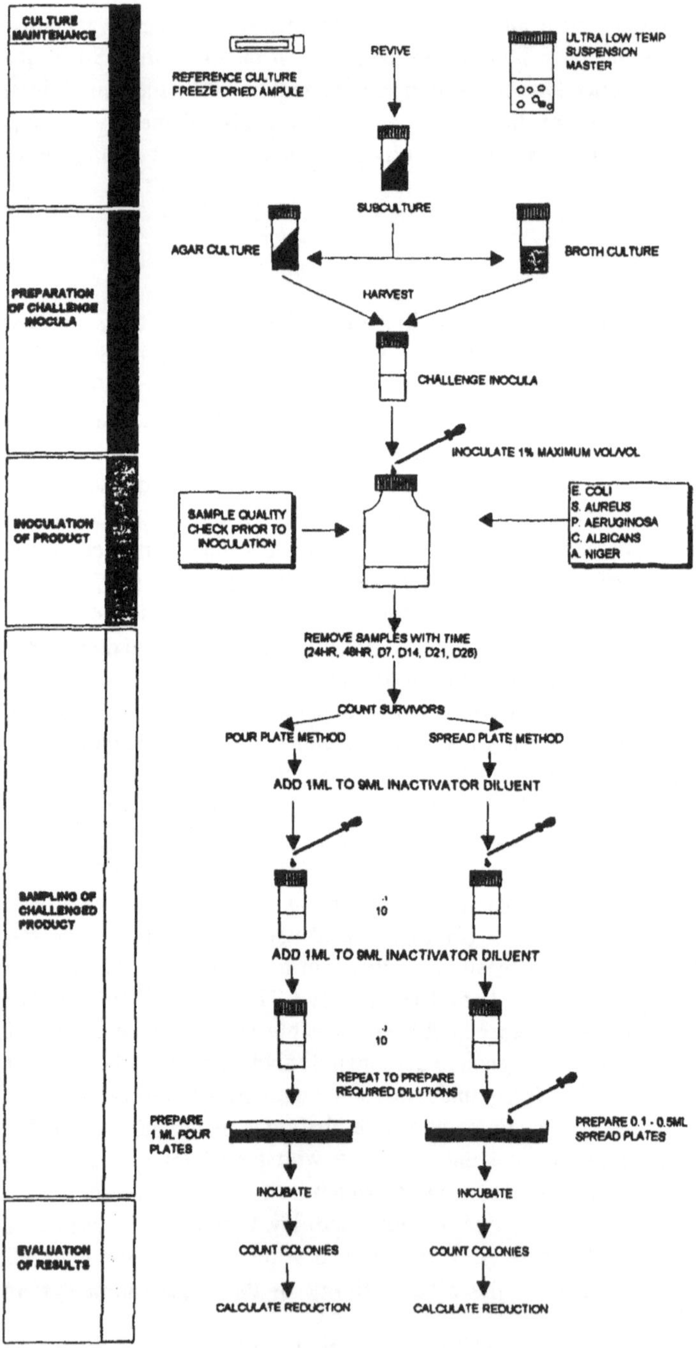

FIGURE 1 A diagrammatical representation of the process involved in preservation efficacy testing. (From CTPA, *Guidelines for Preservation Efficacy Testing*, CTPA, London, 1994. With permission.)

Chemists/CTFA Method (Anon., 1993e). These methods together with a number of other methods have been reviewed recently (see Tables 2 and 3).

PREPARATION OF INOCULA

The objective of a preservation efficacy test (PET) is to determine the fate of challenge microorganisms contacting with the product. The key word is "contact", since without physical contact between the microorganism and a formulation containing an antimicrobial ingredient, e.g., a preservative, the preservative cannot be effective. For this reason all challenge test methodologies involve placing microorganisms in contact with the product. The main differences between PETs can be determined by answering the following questions about the test:

1. Which microorganisms should be selected?
2. How many microorganisms should be added?
3. How is contact made between formulation and microorganism?
4. What number of insults will be introduced into the formulation?
5. When are samples plated?
6. What type(s) of artificial stress(es) is introduced?
7. What are the pass criteria?

It is important to keep this list in mind as different methodologies are discussed. In the following sections, consideration is given to the approaches that may be employed in inoculum preparation and addition to products. The emphasis will be on the need to obtain a repeatable and reproducible challenge and response. Reference will be made to the Pharmacopoeial methods since cosmetic PETs have tended to derive from them.

The physiological status of microbial inocula is one of the major sources of variation in microbiological tests. The variation can be attributed to inoculum size, history, and subsequent preparation (see Chapters 1.1 and 3.1). This variation can only be controlled by the standardization of the preparation procedures so that PETs are both reproducible and repeatable.

The route of preparation of challenge inocula is summarized in the flow chart (see Figure 2, CTPA, 1994). To facilitate testing, once a stock culture is prepared and validated only the final inoculation step is required for routine testing.

Selection of challenge test microorganisms

↓

Establishment and maintenance of master cultures

↓

Preparation of stock cultures

↓

Preparation of challenge inocula from stock/working culture

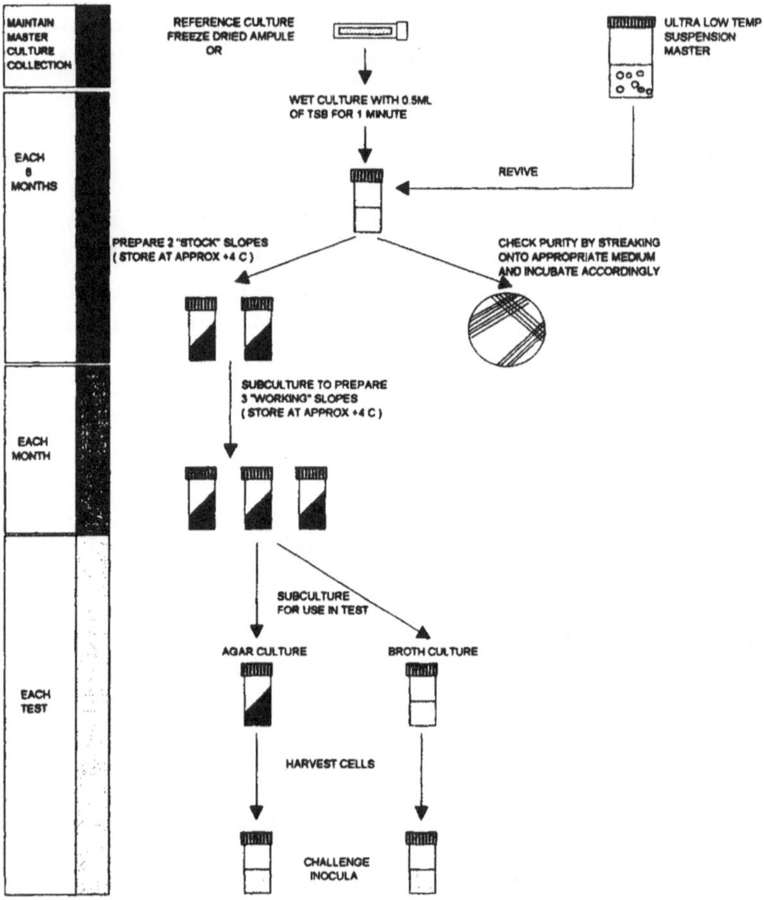

FIGURE 2 A diagrammatical representation of the preparation of challenge inocula from master culture collection. (From CTPA, *Guidelines for Preservation Efficacy Testing*, CTPA, London, 1994. With permission.)

Selection of Preservative Efficacy Test Microorganisms

The types of microorganisms included within a challenge test are a major factor in assessing preservation capacity. To be meaningful, PET strains must be chosen for their known association with cosmetic contamination, resistance to preservatives, opportunistic contamination capacity, low nutrient demand, and adaptability. In addition, it should be standard practice to include a range of primary or potential pathogens which might contaminate and cause infection from an inadequately preserved cosmetic (Perry, 1994; Ringertz and Ringertz, 1982).

Principal isolates from cosmetics, topical pharmaceuticals, and susceptible ingredients rest heavily with gram-negative bacteria, principally the nonfermenters, *Pseudomonas*, *Achromobacter*, and *Flavobacter*; the fermenters *Klebsiella* and *Enterobacter*; and gram-positive cocci and fungi of *Candida*, *Penicillium*, and *Aspergillus*, (Cowen and Steiger, 1976; Sabourin, 1990). *Staphylococcus aureus* does not figure prominently in this list, probably due to its very demanding nutrient requirements. It is, nevertheless, routinely used in PETs because of the threat it poses from consumer contamination of frequently used cosmetics. A selection of the above named microorganisms appears in most of the major PETs (Table 1).

The Pharmacopoeias (USP, BP and EP) rely heavily on five culture collection type strains for the testing of topical products; *Candida albicans*, *Aspergillus niger*, *Staphylococcus aureus*, *Pseudomonas aeruginosa*, and *Escherichia coli* (oral BP and EP). Although opinion is given for the use of other microorganisms, where these may represent likely contaminants to the preparation (Table 2), it should be noted that these are not referenced in USP/EP/BP harmonization proposals (Pharmeuropa, 1993). The proposed CTFA/AOAC PET for non-eye-area, water-miscible cosmetic and toiletry formulations advocates the same range of culture collection strains as do the USP, BP, and EP, but in addition, employs type-strains of *Staphylococcus epidermidis*, *Klebsiella pneumoniae*, *Enterobacter gergoviae*, *Pseudomonas cepacia*, and *Acinetobacter baumanis* (Table 3).

A standard core of challenge microorganisms such as those detailed in the Pharmacopoeias and CTFA/AOAC are recommended to aid development and standardization of a method, since these may be readily retrieved from culture collections and have published data associated with them. Additional microorganisms can be included in the test procedure (Table 1) when

1. Preservation problems have been encountered or may not be predicted with type-strains of microorganism, since these may not necessarily represent strains of typical resistance or growth characteristics. This, therefore, underlines the use of vigorous strains of contaminants which have been isolated either from the cosmetic itself or from the factory environment, in addition to type-strains. These "environmental isolates" broaden the testing spectrum and are selected on the basis of their ability to grow in unpreserved product.
2. Addition of microorganisms which are specific to the product use or type. This may include the use of mixed natural populations, e.g., saliva in intimate-eye-area cosmetic PETs.

Particular isolation techniques may not, in themselves, be sufficient to ensure challenge with an isolate in its most resistant state. Isolates recovered from contaminated product have often failed to establish themselves in noncontaminated samples of the same preserved product even though the isolates in question have only been subjected to a single passage in nutrient medium. The lack of methodologies which maintain, in a reproducible manner,

TABLE 1
Test Organisms Employed in Various Preservative Efficacy Challenge Tests

Pharmacopoeia	CTFA/AOAC	Society of Cosmetic Chemists (UK)	Examples from Various Authors
			Yeasts and Molds
C. albicans ATCC 10231	*C. albicans* ATCC 10231	**Toothpaste**	*A. niger*
		Product contaminants	*A. flavus*
A. niger ATCC 16404	*A. niger* ATCC 16404	**Shampoo**	*Cladosporium* spp.
		Gram negative factory contaminants product spoilage	*Fusarium* spp.
P. aeruginosa ATCC 9027	*P. aeruginosa* ATCC 9027	*Pseudomonas* spp.	*Eupenicillium levitum*
	P. cepacia ATCC 25416	**Creams and Lotions**	*Rhizopus* spp.
	Acin. baumanii ATCC 19606	*S. aureus*	*Trichoderma viride*
S. aureus ATCC 6538	*S. aureus* ATCC 6538	*S. faecalis*	*Penicillium* spp.
		P. aeruginosa	*C. albicans*
	S. epidermidis ATCC 12228	*P. fluorescens*	*C. parasilosis*
E. coli ATCC 8739	*E. coli* ATCC 8739	*E. coli*	*C. tropicalis*
Zygosacch cerevisiae NCYC 381	*Ent. gergoviae* ATCC 33028	*Klebsiella* spp.	*(Zygo) Sacch. cerevisiae* — 4105
	K. pneumoniae ATCC 10031	*Proteus* spp.	*Torula* spp.
		A. niger	**Gram-Positives**
Product contaminants		*Penicillium* spp.	*Bacillus* spp. e.g. *subtilis* ATCC 8478
		Cladosporum spp.	*Corynebacterium* spp.
		Alternaria spp.	*Enterococcus* spp.
		Fusarium spp.	*Micrococcus/Sarcina*

Mucor spp.	**Gram-Negatives**
Rhizopus spp.	*Acin. anitratus*
Phoma spp.	*Alcaligenes* spp.
Trichoderma spp.	*Cit. freundii*
Verticillium spp.	*E. coli*
C. albicans	*Enterobacter aerogenes* ATCC 13048
Saccharomyces cerevisae	*cloacae-gergoviae*
Eye Cosmetics	*K. pneumoniae* ATCC 8303 *oxytoca*
P. aeruginosa	*Morganella morgani*
P. fluorescens	*Moraxella* sp.
Micrococcus luteus	*Proteus* spp. e.g. *vulgaris* ATCC 881
S. faecalis	*Providencia stuartii/rettgeri*
Fresh saliva	*Salm. choleraesius* ATCC 10708
	enteriditis
	Serr. liquefacians/marcescens
	Pseudomonas aeruginosa/cepacia
	Pseudomonas fluorescens/putida
	Pseudomonas maltophila/stutzeri

Data from Cowen, R.A. and Steiger, P., *Journal of the Society for Cosmetic Chemists*, 27, 467–481, 1976; and Sabourin, J.R., *Drug and Cosmetics Industry*, 147, 24–26, 64, 65, 1990.

TABLE 2
Inoculation in Preservative Efficacy Challenge Tests: A Pharmacopoeia

EP (Ph. Eur. 2)VIII.14 1992 and BP Appendix XVIC 1993	USP XXII 1990	International Harmonisation Ph. Eur. 2 and SUP XXII Proposals (Pharmeuropa 5.4 1993)
Candida albicans ATCC 10231	*Candida albicans* ATCC 10231	*Candida albicans* ATCC 10231
Aspergillus niger ATCC 16404	*Aspergillus niger* ATCC 10644	*Aspergillus niger* ATCC 10644
Escherichia coli ATCC 8739 (Oral)	*Escherichia coli* ATCC 8739	*Escherichia coli* ATCC 8739
Pseudomonas aeruginosa ATCC 9027	*Pseudomonas aeruginosa* ATCC 9027	*Pseudomonas aeruginosa* ATCC 9027
Staphylococcus aureus ATCC 6538	*Staphylococcus aureus* ATCC 6538	*Staphylococcus aureus* ATCC 6538
Other microorganisms[a]	Other microorganisms[b]	
From solid medium, suspend in sterile saline-peptone (0.9% NaCl and 0.1% peptone) add 0.05% polysorbate 80 for *A. niger* spores.	From solid or liquid medium, suspend in sterile saline (0.85% NaCl); add 0.05% polysorbate 80 for *A. niger* spores)	Saline (0.9% NaCl) — peptone (0.1%) for bacteria and yeast Saline (0.9% NaCl) polysorbate 80 (0.05%) for mold spores
Single-strain challenges without rechallenge	Single-strain without rechallenge	Single-strain challenges without rechallenge
Maximum 1% inoculum volume in volume of product to give 10^{5-6} cfu/ml or g	0.1 ml into 20 ml product (≤0.5%) to give 10^{5-6} cfu/ml or g	Maximum 1% inoculum volume in volume of product to give 10^{5-6} cfu/ml or g
High sugar oral preparations *Zygosaccharomyces rouxii* NCYC 381		
BP XVIC 1980 and 1988	USP XX 1980 and USP XX1 1985; optional basis, especially those that may represent contaminants likely to be introduced during use	
Specified organisms representative of those found in environment during manufacture, storage and use *also* should be supplemented with other strains especially those likely to be found in manufacture and during use or offer particular challenge to product		
Mix inocula thoroughly to ensure homogeneous distribution	Inocula dispersion not specified but should be homogeneous	Mix thoroughly to ensure homogeneous distribution

[a] Those that may represent likely contaminants to the preparation.
[b] Optional basis, especially those that may represent contaminants likely to be introduced during use.

TABLE 3
Inoculation in Preservative Efficacy Challenge Tests: B Cosmetics and Toiletries

CTFA - AOAC: Draft 6.2 1994 Non-Eye-Area Water-Miscible	CTFA - M3: 1993 Cosmetics and Toiletries - Various	CTFA - M4:1993 Eye Area
I *Gram-Positive Cocci:*	**I *Gram-Positive Cocci***	
Staphylococcus aureus ATCC 6538	*Staphylococcus aureus*	At least one
Staphylococcus epidermidis ATCC 12228	*Staphylococcus epidermidis*	At least one
II *Fermentative Gram-Negative Bacilli*	**II *Fermentative Gram-Negative Rod***	
Klebsiella pneumonia ATCC 10031	*Klebsiella pneumoniae*	At least one
Escherichia coli ATCC 8739	*Enterobacter cloacae*	At least one
Enterobacter gergoviae ATCC 33028	*Enterobacter gergoviae*	At least one
	Escherichia coli	At least one
	Proteus spp.	At least one
III *Non-Fermentative Gram-Negative Bacilli*	**III *Nonfermentative Gram-Negative Rod***	
Pseudomonas aeruginosa ATCC 9027	*Pseudomonas aeruginosa*	At least two (at least one in addition to *P. aeruginosa*)
Pseudomonas cepacia ATCC 25416	*Pseudomonas cepacia*	
Acinetobacter baumanii ATCC 19606	*Pseudomonas fluorescens/putida*	
	Flavobacter spp.	
	Acinetobacter spp.	
IV *Yeast and Mold*	**IV *Yeast***	
Candida albicans ATCC 10231	*Candida albicans*	At least one
Aspergillus niger ATCC 16404	*Candida parasilosis*	At least one

TABLE 3 (Continued)
Inoculation in Preservative Efficacy Challenge Tests: B Cosmetics and Toiletries

CTFA - AOAC: Draft 6.2 1994 Non-Eye- Area Water-Miscible	CTFA - M3: 1993 Cosmetics and Toiletries - Various	CTFA - M4:1993 Eye Area
	V *Mold*	
	Aspergillus niger	At least one
	Penicillium luteum	At least one
	VI ***In-house Isolates*** ***("resistant" strains)***	At least one
	As appropriate	At least one
From solid medium suspend in sterile saline (0.85% NaCl), add 0.05% polysorbate 80 for *A. niger* spores	From solid or liquid medium suspend in sterile saline (0.85% NaCl) or phosphate buffer (pH 7.0) or from similar contaminated product or specific methods for powders or waxed based products, e.g., spraying (aqueous or oil or emulsified or dipping)	
Four groups mixed-strain challenges — see above	Either single-or-mixed strain challenges — see above — with at least one rechallenge	
0.2 ml each pool into 20 ml product to give 10^{6-7} bacterial/ml or 10^{5-6} fungi/ml	0.1–1 ml into minimum 20 ml to give 10^6 bacteria/g or 10^5 yeast and mold spores/g challenge should be larger than the total challenge expected during consumer use	10^5 yeast and 10^5 mold spores/g
Inocula evenly distributed throughout product	Inocula thoroughly mixed manually or mechanically to provide proper even dispersion of inoculum in product	Inocula thoroughly mixed manually or mechanically to provide proper even dispersion of inoculum in product

the genetic and phenotypic characteristics of primary "environmental isolates" has been one of the major criticisms against their use in routine testing.

At this point, it must be emphasized that product preservation is not a substitute for poor GMP and that the choice of test strains must be aligned to the preservation strategy adopted for a particular product. The selection of preservative levels on the basis of the most resistant known organism is not recommended by the author since it may evoke toxicity issues and place an unnecessary burden on the environment. Also, if this approach is used it represents an upward spiral of events since highly resistant microorganisms can often be isolated from the environment but are unlikely to contaminate cosmetic and pharmaceutical products.

Establishment and Maintenance of Master Cultures

Master culture collections should be established from reference cultures by revival and culture (see Figure 2), and maintained and preserved by an appropriate long-term technique (CTFA, 1993; Heckly, 1978). These techniques are covered elsewhere in this book (Chapter 2.1) and are also reviewed admirably in guidelines published by the CTFA (CTFA, 1993). The techniques include

Short-Term Methods

Subculturing
Overlaying

Long-Term Methods

Drying

Sand and soil
Porcelain penicylinders and silica gel crystals
Paper strips or disks
Predried plugs
Gelatine disks

Freezing

Frozen suspension
Glass/bead storage
Ultrafreezing

Freeze Drying

If type-strains only are employed, then master cultures, and subsequently inocula, should be prepared from cultures received from internationally recognized culture collections.

Permanent service culture collections provide customers with an assurance of supply of the "same" strains over time, and use of the "same" cultures in different laboratories makes their work directly comparable. The "same" test organisms validates tests carried out at different times or places. While returning to the same source to replenish master cultures is self-evidently good user practice, it is effective only if the collection can guarantee that "sameness". Yet no collection gives that guarantee, the subject material is biological, methods of QC have their own inherent variability, most tests have elements of interpretative subjectivity, and media and their constituents can change. All a service collection can do, and what the customer can expect, is to apply the best technology science can provide, within the limits of their resources. "Sameness" is dependent on the state of the art (Hill, 1991).

Service culture collections take three policy measures to ensure "sameness" as far as it is practicable with biological materials: conservation, authentication, and development.

Preservation (Conservation)

Freeze-drying and/or immersion in liquid nitrogen are used as preservation methods, with customers usually being supplied with freeze-dried ampoules of culture. The protocols are standard, with regular viability and purity checks being made and recorded. The simplistic view is that microorganisms are viable but neither metabolizing nor dividing, thus obviating genetic drift in the stored culture. In contrast, active subculture is much less secure, with accidental loss, contamination at time of subculture, human error in getting cultures muddled or mislabeled, and the risk of selecting mutant lines as potential risks. The question is often asked: "How many active subcultures can be made from a single freeze-dried ampoule in order to maintain local supplies?" It is impossible to be categorical in reply: as few as can be managed, but the proposed USP PET revision is considering five as a recommendation.

Authentication

A wide set of characterizing tests are used to identify the preserved cultures. These are detailed, and permanent records are kept of the data (e.g., GLP documentation) generated.

Development

Service culture collections generally have significant Research and Development programs to retain and improve standards. In recent years there have been very significant advances in molecular biology that enable a greater degree of certainty with regard to "sameness" fingerprinting of microorganisms. For much-used test strains, the user will find equivalent strains in different service collections. Many original NCTC strains are also held in ATCC under ATCC numbers, but NCTC numbers are also given, and vice versa. If the designated specific culture collection strains (see Tables 2 and 3) are available from say, NCTC, and ATCC, then the respective culture collections do not claim that their versions are identical, but rather that one would have been derived from the other. Both collections, however, perform regular quality control procedures, including reciprocal rechecking of each other's versions. Using modern molecular techniques it is conceivable that versions currently regarded as equivalent in different culture collections may be termed "identical".

A master culture is intended as an internal reference standard to maintain the genetic and phenotypic characteristics of the reference culture. These master cultures should not be used as a stock/working culture. They can be stored as freeze-dried ampoules, low-temperature suspensions, low-temperature suspensions on beads, or through suitably validated alternatives. For those systems which allow repeated retrieval from the same ampoule, care must be exercized so that the errors described under active subculture do not occur.

Microbiologists must have some assurance that the microorganisms used, especially "environmental isolates" will retain their original characteristics. In many cases, conventional methods for maintaining and preserving such isolates may be inadequate. The microbiologist should give special attention to these microorganisms and store them in such a manner that their original attributes are not lost and viability is maintained. This may mean periodic subculture in the product from which the original isolate was obtained (although this does not guarantee success), or devising laboratory media, incorporating a percentage of the product's preservative, that will ensure a culture that retains the desired spoilage/resistance traits.

As with type-strains, environmental isolates should have well-documented profiles. It must be recognized, however, that identifications of bacteria isolated from cosmetics may differ depending on the system used (McLaughlin et al., 1984). In addition to periodic screening, certain key traits unique to the isolate should be monitored frequently to provide an added measure of assurance of strain stability. Any deviation from the expected profile may indicate that the culture is no longer suited for use.

In straightforward terms, reference collections of environmental isolates should be maintained and preserved in an identical manner to service culture collection strains. This is especially true for PETs which employ both types of strains linked to the acceptance criteria. This means establishing a reference culture collection akin to those of the national culture collections or having strains deposited and maintained by them, on a contract basis. The latter could become an expensive option and some degree of compromise generally occurs. The simplest option is, however, to use *only* culture collection type-strains as reference strains, and then to establish a master and stock-working-culture system in-house.

Microorganisms should be revived using the method as recommended by the culture collection. This revived culture can be used to establish a master culture collection. When recovering microorganisms from freeze-dried ampoules it must be recognized that

1. After revival, several subcultures may be needed before the microorganism is in its optimum state.
2. The selection of growth/recovery media is important since minor changes can alter microorganism susceptibility (Chapters 1.2 and 3.1).

Preparation of Stock Culture

Stock cultures must be prepared from master cultures (see Figure 2). To minimize the use of the master cultures, and to keep the number of passages to a minimum, stock cultures should be prepared from them. These may then be stored at low temperatures as cell suspensions or on solid media. When storing stock cultures, it should be remembered that

1. All techniques should be standardized and validated.
2. All isolates should be periodically investigated to confirm identity and resistance. (The USP convention [April 94] has considered phenol coefficients as a means of assessing and validating resistance. This proposal is now to be deleted but still under consideration is an antibiogram or the use of "standard" preservatives as a means of standardizing susceptibility characteristics).
3. It is not recommended to continuously subculture isolates once recovered from stock cultures. Microorganisms should be periodically revived from stock preparations. A maximum of five passages is currently being considered by the USP convention, but at present has not been adopted.
4. In-house (environmental) isolates may present unique storage problems. Storage in the original product or the addition of preservative(s) to a maintenance medium is often the only way to retain viability and continued aggressiveness/resistance. Such methods are especially appropriate where the isolates are subsequently inoculated into similar products.

Stock cultures may be used directly as working cultures, or, alternatively, working cultures can be prepared from stock cultures. Working cultures are defined in this chapter as those used to prepare the challenge inocula.

Preparation of Challenge Inocula from Stock/Working Cultures

A standardized method must be designed for preparation of challenge inocula, using either liquid or solid culture media (Figure 2 and Table 4). The level of reproducibility is, however, significantly greater for liquid than solid cultures and decreases markedly with the density of colonies upon seeded agar plates (Al-Hiti and Gilbert, 1980, 1983).

Soya-bean casein digest medium, Nutrient agar, and Sabouraud agar are commonly used media for storing and propagating bacteria and fungal species for use in PETs (Table 4). The growth of challenge organisms for PETs in conventional microbiological media is, however, open to question. In order to

TABLE 4
Maintenance of Challenge Microorganisms

Source	Culture	Maintenance Media	Storage Conditions	Transfer Frequency
CTFA	Bacteria	Nutrient agar	Refrigeration, 4–8°C	Weekly, biweekly, or monthly
PET Guidelines		Tryptic (Trypticase) soy agar		
M-3, M-4		Eugon agar		
	Yeasts	Tryptic (Trypticase) soy agar	Refrigeration, 4–8°C	Biweekly or monthly
		Potato dextrose agar		
		Mycophil (mycological agar)		
	Molds	Sabouraud dextrose agar	Refrigeration, 4–8°C	Biweekly or monthly
		Potato dextrose agar		
		Mycophil (mycological agar)		
CTFA/AOAC	Bacteria	Nutrient agar		
PET Guidelines	Yeast	Yeast mold agar	No details given	No details given
Draft 6.2	Mold	Potato dextrose agar		
BP 1993 and EP 1992	Bacteria	Tryptone soya agar	No details given	Recently grown stock culture
	Fungi	Sabouraud agar		
USP 1990	Bacteria	Soy-bean-casein digest agar (Tryptone soya agar)		
	Fungi	Sabouraud dextrose agar	No details given	Recently grown stock culture
		Potato dextrose agar		

ensure adequate retention of resistance, the possibility of propagating the organism in a suitable medium to which has been added the preservative or product in low concentration has been proposed (Cowen and Steiger, 1976). Similarly in order to simulate low-nutrient, natural conditions, then isolates can be propagated in conventional media which have been reconstituted at reduced strength, for example at 10% w/v.

Minor changes in growth medium can vary resistance in a marked fashion (Cowen and Steiger, 1976). Standardization and validation of media preparation has to be carefully considered (Post, 1991). Validation of the growth promotion from low levels of inocula of each autoclaved lot of media requires standardized and reproducible inocula. A simple and effective approach is the Quanti Cult system from Chrisope Technologies, Lake Charles, LA. The basic kit delivers a guaranteed specified range of colony-forming units (10 to 100) of a number of ATCC strains. Cell suspensions require no growth periods and have shelf lives of at least 6 months.

Mold inocula should be prepared as mycelial-free spore suspensions derived from solid culture media. If mycelial growth masses are encountered during preparation of, for example, *Aspergillus niger* spore suspensions, these can be removed by filtration through sterile glass wool or gauze.

In order to prevent the inadvertant transfer of growth media, used during revival from stock culture, into product, isolates can be washed (e.g., filtration or centrifugation) in harvesting medium prior to suspension preparation. This will, however, have effects upon susceptibility as many of the prepared cell suspension will have become damaged by the process (see Chapter 2.2). Tables 2 and 3 list the recommended harvesting media from various PETs. Orth, Lutes and Smith (1989) have shown that broth inocula perform differently in PETs than saline inocula prepared from surface growth on agar media. These authors recommend the latter approach.

Harvesting medium is intended as a transport medium; it should neither affect viability nor support growth and should allow the recovery of sublethally damaged microorganisms. *Staphylococcus aureus* may die in saline yet retains its viability in phosphate buffer. Van Doorne (1992) has shown that *E. coli* and *P. aeruginosa* are capable of some growth in phosphate buffer, but in borax buffer at the same pH both organisms die rapidly. This shows that the choice of buffer may affect the potential for microbial survival. The International Commission for Harmonization (ICH) of Pharmacopoeias recommends peptone saline for this purpose. The actual inoculum count of the final standardized preservation efficacy test is established using agar plate count or other alternative validated methods.

Inoculation of Product

Before inoculating the product, a sample quality check of uninoculated product should be performed. This is used to ensure that the product is not contaminated before inoculation with challenge test microorganisms.

All products should be mixed thoroughly after inoculation — manually or mechanically — in order to distribute the challenge microorganisms uniformly. The various mixing methods include vigorous mixing with a stirring rod or by capping and hand-shaking; mixing in a vortex mixer; mixing with a magnetic stirrer, a non-aerating stirrer, or micro blender; gentle mixing with ultrasonic equipment at lower power; mixing with a tissue grinder; or finally, mixing in a plastic bag either by hand or with machines such as a Stomacher. For some solid cosmetics — for example, sticks, soaps, and powders — it is not possible to uniformly distribute the inoculum throughout the product, and some surface inoculation technique is required (CTFA M3 and M4; direct membrane inoculation techniques of Tran and Collier [1992] and Tran, Hitchins and Collier [1990]).

Levels of Inoculum

Samples should be inoculated with freshly prepared inocula to give approximately 10^6 cfu per ml/gram for bacteria, 10^5 cfu per ml/gram yeasts and molds. Cell density standardization can be achieved through rigorously adhering to cultural and harvest conditions; by spectrophotometric adjustment or by reference to McFarland barium sulfate standards. The volume of inoculum added should not exceed 1% v/v so that inocula volume does not significantly change the formulation.

Levels of inocula recommended in various test methods are not intended to represent any real challenge situation. They are, however, intended to ensure that the challenge test can demonstrate an appropriate number of log reductions of the challenge inocula, given detectable limits of the recovery system. This determines whether the preservation system has the capacity to control a large number of inoculum (Van Doorne and De Vringer, 1994).

Ideally, freshly prepared inocula should be used for challenging products. If stored inocula are used, the storage method and conditions must be controlled to ensure continued repeatability. The reproducibility of tests has been much improved by the use of cell suspensions stored at –50°C (Bohannon and Wen, 1978) or at –196°C/liquid nitrogen (Beezer et al., 1976, 1981, 1986; Cosgrove, Beezer and Miles, 1979). Care must be exercised, however, and it is not safe to assume that frozen-recovered inocula perform in the same manner as fresh inocula (Gay, 1983, 1991; Chapter 2.1).

The product should be inoculated with a single species to obtain specific data on each microorganism used in the test. If appropriate, however, a pooled culture approach can be used, for example, as a means of reducing experimental effort or in preference to limiting the range of test strains and/or the number of tests. If mixed pools of microorganisms are used it is recommended that, broadly speaking, related types of microorganisms be separately pooled — for example, gram-positive species, gram-negative species, fungal spores, and yeast cells (Table 3). It must be noted, however, that competitive effects of pooled microorganisms may affect test reproducibility.

Number of Challenges

No fixed pattern for rechallenging products with microorganisms has so far emerged for challenge tests of cosmetics and toiletries. The Pharmacopoeias and proposed AOAC use a single challenge. At least one rechallenge is recommended by the CTFA, particularly for those cosmetics subjected to multiple challenge during use. A rechallenge is useful for determining whether or not a formulation is marginally preserved, and in indicating which types of microorganisms may represent potential problems for that particular formulation. Some authors favor a multichallenge test (Cowen and Steiger, 1976) because it seems to represent more closely the in-use situation and is a means of predicting that preservation is likely to be adequate over the life of the product. While it could be argued that such testing could lead to excessive preservation of a product, it is not necessarily the case. This will depend upon the preservation system and the number of challenges made. The frequency of rechallenge will dictate the ease with which inoculum preparation can be standardized.

In-Use Tests

The purpose of a preservation system is to protect the consumer and product formulation from microbial contamination. The most significant product contamination can occur through consumer misuse. The ultimate test is, therefore, to have the product evaluated in the hands of the consumer.

Consumer-use testing should be regarded as confirmatory testing while laboratory challenge tests are intended to be predictive of preservation during use. Predictive-challenge tests can only be developed by comparing the results of laboratory tests with the experience of manufacturing the products, and with the results of consumer tests (Brannan, Dille and Kaufman, 1987). Often challenge tests' conditions are stressed in order to simulate in-use conditions, e.g., dilution of product and "dirty" inoculations by the addition of glucose and peptone. The ultimate aim of this approach is to simulate in the laboratory that which might occur during use. Clearly any stressed PETs must be also carefully standardized.

SPECIAL CONSIDERATIONS

The structure, composition, and physiology of microorganisms changes markedly according to their rate of growth, the availability of nutrients, and whether or not they are free-living or growing as biofilms in association with surfaces (Brown, 1975; Gilbert, 1991; Gilbert and Wright, 1987). In this respect, envelope structures are generally optimized towards the acquisition of the least available nutrient, and cellular metabolism is geared towards the most efficient utilization (see Chapters 1.2, 1.4, and 3.1). Typical nutrient-broth-grown cells do not represent the phenotypes expressed in nature. Since antimicrobial preservatives must interact with the cell envelope either directly, as part of their mechanism of action, or indirectly, in order to access cytoplasmic targets, then it is not surprising that such envelope changes are accompanied by marked changes in preservative susceptibility (Gilbert and Wright, 1987). In addition, growth of bacteria, as biofilms in association with surfaces, causes them to express unique physiologies which are notably recalcitrant towards chemical treatment. Preservative susceptibility of microorganisms in relation to their mode of growth and nutritive environment must be given due consideration (Chapter 3.1). It may be more appropriate to grow inocula under the defined conditions of a chemostat (Chapter 1.4) than in batch cultures (Chapter 1.2). However, such an approach would be very demanding on laboratory operations if a large number of strains are to be employed in PETs.

CONCLUDING REMARKS

Microorganisms are used by the cosmetics industry in order to obtain a variety of important data. Consequently, the microbiologist must have some assurance that the organisms used retain their original genotype. This is especially true of microorganisms isolated from contaminated cosmetics or raw materials, and/or isolates known to have resistance to preservatives (Favet et al., 1987; Levy, 1987; Orth and Lutes, 1985; Russell, 1991). In many instances, conventional methods for maintaining such organisms may not be adequate. Spontaneous reversion may occur if resistant strains are removed from their original environment to artificial media or subjected to physical stresses. The microbiologist should give special attention to spoilage and environmental isolates intended for use in the PET, and store them in such a manner that their *original* characteristics are not lost and viability is maintained. This may mean periodic subculturing into the material from which the original isolate was obtained or devising an artificial medium that will ensure that a culture has retained the desired trait (for example, by incorporating preservatives into artificial media). Due to the problems associated with the maintenance and preservation of "environmental" isolates as challenge strains, "standardized" culture collection strains may be more appropriate for reproducibility. Future editions of the Pharmacopoeias will undoubtedly clarify and define inocula maintenance and preparation regimens.

Challenge tests strains should also have well-documented profiles. In addition to periodic screening, certain key traits, unique to the isolate, should be monitored frequently in order to provide added assurance of strain stability. Any deviation from the expected profile may indicate that the strain is no longer suited for use. The laboratory testing conditions aim for reproducibility and repeatability through standardization, rationalization, and validation of the parameters and techniques which appear critical in routine PETs.

Beside the validation of individual parameters the overall reproducibility of the standard PET has to be observed in routine use and periodically verified in a selected reference system for Good Laboratory Practices (GLP).

For the assessment of safety, confidence evolves from validated laboratory work conforming with GLP as well as from the specialized experience of the cosmetic manufacturer with the efficacy criteria within an overall MQM strategy. Even when a control agency, in part, takes a responsibility in accepting the preservation documentation provided to it, the prime responsibility resides with the manufacturer who overviews all the circumstances for his assessment of safety and quality.

ACKNOWLEDGMENTS

The author wishes to express his thanks to the CTPA for kind permission to reproduce parts of their PET guidelines.

REFERENCES

Al-Hiti, M.M.A. and Gilbert, P. 1980. Changes in preservative sensitivity for the USP antimicrobial agents effectiveness test microorganisms. *Journal of Applied Bacteriology* **49**, 119-126

Al-Hiti, M.M.A. and Gilbert, P. 1983. A note on inoculum reproducibility: a comparison between solid and liquid culture. *Journal of Applied Bacteriology* **55**, 173-175

Anon. 1990. Microbiological tests, antimicrobial preservatives — effectiveness. *United States Pharmacopoeia* XXII 51. pp 1478–1479. *United States Pharmacopoeial Convention*, Rockford MD.

Anon. 1993a. Preservative frequency of use: FDA data. June 1993 update. *Cosmetics and Toiletries* **108**: 47-48.

Anon. 1993b. Efficacy of antimicrobial preservatives in pharmaceutical products. *British Pharmacopoeia* 1993 Vol II. Appendix XVIC A191-192. Her Majesty's Stationery Office, London.

Anon. 1993c. M-3 The determination of preservation adequacy of water-miscible cosmetic and toiletry formulations. *CTFA Microbiology Guidelines,* Washington DC: CTFA.

Anon. 1993d. M-4 Method for preservation testing of eye area cosmetics. *CTFA Microbiology Guidelines,* Washington DC: CTFA.

Anon. 1993e. *CTFA Proposed AOAC Challenge Test.* Draft 6.2 of January 4, 1994. 94-MICRO-16 Washington DC: CTFA.

Anon. 1994. *Cosmetic Good Manufacturing Practices: A Guideline for the Manufacturer of Cosmetics and Toiletries.* The European Cosmetic Toiletry and Perfumery Association, Brussels.

Anon. 1994. *Proposed Guidelines of Good Manufacturing Practices of Cosmetic Products.* Council of Europe, Brussels.

Beezer, A.E., Newell, R.D. and Tyrrell, H.J.V. 1976. Application of flow microcalorimetry to analytical problems: the preparation, storage and assay of frozen inocula of *Saccharomyces cerevisiae*. *Journal of Applied Bacteriology* **41**, 197-207.

Beezer, A.E., Hunter, W.H. and Storey, D.E., 1981. The measurement of phenol coefficients by flow microcalorimetry. *Journal of Pharmacy and Pharmacology* **33**, 65-58.

Beezer, A.E., Volpe, P.L.O., Gooch, C.A., Hunter, W.H. and Miles, R.J. 1986. Quantitative structure–activity relationships: microcalorimetric determination of a group additivity scheme for biological response. *International Journal of Pharmaceutics* **29**: 237-242.

Bloomfield, S.F., Baird, R., Leak, R.E. and Leech, R. 1988. *Microbial Quality Assurance in Pharmaceuticals, Cosmetics and Toiletries*. Chichester, England: Ellis Horwood.

Bohannon, T.E. and Wen, L-F. L. 1978. Methods for storage of antimicrobial effectiveness test inoculum suspensions below freezing. *Journal of Pharmaceutical Science* **67**, 815-818.

Brannan, D.K., Dille, J.C. and Kaufman, D.J. 1987. Correlation of *in-vitro* challenge testing with consumer use testing for cosmetic products. *Applied and Environmental Microbiology* **53**, 1827-1832.

Brown, M.R.W. 1975. The role of the cell envelope in resistance. Resistance in *Pseudomonas aeruginosa* Brown M.R.W. (ed.) pp 71-107. London: John Wiley & Sons.

Bryan, W.L., Fizer, E.R. and Farrington, J.K. 1980. A review of methods for determining preservative efficacy in cosmetic products. *Developments in Industrial Microbiology* **21**, 273-276.

Corbett, R.J. 1992. Preservation of cosmetics and toiletries: a microbiological overview. *Parfumerie und Kosmetik* **73**, 22-27.

Cosgrove, R.F., Beezer, A.E. and Miles. R.J. 1979. The application of cryobiology to the microbiological assay of nystatin. *Journal of Pharmacy and Pharmacology* **31**, 83-86.

Cowen, R.A. and Steiger, B. 1976. Antimicrobial activity — a critical review of test methods of preservative efficacy. *Journal of the Society for Cosmetic Chemists* **27**, 467-481.

Cowen, R.A. and Steiger, B. 1977. Why a preservative system must be tailored to a specific product. *Cosmetics and Toiletries* **92**, 15-20.

CTFA. 1986. Microtopics. Wheaton Illinois: Allured Pub. Corp.

CTFA. 1993. *C.T.F.A. Microbiology Guidelines*. Washington D.C: Cosmetic, Toiletry and Fragrance Association.

CTPA. 1990. *MQM Microbial Quality Management. C.T.P.A. Limits and Guidelines* London: Cosmetic Toiletry and Perfumery Association.

CTPA. 1990b. BS 5750: Part 2: 1987 (ISO 9004) *Guidelines for Use by the Cosmetics Industry 1990*. London: Cosmetic Toiletry and Perfumery Association.

CTPA. 1993. *C.T.P.A. Guidelines on Preservation*. London: Cosmetic Toiletry and Perfumery Association.

CTPA. 1994. *C.T.P.A. Guidelines on Preservation Efficacy Testing*. London: Cosmetic Toiletry and Perfumery Association.

Curry, J. 1987. Thoughts on preservation testing of water-based products. *Cosmetics and Toiletries* **102**, 93-95.

Denyer, S. and Baird, R. 1990. *Guide to Microbiological Control in Pharmaceuticals,* Chichester, England: Ellis Horwood.

Diehl, K-H. 1992. The key to microbiological quality assurance. *Seife Oele Fette Wachse* **118**, 136-146.

Durant, C. and Higdon, P. 1987. Preservation of cosmetic and toiletry products. In *Preservatives in the Food Pharmaceutical and Environmental Industries,* Board, R.G., Allwood, M.C. and Banks, J.G. (eds.) Society of Applied Bacteriology Technical Series 22, pp 231-235. Oxford: Blackwell.

Favet, J. Fehr, A., Griffiths, W., Amacker, P.A. and Schorer, E. 1987. Adaptation of *Escherichia coli, Pseudomonas aeruginosa* and *Staphylococcus aureus* to Kathon CG and Germall II in an O/W Cream. *Cosmetics and Toiletries* **102**, 75-85.

Gay, M. 1983. Testing the efficacy of antimicrobial preservation: pharmacopoeial requirements, practical experience and drug safety. *Pharmaceutical Technology* **7**, 58-70.

Gay, M. 1991. Preservative efficacy testing of non-sterile pharmaceuticals: Laboratory practice and predicting safety assessment. *Proceedings in Preservative Efficacy Testing.* Nov. 1991. London: Scientific Symposia.

Geis, P.A. 1988. Preservation of cosmetics and consumer products: rationale and application. *Developments in Industrial Microbiology* **29**, 305-315.

Gilbert, P. 1991. Preservative susceptibility of microorganisms in relation to their mode of growth and nutritive environment. *Proceedings in Preservative Efficacy Testing.* Nov. 1991. London: Scientific Symposia.

Gilbert, P. and Wright, N. 1987. Non-plasmidic resistance towards preservatives of pharmaceutical products. In *Preservatives in the Food, Pharmaceutical and Environmental Industries.* Board, R.G., Allwood, M.C. and Banks, J.G. (eds.). Society of Applied Bacteriology Technical Series 22, pp 255-279. Oxford: Blackwell.

Heckly, R.J. 1978. Preservation of microorganisms. *Advances in Applied Microbiology* **24**, 1-53.

Hill, L.R. 1991. National collection of type cultures — test organisms. *Proceedings in Preservative Efficacy Testing:* Nov. 1991. London: Scientific Symposia.

IFSCC 1992. *The Fundamentals of Stability Testing.* New Jersey: Micelle Press

Kabara J.J. 1984. *Cosmetic and Drug Preservation: Principles and Practice,* New York and Basel: Marcel Dekker.

Levy, E. 1987. Insights into microbial adaptation to cosmetic and pharmaceutical products. *Cosmetics and Toiletries* **102**, 69-74.

McLaughlin, J.K., Zuckerman, B.D., Tenenbaum, S. and Wolf, B.A. 1984. Comparison of the API 20E, Flow and Minitek systems for the identification of enteric and nonfermentative bacteria isolated from cosmetic raw materials. *Journal of the Society for Cosmetic Chemists* **35**, 253-263.

Muscatiello, M.J. 1993. CTPA's preservation guidelines: an historical perspective and review, *Cosmetics and Toiletries,* **108:** 53-59.

Orth, D.S. 1993. *Handbook of Cosmetic Microbiology.* New York: Marcel Dekker.

Orth, D.S. and Lutes, C.M. 1985. Adaptation of bacteria to cosmetic preservatives. *Cosmetics and Toiletries* **100**, 57-64.

Orth, D.S., Lutes, C.M. and Smith, D.K. 1989. Effect of culture conditions and method of inoculum preparation on the kinetics of bacterial death during preservative efficacy testing. *Journal of the Society for Cosmetic Chemists* **40**, 193-204.

Perry, B.F. 1994. Adverse health consequences associated with the microbiological contamination of cosmetics: role of microbial quality management strategy in reducing risk. In *Proceedings of the Third International Symposium on Microbiology of Food and Cosmetics in Europe 1993: Yesterday's Projects and Today's Reality*, Marengo, G. and Pastoni, F. (eds.). pp 163-175. Joint Research Centre. EUR 15601 EN. Luxembourg: European Commission.

Pharmeuropa 1993. *Efficacy of Antimocrobial Preservation*, 5.4. Strasbourg: Council of Europe.

Post, P.E. 1991. Quality assurance of industrially manufactured culture media. *Proceedings in Preservative Efficacy Testing*. Nov. 1991. London: Scientific Symposia.

Ringertz, O. and Ringertz, S. 1982. The clinical significance of microbial contamination in pharmaceutical and allied products. *Advances in Pharmaceutical Science* **5**, 201-225.

Russell, A.D. 1991. Mechanisms of bacterial resistance to non-antibiotics: food additives and food and pharmaceutical preservatives. *Journal of Applied Bacteriology* **71**, 191-201.

Sabourin, J.R. 1990. Evaluation of preservatives for cosmetic products. *Drug and Cosmetics Industry* **147**, 24-26, 64, 65.

Spooner, D.F. 1989. Problems in the preservation testing of topical products. *Journal of Applied Cosmetology* **7**, 93-101.

Tran, A.T. and Collier, S.W. 1992. Direct contact membrane inoculation of yeasts and moulds for evaluating preservative efficacy in solid cosmetics. *International Journal of Cosmetic Science* **14**, 163-172.

Tran, A.T., Hitchins, A.D. and Collier, S.W. 1990. Direct contact membrane method for evaluating preservative efficacy in solid cosmetics. *International Journal of Cosmetic Science* **12**, 175-183.

Van Doorne, H. 1992. Fundamental aspects of preservation of cosmetics and toiletries. *Parfumerie und Kosmetik* **73**, 84-92.

Van Doorne, H. and De Vringer, T. 1994. Effect of inoculum size on survival rate of *Candida albicans* and *Aspergillus niger* in topical preparations. *Letters in Applied Microbiology* **18**, 289-291.

Warwick, E.F. 1993. Preventing microbial contamination in manufacturing. *Cosmetics and Toiletries* **108**, 77-92.

3.4 Reproducibility and Predictivity of Disinfection and Biocide Tests

Sally F. Bloomfield

INTRODUCTION

Disinfectant products containing chemical agents, and disinfection processes involving heat or other agents, are marketed worldwide for use in a variety of situations which include medicine, agricultural and veterinary practice, and food, domestic, and public hygiene. For development and official approval of these products and processes, laboratory tests which are robust, relevant to practical conditions, and acceptable on an international basis are required to ensure optimum standards of efficacy under conditions of practical use.

Although there is no universally agreed definition of the term "disinfection", there is general acceptance that the implied purpose is to reduce or prevent transfer of contamination and/or infection. Compared with the process of sterilization, in which the requirement is total absence of microorganisms, the requirements of a disinfection procedure (i.e., the extent to which destruction of microbes is achieved) may vary considerably according to what is acceptable and achievable, and what is considered to be cost-effective for the particular situation. Whereas, in the chemical disinfection of critical surfaces such as endoscopes, sterility (it is only in the United States that the term "chemosterilization" is accepted at the present time) is desirable, the requirements for disinfection of surfaces in the domestic environment may be less stringent.

In laboratory testing of disinfectants the ultimate purpose is to establish whether products meet specified requirements under "in use" conditions. Ideally disinfectant testing should comprise a number of phases.

0-8493-4752-1/95/$0.00+$.50

Phase 1: Suspension tests evaluate the activity against relevant microbial species under conditions which simulate practical use, e.g., in the presence of hard water or organic soil, at varying temperatures and contact times.

Phase 2: Test inocula are applied to skin or inanimate surfaces (or other conditions relevant to use, e.g., bacterial aerosols) under controlled laboratory conditions and the efficacy of the product determined.

Phase 3: The product is evaluated in "field" tests against contamination generated under use conditions, using processes simulating the intended use of the product.

Over the years various laboratory methods have been developed for biocide testing. Although these methods differ in experimental detail, all are based on the principle of adding a test inoculum to disinfectant, and removing samples at specified times. The biocide in each sample is neutralized and levels of survival of the organisms assessed. In practice the methods can be classified into three groups according to how the end point is determined:

End-point tests: The sample of biocide-treated cells is transferred to nutrient medium and incubated to determine the presence or absence of survivors. The result is expressed as the concentration of biocide producing kill (i.e., no detectable survivors) within a specified contact period, or the time required to achieve kill using a given concentration.

Quantitative tests: Samples of untreated and biocide-treated cells are plated on nutrient medium. After incubation the number of colony-forming units is determined and the log reduction in viable counts determined.

Capacity tests: The biocide is challenged successively with bacteria, at defined time intervals. Following each inoculation, samples are taken after a suitable contact period has elapsed, the biocide is neutralized, and the suspension incubated in medium to determine the presence or absence of detectable survivors. The result is expressed as magnitude of the accumulated inoculum that was required to produce the "failure".

Originating from the beginning of the twentieth century various standard test methods have been developed, in different countries worldwide, which are used for approval of disinfectants as part of national official and unofficial registration or approval schemes. The historical background to these methods and the methods themselves are described in more detail by Sykes (1965), Cremieux and Fleurette (1991) and Reybrouck (1992a). In the United States standard methods used by the EPA and FDA for registration and approval of disinfectants are predominantly end-point tests such as the AOAC Use-Dilution Method, but quantitative tests are also applied in some situations. By contrast Europe has favored the use of quantitative tests, although British Standard Methods, some of which were developed at the turn of the century, include end-point tests such as the Rideal-Walker test, and capacity tests such as the Kelsey-Sykes test. With current moves towards European harmonization, it is unclear whether these latter tests will retain official status.

In developing these standard tests, considerable attention has been paid to ensure that they are carried out against relevant challenge inocula under conditions which, as far as possible, are indicative of use conditions. In writing the standards an increasingly rigorous approach has been adopted in specifying the method for preparation and standardization of challenge inocula, together with all aspects of test methodology. In both Europe and the United States, experience gained in the application of these tests has been incorporated into successive second- and third-generation versions of the tests. Despite this, the precision of these methods and their ability adequately to predict efficacy under use conditions remains a matter for concern (Gilbert et al., 1987; Brown et al., 1991). In the following sections various studies of the precision of such standard tests, together with that of other laboratory investigations, are reviewed in order to draw conclusions about the effect of the test inoculum on the reproducibility of the laboratory-based disinfectant tests and their relevance or otherwise to practical use conditions. Details of the disinfectant products cited in this text are given in the appendix.

REPEATABILITY AND REPRODUCIBILITY OF DISINFECTANT TESTING

END-POINT TESTS

One of the earliest standard methods to be developed was the Rideal-Walker (RW) Test (Anon., 1985). This and the AOAC phenol coefficient test (Anon., 1984), are end-point methods in which the concentration of disinfectant, relative to that of phenol, required to kill an inoculum of *Staphylococcus aureus, Salmonella typhi,* or *Pseudomonas aeruginosa* within a given contact time is determined. The use of phenol as an internal standard is intended to eliminate variability associated with changes in resistance of the test suspension, since the same suspension is used for the disinfectant and standard. The main disadvantage of end-point tests is that, as total kill is approached, there is a period of time where the number of surviving organisms is less than one per sample and the end point depends on the probability of removing a sample containing a viable cell from the test mixture. This means that the result is reproducible as long as hypereffective or ineffective dilutions are tested, but, for intermediate concentrations, negative and positive subcultures will occur by chance. This problem can be partly overcome by increasing the number of replicate samples and calculating the proportion of positives, from which a semi-quantitative estimate of viability similar to that determined by the Most Probable Number Method can be obtained.

Despite successive attempts to improve precision, it is estimated that the RW test can produce variations up to 25 to 30% in the results (Hayter, 1973; Croshaw, 1981). There is, however, little indication of the extent to which lack of standardization of the test inoculum contributes to this variability. Test suspensions used for these tests are overnight broth cultures, but the number of cfu/ml is not standardized.

In 1953 the AOAC adopted an end-point test, the Use-Dilution Method (UDM) (Anon., 1984), in which test inocula are dried onto stainless steel penicylinders before exposure to disinfectant. The product passes the test if no more than one sample out of 60 replicates gives a positive result. Since its introduction, the test has been under constant criticism for inconsistent results and lack of reproducibility. A study by Rhodes (1983) indicated that up to 22% of EPA registered claims for hospital disinfectants could not be reproduced. Major deficiencies identified in the method include inconsistency of the surface and of the inoculum level, and ambiguity in the written method. A collaborative study of a modified UDM which incorporated 32 technical changes (Rutala et al., 1987; Cole et al., 1988) indicated no improvement in test precision. In this study failure rates for six EPA registered disinfectants tested in 18 laboratories ranged from 20% for *Salmonella choleraesuis* up to 62% for *P. aeruginosa*. These workers concluded that the interlaboratory variability of the UDM is primarily a methodologic problem (i.e., related to the problems of end-point sampling) and that the method should be abandoned in favor of a quantitative method.

More recently a modified UDM, the hard surface carrier test (HSCT), has been developed by Rubino et al. (1992). Major differences for this method are that glass cylinders are used, the test suspension is harvested from a synthetic medium (Bacto synthetic broth AOAC) and standardized before use, and the method is more rigorously described. A collaborative study of this test in ten laboratories with five disinfectants (Rubino et al., 1992) indicated some improvement in precision of the HSCT, although inconsistencies were still observed. For the HSCT, failure rates for three EPA registered disinfectants tested in 18 laboratories ranged from 15% for *S. choleraesuis* up to 30% for *P. aeruginosa*. Rubino et al. (1992) suggested that reproducibility of the test could be improved by increasing the performance standard to ≤2/60 positive carriers for *S. aureus* and *S. choleraesuis* and ≤3/60 positive carriers for *P. aeruginosa*. Although these studies quantified the intra- and interlaboratory variation of the method, again there was little attempt to identify the extent to which the test inoculum may contribute to the lack of precision.

Quantitative Tests

During the 1970s quantitative suspension tests for evaluation of disinfectants were developed in a number of European countries including France (Anon., 1981), Germany (Beck et al., 1977), UK (Anon., 1984), and Holland (Van Klingeren and Mossel, 1978). These tests are similar in design and determine those disinfectant concentrations which produce a given microbicidal effect or ME value (usually a 5 log reduction in cfu in 5 min). In 1978 the Council of Europe set up an initiative to harmonize disinfectant test methods for applications in food hygiene. A suspension test based on the Dutch test (the so-called "5-5-5" test) was drawn up and a collaborative trial carried out in ten European countries. It was concluded (Van Klingeren et al., 1981) that precision might be improved by more rigorous standardization of methods for preparation of disinfectant solutions and bacterial suspensions. The test was modified accordingly and republished in 1987 (Anon., 1988) as the *Method of Test for the Antimicrobial Activity of Disinfectants in Food Hygiene* usually referred to as the 1987 EST.

Over the last 20 years quantitative surface tests have also been introduced in various European countries including Germany, France, Holland, and Belgium. These tests are described in more detail by Reybrouck (1990, 1992a) and involve determining the log reduction in recoverable organisms (recovery is achieved by a rinsing procedure followed by a colony-counting technique) by application of the disinfectant to a bacterial inoculum dried onto a specified surface. ME values are expressed as log reduction in cfus over and above that produced by drying alone.

Most recently, in 1989, a European committee, Comité Européen de Normalisation (CEN) TC 216 has been established to produce harmonized European test methods for antiseptics and disinfectants used, not only in food hygiene, but also public hygiene, medicine, agriculture, and veterinary practice. Suspension tests for bactericidal, fungicidal, and sporicidal activity have been drafted (Anon., 1993, 1994a,b,c,d) which are closely based on the 1987 EST. Collaborative studies of the method are currently in progress and, if acceptable, the tests will be introduced over the next five or so years to replace existing European standards. A quantitative European "surface test" in which the test inoculum is dried onto a stainless steel surface before exposure to the disinfectant is also being developed.

Since 1989, as part of our involvement in development of the CEN tests, we have had the opportunity to study the precision of the 1987 EST and the proposed European surface test (Bloomfield and Looney, 1992; Bloomfield et al., 1991, 1993, 1994). In effect the 1987 EST is the grandchild of the Dutch 5-5-5 test on which it was based. From a statistical evaluation of the 1987 test compared with the 5-5-5 test, Van Klingeren (1983) noted a reduction in the between-day variance of the ME values, indicating that better standardization had been obtained. Our most recent studies of the 1987 EST as described below, indicate that despite further modification, the precision of these tests still represents a problem. The results of these studies (Bloomfield et al., 1991; Bloomfield and Looney, 1992), together with previous studies of the 5-5-5 test and the 1978 EST (Van Klingeren et al., 1977, 1981) and studies of the surface test method (Bloomfield et al., 1993, 1994) give useful insights into the sources of variability and the extent to which the test inocula contribute to this variability.

Intralaboratory Variability of Test Results

The 1987 EST was conducted in two laboratories (the Chelsea laboratory at Kings College London and the Hospital Infection Research Laboratory, Birmingham [HIRL]) (Bloomfield and Looney, 1992). In each laboratory, tests were carried out on five separate, but not necessarily consecutive, days with three replicate tests on each day, using five disinfectant products at concentrations producing ME values between 0 and 5 log reductions. Tests were carried out in each laboratory by an experienced operator, but in one the testing was repeated by an inexperienced student. An additional surface test was carried out in the Chelsea laboratory, supplementing *Enterococcus faecium* for *P. aeruginosa* as the test organism (Bloomfield et al., 1994). Mean ME values and 95% limits of error, calculated for each product and each test organism, are shown in Tables 1 and 2.

In repeating this study with the surface test it was anticipated that the presence of a carrier surface would further reduce the precision of the test. One of the major problems with surface testing is achieving consistent recovery of survivors from surfaces. This is usually carried out by a rinsing method involving sonication or vortexing the sample with glass beads. Another problem incurred was the substantial loss of viability during drying of the inoculum onto the test surface. Surprisingly, however, the results (Table 2) indicated that the precision of surface tests was better than that of suspension tests. Whereas suspension tests indicated limits of error for a single ME value of up to ±3.38 to ±4.10, by contrast the results of the surface tests indicate limits of error of ±0.38 for 35% alcohol against *E. faecium*, up to not more than ±1.64 for Dettol against *P. aeruginosa*. From similar studies of suspension and surface tests Van Klingeren (1983) also reported that the precision of the Dutch surface test was at least as good as that of the suspension test. From a more detailed examination of the results a number of observations can be made regarding the sources of variability.

Standardization of Test Suspensions

Results (Table 1) show that, in the Chelsea laboratory where cell density in the test suspensions were standardized spectrophotometrically, the variability in initial viable count (log Nc) was relatively small (±0.30 to 0.34). By contrast, variability of log Nc obtained at HIRL, where test suspensions were harvested and prepared in the prescribed manner but were not standardized photometrically, was much higher (±0.74 to 0.82). Table 2 shows that the process of drying and recovery of test organisms from surfaces was associated with limits of error for Nc values four times higher than for the suspension test. Since the variation in suspension test ME values for the HIRL skilled operator were no greater than those from Chelsea, and the variability of the ME values for the surface test was actually less than for the suspension test, it is concluded,

TABLE 1
Mean Microbial Effect (ME Values), Initial Log Counts (N_c) and 95% Limits of Error Using Disinfectant Products Against Bacterial and Fungal Suspensions Using the 1987 European Suspension Test

		Mean ME Values from 15 Replicate Tests: 95% Limits of Error for a Single ME Value				
Study Center	**Initial Count (N_c)**	**Dettol 0.35% v/v**	**Hypochlorite 50 ppm**	**CHDNE 0.005% w/v**	**BAK 0.0025% w/v**	**Alcohol 40% v/v**
Staphylococus aureus						
Chelsea (Study 1)	7.10 ± 0.30	2.43 ± 2.73	1.96 ± 4.10	3.22 ± 2.07	1.10 ± 0.56	2.46 ± 2.70
Chelsea (Study 2)	6.97 ± 0.34	1.37 ± 0.92	4.76 ± 3.64	1.71 ± 3.38	1.11 ± 0.74	3.74 ± 4.27
HIRL skilled	7.35 ± 0.82	0.74 ± 0.62	5.5 ± 0.92	2.12 ± 0.73	0.05 ± 0.45	3.07 ± 3.48
HIRL non-skilled	6.85 ± 0.74	1.28 ± 4.28	4.80 ± 3.06	1.73 ± 1.01	0.07 ± 0.85	3.10 ± 6.28
Pseudomonas aeruginosa		**Dettol 4% v/v**	**Hypochlorite 50 ppm**	**CHDNE 0.01% w/v**	**BAK 0.03% w/v**	**Alcohol 37% v/v**
Chelsea (Study 1)	7.07 ± 0.14	2.08 ± 1.03	2.31 ± 1.33	1.27 ± 1.39	3.50 ± 1.24	3.11 ± 2.19
Candida albicans		**Dettol 0.35% v/v**	**Hypochlorite 50 ppm**	**CHDNE 0.025% w/v**	**BAK 0.015% w/v**	**Alcohol 27.5% v/v**
Chelsea (Study 2)	5.97 ± 0.13	2.29 ± 1.53	1.01 ± 1.37	2.48 ± 2.51	2.86 ± 1.34	2.08 ± 1.91

Note: CHDNE = Chlorhexidine acetate; BAK = Benzalkonium chloride.

From Bloomfield, S.F. and Looney, E., *Journal of Applied Bacteriology*, 1992.

TABLE 2
Mean Microbial Effect (ME Values), Initial Log Counts (N_c) and 95% Confidence Limits for Disinfectant Products Tested on Stainless Steel and Formica Surfaces Using the Proposed European Surface Test

		Mean Me Values from 15 Replicate Tests: 95% Limits of Error for a Single ME Value							
Organism	**Surface**	**Initial Count (N_c)**	**Dettol 5% v/v**	**NaOCl 500 ppm**	**NaOCl 1000 ppm**	**CHDNE 0.1% w/w**	**BAK 0.2% w/v**	**Alcohol 35% v/v**	**Alcohol 50% v/v**
Enterococcus	Stainless steel	7.67 ± 0.26	1.33 ± 0.87	1.68 ± 0.83	2.43 ± 0.40	0.54 ± 0.58	1.27 ± 1.14	1.02 ± 0.4	4.24 ± 0.57
faecium	Formica	7.63 ± 0.27	1.29 ± 1.61	1.69 ± 0.72	2.36 ± 0.78	0.43 ± 0.24	1.14 ± 0.87	1.09 ± 0.38	4.18 ± 0.65
Pseudomonas	Stainless steel	6.55 ± 0.59	2.95 ± 0.61	1.86 ± 0.43	4.08 ± 0.59	1.36 ± 1.24	1.50 ± 0.72	1.14 ± 0.76	2.54 ± 0.57
aeruginosa	Formica	6.45 ± 0.57	2.57 ± 1.64	1.93 ± 0.58	4.15 ± 0.45	1.22 ± 1.19	1.37 ± 1.13	1.05 ± 0.35	2.17 ± 0.62

Note: CHDNE = chlorhexidine gluconate; BAK = benzalkonium chloride.

From Bloomfield, S.F., et al., *Journal of Applied Biology*, 1994.

as was also found by Van Klingeren et al. (1981), that variability in inoculum size contributes little to the variance in ME.

Operator Skill

Table 1 shows that the confidence limits for ME values obtained by the untrained operator were significantly wider (e.g., up to as much as ±6.3) than for the experienced operators, indicating that test precision is related to operator skill. By contrast, limits of error for the original tests carried out at Chelsea were not significantly greater than for the repeat test period, indicating that the precision of this operator did not alter over a prolonged period of testing.

Variations Between Test Organisms and Products

Results of suspension test studies (Table 1) indicate, as found in previous studies of the 5-5-5 and 1978 EST (Van Klingeren et al., 1977, 1981), that the limits of error for some organisms and some products were greater than for others. On further consideration, however, it was realized that if confidence limits for all the products were plotted against the relevant ME values (Figure 1), there was some correlation, which suggests that the precision of the test decreases as the ME values increase. Al-Hiti and Gilbert (1983) also showed that increasing sensitivity of *E. coli* and *P. aeruginosa* to (chlorhexidine) CHDNE was associated with a decrease in the reproducibility of the log mean percentage survival.

Within-Day and Between-Day Variability

If individual ME values from these initial studies are considered and the probability of obtaining a significantly different result within and between days is calculated as shown in Table 3 (study period 1 for the suspension and surface tests), results indicate that, where 3 replicate tests are carried out on the same day by the same operator, consistent ME values are obtained which differ from one another by no more than 1 log. These studies indicate that the probability of one ME value being significantly different from the others was less than 1% for the surface test and about 5% for the suspension test (although a value of 9.3% was recorded for the relatively limited study [75 trials] at HIRL). Since the same batches of test suspension and biocide solution were used in tests carried out on any single day, it is concluded that this occasional event must be due to operator error.

For day-to-day maintenance of test strains the 1987 EST and CEN standard methods, as currently written, prescribe that stock cultures, on TSA (Tryptone Soya Agar, prepared from freeze dried cultures), are stored in a refrigerator at 5°C and subcultured monthly. Test suspensions are prepared from TSA subcultures from the stock culture, and are grown at 37°C on at least two but not more than three successive occasions. Overnight incubation on TSA produces a

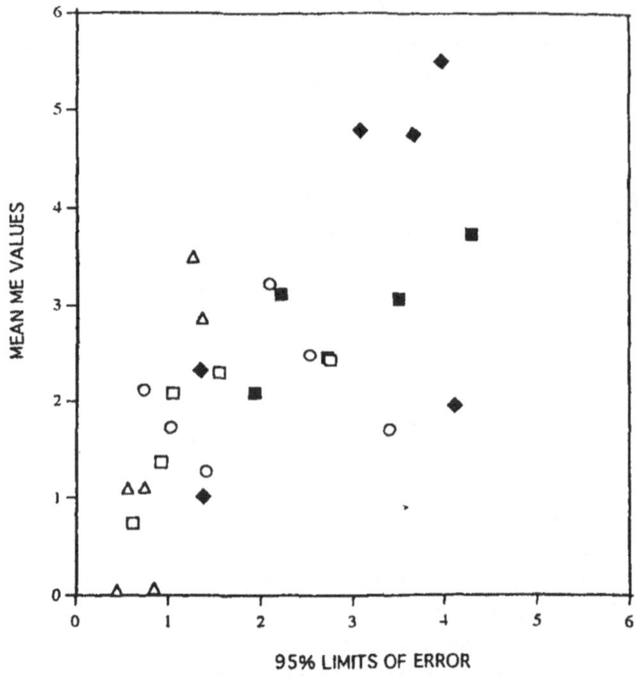

FIGURE 1 Relationship between mean ME values and 95% limits for (□), Dettol; (◆), NaOCl; (○), chlorhexidine; (△), benzalkonium chloride and (■), alcohol tested against *Staphylococcus aureus, Pseudomonas aeruginosa,* and *Candida albicans* using the 1987 European Suspension Test. (From Bloomfield, S.F. and Looney, E., *Journal of Applied Bacteriology,* 1992.)

population of cells in the stationary growth phase, which is harvested and standardized to contain a specified number of cfu/ml. The results of our studies showed that where tests were carried out daily for five days, using suspensions prepared from sequential subcultures prepared in this way, consistent ME values were mostly obtained on each of the 5 days, but a mean ME value which was significantly different (>1.3 logs) from the other mean values was sometimes obtained. For the suspension test the frequency occurence of a significantly different ME value was about 15% and, for the surface test, up to about 5%. Since ME values were generally consistent within a given day it must be concluded that these variations are due to day-to-day changes in the resistance of the test suspension, although they may be partly due to variations in preparing the disinfectant solution.

Inter- and Intralaboratory Variability of Test Results

In assessing the precision of standard disinfectant tests it is often assumed that variability between laboratories comes largely from variations in

TABLE 3
Repeatability and Reproducibility of Surface Test Results

Study Center	Study No.	Probability of Obtaining a Significantly Different Result: Within Days a	Within Days b	Within Days c	Between Days a	Between Days b	Between Days c	Between Periods Within Labs a	Between Periods Within Labs b	Between Periods Within Labs c	Between Labs Between Periods a	Between Labs Between Periods b	Between Labs Between Periods c
					Suspension tests								
Chelsea[a]	1	375	17	5%	90	14	15%	90	11	12.5%	15 (Chelsea[a] and HIRL)	2	13.3%
HIRL	1	75	7	9.3%	25	2	8%	—	—	—			
Chelsea[b]	2	—	—	—	297	14	4.7%	50	8	16.0%	—	—	—
					Surface tests								
Chelsea[c] study 1	1	420	1	0.24%	140	6	4.2%	—	—	—	—	—	—
Chelsea[d] study 2	2	—	—	—	369	25	6.8%	84	15	17.8%	—	—	—
Chelsea/[c]	3 (Chelsea, Bilthoven, Campden)	108 (Chelsea, Bilthoven, Campden)	0	0%	54	1	1.9%	—	—	—	24 (Chelsea, Bilthoven, Campden)	4	16.6%
Bilthoven													
Campden					18	4	22.0%	—	—	—			

Note: a = number of tests or trials; b = number of tests or trials which were significantly different; c = % frequency of occurrence of a different result.

[a] From Bloomfield, S.F. and Looney, E., *Journal of Applied Bacteriology*, 1992.
[b] From Bloomfield, S.F., et al., *Letters in Applied Microbiology*, 1991.
[c] From Bloomfield, S.F., et al., *Journal of Applied Bacteriology*, 1994.
[d] From Bloomfield, S.F., et al., *Letters in Applied Microbiology*, 1993.

methodology between the different operators. In further studies (study periods 2 and 3) as summarized in Table 3, the activity of a larger range of biocides was determined using suspension and surface test methods as described previously (Bloomfield et al., 1991, 1993). In the initial test period, products were tested daily for not less than three days. For a proportion of products, testing was repeated in a second test period in the same or another laboratory and, in some cases, a third and fourth testing period within the same laboratory. The interval between testing periods was never less than 2 weeks, and, for both the suspension and surface tests, each total study took place over not more than 9 months. In total, 1500 ME values were determined in five different laboratories.

From this study it was found that where a sequential series of daily tests (i.e., a test period) was repeated at a later date, either in the same laboratory by the same operator or in another laboratory by a different operator, the two sets of ME values were, for the most part, internally consistent, but sets of values, consistent within the given test period but significantly different from the previous test period were sometimes obtained (difference between mean ME values >1.3 logs up to as much as 3.0 logs). For this situation the probability of obtaining a consistently different result between test periods (either within or between laboratories) was 16 to 18% for the surface test, and (although only limited data was available) 13% for the suspension test. What was particularly significant was the finding that the differences in ME values between test periods occurred as frequently within a laboratory (17.8% for the surface test) as between different laboratories (16.6% for the surface test). This suggests that the variability does not result primarily from differences in methodology between different laboratories. Since to a large extent the results were consistent within each test period, it suggests that this variability derives from changes in the resistance of the test strain.

For all of these studies test strains were maintained as specified in the European standards. The strains were obtained as freeze dried cultures from a national collection and cultivated on agar slopes. The agar slopes, which were used for preparation of test suspensions, were stored in the refrigerator at 4°C for up to, but not more than, 1 month, at which time the slope was subcultured onto a fresh agar slope. The variability observed in these studies suggests that random but significant changes in resistance of the stock agar slopes may occur within refrigerated microbial populations normally considered to be in a stationary condition.

Further observations can be made if the results are considered in chronological sequence. From a typical set of surface test results with *P. aeruginosa* (Table 4) where a number of different biocides were tested in each of a number of study periods, it can be seen that increased resistance to one biocide was not accompanied by a general increase in resistance to the other biocides and that the increased resistance was sometimes maintained in a third testing period but in some cases reverted to its original level. It can also be seen that these variations occurred mainly with biocides which produced ME values greater than 2.0 but not total kill.

TABLE 4

Mean Microbicidal Effect (ME Values) Obtained in Successive Test Periods for Disinfectants Tested Against *Pseudomonas aeruginosa* on Stainless Steel Surfaces Using the Proposed European Surface Test Method

	Mean ME Values from 3 or 5 Replicate Tests on Successive Days in Test Period								
	1	2	3	4	5	6	7	8	9
	Nov 1991	Jan 1992	Mar 1992	Apr 1992	July 1992	Oct 1992	Jan 1993	Feb 1993	Apr 1993
Iodophor 10,000 ppm	—	>5.5	—	—	—	—	4.1	—	—
Alcohol 70% v/v	—	>6.0	—	—	—	3.7	>6	—	—
Peratol 0.5% v/v	—	>5.5	—	—	—	>6.0	>6	—	>6.1
Clearsol 0.625% v/v	—	>5.5	—	—	—	—	>6	—	—
Hibisol 100%	—	>5.5	—	—	—	—	>6	—	—
NaOCl 2500 ppm	—	>6.0	—	—	—	>6.0	>6	—	—
NaOCl 500 ppm	1.9	—	3.7	—	—	—	—	—	—
NaOCl 250 ppm	—	4.3	—	3.7	—	—	—	>6.0	>6.1
Dettol 2.5% v/v	—	1.6	—	>6.0	>6.0	—	—	4.9	>6.1
Dettol 5.0% v/v	2.9	3.8	>6.0	—	>6.0	—	—	—	—
Dettox 100%	—	2.9	3.2	>6.0	3.5	—	—	—	—
Dettol fresh 2.5% v/v	—	1.7	—	2.1	2.8	—	—	1.7	—
Dettol fresh 5.0% v/v	—	2.5	1.6	—	2.9	—	—	—	—
BAK 0.2% w/v	1.5	1.5	—	2.0	—	—	—	2.5	1.9
CHDNE 0.1% w/v	1.4	1.4	—	1.3	—	—	—	1.5	2.8
Savlon 5%	—	1.5	—	1.2	—	—	—	—	—

Note: CHDNE = chlorhexidine gluconate; BAK = benzalkonium chloride.

From Bloomfield, S.F., et al., *Letters in Applied Microbiology*, 1993.

From the results of these various investigations, it is concluded that, in the short term, if the European tests are not to be abused, adequate replication must be specified in order to determine whether products comply with the specified criteria. At present there has been little discussion of test replication by committee CEN TC 216 but it would seem that, as in the 1987 EST, a minimum of two replicate tests should be performed with freshly prepared test suspensions on two separate (and preferably not sequential) days — with provision for one or more further replicates in the case of conflicting results. All test suspensions should be sourced directly from the stock culture. For the longer term, however, there is a need to develop alternative methods for maintaining test strains and preparing test suspensions which achieve a more consistent resistance to biocides. The ways in which the method of stock culture preservation and the method of preparation of test suspensions can affect the resistance of challenge inocula to antimicrobial agents are currently under investigation in several centers.

CAPACITY TESTS

The only official capacity test method, the Kelsey-Sykes test (Anon 1987) was originally introduced as a potential alternative to the much criticized Rideal-Walker test. In this test, the disinfectant concentration with a capacity to kill three successive inocula of the test organism added at 10 min intervals is determined. The intention was to provide a method for determining disinfectant concentrations for use in hospitals under clean and dirty (simulated by the presence of 5% yeast) conditions, which, unlike the RW test (which is only applicable to phenolic disinfectants), could be applied to all disinfectant types. Studies of the method by Werner et al. (1975), Coates (1977), and Cowen (1978) indicate, however, that, as with other methods, precision represents a significant problem, although again there is little data to indicate the extent to which the test inoculum contributes to this variability.

RELEVANCE TO PRACTICE

In laboratory testing of disinfectants, the ultimate purpose is to establish whether products meet agreed efficacy requirements under practical conditions, whether this be a critical hospital situation, processing surfaces in the food or pharmaceutical industries, or the decontamination of hands. Whereas antibiotics and other agents used in treatment of infections are invariably subjected to clinical trials, for disinfectants, because of economic constraints, there is a tendency to rely largely on laboratory tests. In view of the reliance placed on these tests it is particularly important that they be adequately designed.

Since their introduction at the turn of the century, suspension tests, using bacterial inocula harvested from complex media, have been mainly used for evaluating disinfectants and determining their use concentrations. Although these tests give valuable information, the evidence, as reviewed in this text, suggests that such tests have the potential to mislead if rigidly applied. In an attempt to address this problem surface tests using inocula dried onto test surfaces were developed during the 1960s and 1970s. Experience with these tests, however, increasingly indicates that they may be equally misleading in terms of predicting activity against naturally occurring resident and transient biofilms, and that test models which more closely simulate use situations are urgently required if effective evaluation is to be achieved in the laboratory. At the same time it is realized that repeatability and reproducibility often suffer when generating tests which are more relevant to practice. Laboratory tests must always, therefore, represent a compromise between the requirement for test precision and practical relevance.

In setting up disinfectant testing protocols it is recognized that careful consideration of all aspects of the test conditions — including the choice of test strains, preparation of test inocula, and introduction of test surfaces — can do much to optimize the value of the information gained (Gilbert et al., 1987). In the following section the various aspects to be taken into account in designing disinfectant test methods are reviewed.

Choice of Test Organism

In designing any program of disinfectant testing, one of the first considerations must be the choice of test strains. In the majority of practical situations, the range of microbial species encountered will be quite extensive and largely unpredictable. For this reason it is desirable that chemical agents used as disinfectants should have good activity against a wide range of microbial species, including gram-positive and gram-negative bacteria and fungi (including yeasts), and in some cases also have virucidal and sporicidal action. Although, because of their generally nonspecific action, agents used as disinfectants tend to have a broad spectrum of action, relative to antibiotics, variability in resistance between different species and different strains of the same species are quite frequently reported. Where investigated, biocide resistance is usually found to be associated with phenotypic changes to cell outer layers which inhibit access of the biocide to its site of action on the underlying cell protoplast. Reports of plasmid-mediated resistance to biocides are confined to a small range of agents, e.g., mercury compounds. The inherent resistance patterns and the mechanisms of development of resistance to biocides is reviewed in more detail by Russell and Gould (1988) and Russell and Chopra (1990). Since differences in resistance between laboratory and wild strains and between different wild strains is a significant possibility, tests against "clinical/environmental isolates" must always be considered.

As far as the European methods are concerned, it is agreed that, for a product to be designated as bactericidal or fungicidal, it must be active against *S. aureus* and *P. aeruginosa* (5 log reduction within 60 min), and against

C. albicans and *Aspergillus niger* (4 log reduction within 60 min). To be recommended as a disinfectant for a specified use, it is required that the use-dilution of the product must also be active against additional test strains (up to 5 log reduction within the designated use contact time), under test conditions relevant to the intended use. Test organisms and test conditions, agreed for different uses are summarized in Table 5. As yet no agreement has been reached on test strains for sporicidal and mycobactericidal tests.

In the United States, a different approach is adopted whereby EPA-registered products must be tested against specified organisms as shown in Table 6 in order to support limited activity, broad spectrum, hospital or medical environment disinfectant efficacy claims, and fungicidal, mycobactericidal, and sporicidal claims.

Cultivation of Test Organisms

Inocula for disinfectant testing are typically grown by overnight incubation on nutrient-rich complex media. In this situation, maximal cell division rates up to 3.0/h are attained, giving a population which at 16 h is in its stationary phase of growth (Chapters 1.4 and 1.2). As with all types of antimicrobial agents, test inocula cultivated under these conditions may have different sensitivity to disinfectants compared with the same organisms occurring naturally in the environment. As far as growth rate and growth phase are concerned, indications are that the resistance of exponentially growing cells to agents used as disinfectants, as with other types of antimicrobial agents, mostly increases as the growth rate is decreased, and that cell populations in the stationary phase of growth are generally more resistant than cells in exponential growth. Inocula grown under conditions of various nutrient depletions in batch culture, or nutrient limitation in continuous culture (carbon, nitrogen, iron, phosphate, or magnesium), may show either relative increases or decreases in resistance according to the nature of the nutrient limitation, the type of biocide, and the test organism. A number of studies have been carried out which indicate the nature and extent of these effects.

Effect of Nutrient Depletion in Batch Culture

Using *E. coli*, *P. aeruginosa*, *S. aureus*, and *C. albicans* grown under conditions of nitrogen (N), carbon (C), and phosphate (P) depletion, Al-Hiti and Gilbert (1980) showed variations in the concentrations of thiomersal, chlorhexidine (CHDNE), and benzalkonium chloride BP (BAK), incorporated into nutrient agar, required to produce a 1 log reduction in colony-forming ability, but there was no identifiable pattern among the systems tested. By contrast, *A. niger* showed little change in biocide sensitivity under these conditions. In this study the greatest variation occurred with *P. aeruginosa* for which isoeffective concentrations of BAK increased from 0.0025% for

TABLE 5
Proposed Test Conditions for Harmonized European Suspension Tests

	Medical Hard Surface	Medical Hand Disinfection	Agriculture and Veterinary	Food, Domestic and Institutional
Test Strains	*Staphylococcus aureus* *Pseudomonas aeruginosa* *Acinetobacter* spp. *Escherichia coli*	*Staphylococcus aureus* *Pseudomonas aeruginosa* *Escherichia coli* *Enterococcus faecium*	*Staphylococcus aureus* *Pseudomonas aeruginosa* *Proteus mirabilis* *Enterococcus faecium*	*Staphylococcus aureus* *Pseudomonas aeruginosa* *Escherichia coli* *Enterococcus faecium* (and other spp)
Diluent	WSH or SDW	WSH or SDW	WSH	WSH
Organic load (clean and/or dirty)	Albumin and/or Yeast extract	None	0.3% albumin or 1% albumin + 1% yeast extract	0.03% albumin or 0.3% albumin and other materials
Temperature	20°C	20°C	4°C, 10°C, 20°C, 40°C	4°C, 20°C, 40°C
Contact time	5 or 60 min	**Handrub** 5 log reduction in 1 min	5, 30, 60 min	5, 15, 60 min
Requirement	5 log reduction	**Handwash** 3 log reduction in 1 min 5 log reduction in 5 min	5 log reduction	5 log reduction

Note: WSH = Water of standard hardness; SDW = Sterile distilled water.

TABLE 6
Test Organisms as Specified in AOAC Disinfectant Test Methods

Test	Objective	Test Strain
AOAC Use-Dilution Test	Disinfectant (limited efficacy)	*Salmonella choleraesuis* or *Staphylococcus aureus*
AOAC Use-Dilution Test	Disinfectant (general efficacy)	*Salmonella choleraesuis* and *Staphylococcus aureus*
AOAC Use-Dilution Test	Disinfectant (hospital or medical)	*Salmonella choleraesuis*, *Staphylococcus aureus* and *Pseudomonas aeruginosa*
AOAC Fungicidal Suspension Test	Fungicidal activity	*Trichophyton mentagrophytes*
AOAC Sporicidal Carrier Test	Sporicidal activity	*Bacillus subtilis* and *Clostridium sporogenes*
AOAC Tuberculocidal Carrier Test	Presumptive tuberculocidal activity	*Mycobacterium smegmatis*
AOAC Germicidal and Detergent Sanitizing Test	Bactericidal activity of product in hard water	*Escherichia coli* *Staphylococcus aureus*

P-depleted cells to 0.01% for N-depleted cells. Among the different biocides, the least variation occurred with thiomersal and greatest variation with BAK. Using a similar method Klemperer et al. (1977, 1980) showed that resistance of *E. coli* and *Proteus mirabilis* to phenol, cetrimide and CHDNE, varied according to whether cells were grown under conditions of C, P, or Mg depletion, but again there was no particular correlation between the nature of the biocide, nutrient depletion, and change in resistance. Similar studies by Gilbert and Brown (1978a) indicated that *E. coli*, grown in Mg- or P-depleted media, was more resistant to 3- and 4-chlorophenol and 2-phenoxyethanol (2-PE), compared with C-limited cells.

Carson et al. (1972) demonstrated that suspensions of naturally occurring *P. aeruginosa* cultivated in hospital distilled water were more resistant to chlorine dioxide, a quaternary ammonium compound, and glutaraldehyde, compared with the same strain cultivated at 37°C on TSA, and that resistance was further increased by reducing the growth temperature of the test inoculum from 37°C to 25°C. In general the resistance of stationary-phase cells was greater than that of growing cells, although for glutaraldehyde the reverse effect was observed. In a similar study Leak (1983) showed that overnight cultures of wild and laboratory strains of *Enterobacter cloacae* and *P. aeruginosa*, "freshly" harvested from TSA, were more sensitive to CHDNE, bronopol, and methylparabens than cells and harvested after storage at 4°C for more than 4 days. Results showed increased recovery of the "aged" compared with the "fresh" cells of the order of 0.5 to 2 logs over a 4 to 5 h contact period.

Effect of Growth Rate and Nutrient Limitation in Continuous Culture

This area has recently been reviewed (Brown et al., 1990; Gilbert et al., 1990). Gilbert and Brown (1980) showed that the growth rate of P-limited vegetative cells of *Bacillus megaterium* had little effect on their sensitivity to CHDNE or 2-PE, but that Mg- and C-limited cultures showed increased resistance with decreasing growth rate. Over dilution rates from 0.45 down to 0.13/h, concentrations required to produce a 1 log reduction within 15 min increased from 0.2 to 0.5% for 2-PE, and 0.02 to 0.001% for CHDNE. These workers also showed that, over a range of growth rates from 0.3 to 0.05/h, the isoeffective concentrations of 3- and 4-chlorophenol against C- and Mg-limited *P. aeruginosa* cells was increased from less than 1 m*M* to between 3 and 5 m*M* (Gilbert and Brown, 1978b). Wright and Gilbert (1987a) on the other hand showed that whereas sensitivity of N- and C-limited *E. coli* to CHDNE increased with increasing growth rate, the sensitivity of Mg- and P-limited cells decreased. In a related study, Wright and Gilbert (1987b) showed that resistance of *E. coli* to a series of n-alkyltrimethylammonium bromides was increased at low growth rates (0.1 to 0.2/h) but then successively decreased at increasing growth rates from 0.2 to 0.5/h, the extent of this effect varying with the type of nutrient limitation and the agent.

Recent studies, as reviewed by Gilbert (1988), indicate the ways in which growth under conditions of nutrient limitation gives rise to cells with different cell structures which in turn may account for the observed differences in biocide sensitivity. Other studies, as reviewed by Kolter (1992), suggest that the reductions in growth rate at the onset of the stationary phase is associated with the expression of genes which confer resistance to environmental stress, including that associated with the action of biocides. These aspects are discussed in more detail in Chapters 1.2, 1.4, and 1.5 respectively.

Although, for antibiotic treatment, it may be possible to assess the nature of the growth environment of infecting organisms for disinfectants, the nutritional status of cells growing in their natural habitat is infinitely variable and almost impossible to assess. On this basis it is suggested that disinfectants should be assessed under likely "worst-case situations" which include relevant species cultivated at slow growth rates under a range of nutrient-limited conditions.

Preparation of Test Suspensions

Challenge inocula for disinfectant testing may be grown on solid or liquid media. Investigations by Al-Hiti and Gilbert (1983) showed that the choice of medium may affect sensitivity to biocides. These workers found that solutions of 0.005 and 0.015% CHDNE used to treat *E. coli* and *P. aeruginosa* cells harvested from liquid culture produced 0.3 and 0.8 log reductions within 15 min. Where cells were harvested from agar plates inoculated to give successive increases in colony density, sensitivity to CHDNE increased progressively up to 3 log reduction in 5 min for cells harvested from plates containing 10 cfu/plate.

Where cells are grown in nutrient-rich media, an effective means is required to remove residual media which might otherwise affect biocide activity. Experience has shown that relatively small amounts of organic material may be sufficient to reduce the activity of some biocides — most particularly the chlorine-releasing agents (CRAs). For the majority of European tests, inocula are prepared by dilution of cell harvests by a factor of ten or more, on the assumption that this is sufficient to dilute out interfering substances. In some laboratories it is customary to wash cells by successive centrifugation and resuspension in fresh diluent. Investigations by Gilbert et al. (1991), however, indicate that sublethal injury induced by centrifugation may increase sensitivity to biocides. These workers showed that the concentrations of cetrimide and CHDNE that were required to be incorporated into the plate count agar in order to produce a 50% reduction in colonies formed were 60 and 1.5 mg/ml^{-1} for cells centrifuged at 20,000 **g** for 15 min, compared with 140 and 3.5 mg/ml^{-1} for uncentrifuged cells.

Contact Time and Temperature

For disinfectant testing it is important that the contact time over which products are tested be equivalent to recommended use conditions. For slow-acting disinfectants such as the aldehydes, where contact times of 60 mins or more may be required, it is suggested that products should be tested at additional times in order to indicate the margin of safety inherent in the contact time stated on the label.

Since biocide activity is strongly affected by temperature, products should be tested at temperatures relevant to use conditions. For European tests, as shown in Table 5, it is required that products should not only meet specifications when tested at 20°C, but also at additional temperatures as relevant to specified use conditions. It is therefore proposed that products for agricultural use under outdoor conditions products must meet required standards when tested additionally at 4°C. While domestic and food hygiene products specified for use at elevated temperatures, must satisfy the test at 20°C, use concentrations are determined from tests carried out at 40°C.

Presence of Interfering Substances

Where disinfectants are applied to either inanimate surfaces or the hands, any number of substances may be present which may affect activity. The effects of interfering substances on the activity of disinfectants are reviewed in more detail by Russell (1992).

Since there is evidence that the activity of some disinfectants may be affected by metal ions such as Ca^{2+} and Mg^{2+}, European and EPA test systems require that products destined for dilution with potable water must be diluted for testing in water of standard hardness.

Since organic soiling is generally associated with inactivation of disinfectants it is recommended that, wherever possible, surfaces should be cleaned before disinfection. In many cases, however, residual contamination must be anticipated, and in some situations (e.g., in the treatment of blood spillages), disinfectants are used specifically to decontaminate soil to prevent infection transfer and to assist in safe disposal. Both EPA and European testing systems state that, where claims are made for use under soil conditions, use concentrations must be determined from tests carried out in the presence of suitable soil. Soil materials commonly used include albumin, serum, blood, yeast, and yeast extract. For European suspension tests albumin at concentrations of 0.03 to 1% w/v is specified according to the area of use (Table 5) while EPA regulations state that products bearing a claim as a "one step" cleaner-disinfectant or disinfectant/sanitizer must be tested in the presence of a "representative soil" such as 5% v/v blood serum.

Suspension tests conducted under conditions as proposed for European tests (Bloomfield et al., 1991), with a range of UK disinfectants, indicated that the majority of products at recommended use dilution in water of standard hardness showed satisfactory activity in the presence of 1.0% w/v albumin. This was with the exception of the CRAs, for which a concentration of 2500 ppm, as compared with 250 ppm in the absence of soil, was required to achieve consistent activity across all test organisms. By contrast, where a number of these products were tested against *S. aureus* suspended in diluent containing 3% w/v albumin (equivalent to 1% w/v albumin in a suspension test) and dried onto stainless steel surfaces (Bloomfield et al., 1993) activity was very significantly reduced, giving ME values of 2.5 or less for the majority of products.

Although albumin may be appropriate for testing products to be used in the presence of blood or body fluids, there is evidence to suggest that this may be entirely inappropriate in other situations. Results of some recent studies in our laboratory (Table 7) indicate that for chlorine- and oxygen-releasing agents, 0.8% w/v albumin produced significantly greater neutralization compared with 1% w/v yeast, while for CHDNE and BAK, yeast produced greater neutralization than albumin.

In view of current concerns about the ecotoxicological effects of disinfectants it would seem important that disinfectant concentrations for use under soil or other conditions not be increased unnecessarily to meet the requirements of tests which are inappropriate to the particular use situation.

Surface Testing

Although suspension tests allow us to single out products likely to be most effective under use conditions, they give no information on how products actually perform on contaminated surfaces. As stated previously, standard surface tests now operate in various European countries which involve

TABLE 7
Mean Microbicidal Effect (ME Values) for Disinfectant Products Tested in the Presence of Either 0.8% w/v Albumin or 1% w/v Yeast Using the 1987 EST Method

		Mean ME Values From 3 Replicate Tests			
		Staphylococcus aureus	***Pseudomonas aeruginosa***	***Proteus mirabilis***	***Enterococcus faecium***
Alcohol 70% v/v	albumin	>6	>6	>6	>6
	yeast	>6,>6	>6,>6	>6	>6,>6
Stericol 2.0% w/v	albumin	>6	>6	>6	>6
	yeast	>6,>6	>6	>6	>6,>6
Peratol 5% v/v	albumin	>6	>6	>6	>6
	yeast	>6,>6	>6	>6	>6,>6
Dettol 5% v/v	albumin	>6	3.4	>6	>6
	yeast	>6,>6	2.4,2.7	>6	>6,>6
NaOCl 250 ppm	albumin	1.4	1.1	0.5	0.4
	yeast	>6	>6	>6	3.5
NaOCl 1000 ppm	albumin	>6	>6, 4.3	>6	4.3, 4.1
	yeast	>6, >6	>6, >6	>6	5.4, >6
NaDCC 250 ppm	albumin	2.6	>6, 4.6	2.9	0.7
	yeast	>6	>6, >6	>6	4.0
NaDCC 1000 ppm	albumin	>6, >6	>6	>6, 4.6	4.3, 4.1
	yeast	>6, >6	>6	>6	>6, 4.4
Virkon 0.5% w/v	albumin	0.5	2.3, 2.4	>6	1.8
	yeast	>6, 4.9	>6	>6, >6	4.8, 5.2
Virkon 1.0% w/v	albumin	1.3	>6, 5.2	>6	1.81
	yeast	>6, >6	>6, >6	>6	>6, >6
CHDNE 0.05% w/v	albumin	3.2, 5.1	5.2, 3.9	0.9, 1.1	1.5, 0.6
	yeast	0.8, 0.9	0.4, 0	0	0.2, 0.3
BAK 0.1% w/v	albumin	>6	5.8	>6	>6
	yeast	>6, >6	2.8, 4.0	4.6	>6

Note: CHDNE = chlorhexidine gluconate; BAK = benzalkonium chloride.

determining the log reduction in recoverable organisms (recovery is achieved by a rinsing procedure followed by a colony-counting technique) produced by application of the disinfectant to a bacterial inoculum dried onto a specified surface. In the USA the AOAC use-dilution method (AOAC, 1984) is used, in which test organisms are dried onto stainless steel penicillin cups before immersion in the disinfectant, and activity is determined using the end-point method.

Since suspension and surface tests are frequently used in conjunction with one another for approval of disinfectants, it is important to know the extent to which the results of these tests correlate. From our recent studies of the activity of a range of UK disinfectants at recommended use concentrations as determined by the 1987 EST, and the proposed European surface test method (Bloomfield et al., 1991, 1993), it was found that, whereas in the suspension tests the majority of products produced consistent ME values above six (i.e., no detectable survivors), ME values for the surface test were different for different products. Mean ME values from surface tests as shown in Table 8 indicate that in general the products divided into three groups. For products containing alcohol, peracetic acid/hydrogen peroxide, glutaraldehyde, CRAs (at 2500 ppm $AvCl_2$) and iodine (at 10,000 ppm AvI_2) a consistent >3.5 log reduction was achieved in all tests, and in many cases there were no detectable survivors, indicating ME values >5.4 to 6.6 log reduction. For the CRAs at 250 ppm, Virkon 0.5% w/v, Domestos multisurface 500 ppm $AvCl_2$, Dettol 2.5 and 5% v/v, and Dettol fresh 2.5 and 5% v/v, ME values were mostly between 2 and 5. For products such as BAK 0.2% w/v, CHDNE 0.1% w/v, and CHDNE 0.015% w/v + cetrimide 0.15% w/v (Savlon), ME values were generally lower, giving mean ME values varying from 0.5 to 3.0 logs only. Although, as reported by Van Klingeren (1990), there was some evidence of increased resistance of *E. faecium* compared with the other test organisms, in general the order of activity was not greatly different for the different test organisms; products ranked consistently as high, intermediate, and low activity against all four organisms.

Thus, whereas suspension tests indicated that virtually all products showed the same activity, surface tests indicated that some products were more effective than others. Studies by Van Klingeren (1983) and Reybrouck (1992b) also indicate that disinfectant concentrations required to produce a given log reduction are generally higher for the surface tests than for suspension tests and the AOAC use-dilution test. Van Klingeren (1983) suggested that these differences may relate to a reduction in the total amount of biocide per bacterial cell in the surface test as compared with the suspension test, but results from our laboratories (unpublished) indicate that this may not be the case.

In using combined data from suspension and surface tests to set performance standards and to determine use concentrations, the most important consideration is whether the methods adequately predict use conditions. Unfortunately there is little data to confirm whether this is the case in practice. Assessment of products under practical conditions are sometimes reported in the literature (Babb et al., 1981; Scott et al., 1984) but, once activity under

TABLE 8
Mean Microbicidal Effect (ME Values) for Disinfectant Products Tested on Contaminated Stainless Steel Disks Using the Proposed European Surface Test Method

		Mean ME Values from 3 or 5 Replicate Tests			
Product	**Concentration**	***Staphylococcus aureus***	***Pseudomonas aeruginosa***	***Enterococcus faecium***	***Candida albicans***
NaOCl	2500 ppm	>6.0*, >6.4	>5.5, >6.0	>6.6, >6.4	>5.4
NaDCC	2500 ppm	>6.3, >6.4	>5.5, >6.0	>6.6, >6.4	>5.4
Alcohol	70% v/v	>6.3, >6.4, >6.4	>5.5, >6.0	>6.6, >6.4	>5.4
Hibisol	Undiluted	>6.3, >6.4, >6.4	>5.5, >6.0	>6.6, >6.4	>5.4
Clearsol	0.625% v/v	4.6, 3.8, 3.6	>5.5, >6.0	5.2, 5.4, >6.6	>5.4
Peratol	0.5% v/v	>6.3, >6.4, 5.2	>5.5, >6.0, >6.0	>6.6, >6.4	4.3
Iodophor	10 000 ppm	4.5, >6.4	>5.4, 4.1	>6.6, >6.4	3.6
Cidex	Undiluted	5.4, >6.4	>5.5, >6.0	>6.6, >6.4	4.5
NaOCl	250 ppm	4.9, 5.5, >6.4	4.3, 3.7	3.1	2.8
NaDCC	250 ppm	2.4	3.7	2.2	1.9
Virkon	0.5% w/v	3.9	4.1	3.2	2.0
Dettol	5% v/v	4.5, 3.5, 3.9	2.9, 3.8, >6.0	1.3, 4.8, 3.9	>4.2
Dettol F	5% v/v	3.5, 2.8, 3.7	2.5, 1.6, 2.9	3.9, 4.1	2.6
Dettox	Undiluted	2.5, >6.1, 2.6	2.9, 3.2, >5.8	3.1, 3.8	1.7
CHDNE	0.1% vw/v	1.2, 1.5, 1.8	1.4, 1.4, 1.3	0.6, 0.5	2.7
BAK	0.2% vw/v	3.3, 3.3, 3.0	1.5, 2.0, 1.5	1.3, 1.3	1.9
Savlon	5% v/v	2.6, 1.1	1.5, 1.2	3.0	1.9

Note: Dettol F = Dettol Fresh; CHDNE = chlorhexidine gluconate; BAK = benzalkonium chloride.

From Bloomfield, S.F., et al., *Letters in Applied Microbiology*, 1993.

practical conditions has been established, there is rarely any attempt to relate this back to the results of laboratory tests. For this reason pass/fail criteria for standard suspension and surface tests are traditionally agreed to on an empirical basis which reflects what is achievable rather than that which is considered desirable from a clinical perspective. In agreeing to acceptance criteria for the proposed European tests, although a pass/fail requirement of 5 log reduction in 5 min would seem appropriate for the suspension test, the results of surface tests as described above (Bloomfield et al., 1993), indicate that, if the same requirement were applied to surface tests, the use concentrations of many UK products used for general environmental disinfection would have to be increased. In view of this, committee CEN TC 216 has agreed in principle that it may be necessary to set lower pass/fail criteria in the surface test for certain situations (e.g., 5 log reduction for products used for immersion disinfection of critical surfaces of medical equipment, but only 2 or 3 log reduction for surfaces in domestic and institutional situations where decontamination is achieved by mechanical cleaning applied with the disinfectant).

In the current climate of concern about environmental issues, there is increasing pressure to ensure not only that use concentrations of disinfectants are kept to a minimum, but also that effectiveness under use conditions and genuine benefits can be demonstrated for the product. Increasingly there is evidence to suggest that, in a large proportion of practical situations, surfaces provide a site for the adhesion and/or development of biofilms (Chapter 1.7). There is evidence, as follows, which suggests that the resistance of both resident or transient microbial biofilms to biocides currently used for disinfection purposes may be very different from that of organisms in suspension or surface dried films prepared using laboratory-grown inocula as used in standard laboratory tests.

Investigations by Pallent et al. (1983), Marrie and Costerton (1981) and Costerton and Lashen (1984) suggested that adsorption of *P. cepacia* and *Serratia marcescens* onto glass surfaces enabled cells to survive in the presence of high concentrations of chlorhexidine (0.5 up to 2%), isothiazolones and quaternary ammonium compounds. Investigations by Stickler et al. (1989, 1991) showed that mixed biofilms of *Citrobacter diversus, P. aeruginosa,* and *Enterococcus faecalis*, and biofilms of *E. coli* grown on silicone disks (representing urinary catheter surfaces) showed increased resistance to chlorhexidine 0.2% and povidone iodine (1% w/v AvI_2) compared with planktonic cells grown in samples of urine. These workers suggested that increased biocide resistance associated with biofilm formation *in vivo* may contribute to the chronic recurrent nature of urinary tract infections in catheterized patients. Studies by Berkelman et al. (1984) and Panlilio et al. (1992) indicated that contamination of povidone iodine formulations implicated in bacteremias following the use of such contaminated products resulted from formation of resistant *Pseudomonas* biofilms on the surfaces of the pipework of the equipment used in manufature of the product. In further studies (Vess et al., 1993) demonstrated that spp. of *Pseudomonas, Mycobacteria,* and *Acinetobacter* produced colonization of PVC surfaces which survived treatment wth a range of biocides including phenolics, quaternaries, aldehhydes, and halogen-releasing agents. They suggested that these biofilms represent a continuous reservoir of potentially protected organisms adhering to and shedding from the interior surfaces of PVC pipework. Holah et al. (1990) evaluated a surface disinfection test in which biocidal action was assessed against biofilms of *P. aeruginosa, S. aureus,* and *P. mirabilis* on stainless steel. In all cases the surface-attached organisms were more resistant than the same organisms grown in suspension.

Using a modified Robbins device, Wright et al. (1991) demonstrated increased resistance of biofilms of *Legionella pneumophila* cultivated on PVC surfaces to Bronopol and Kathon as compared with planktonic organisms. Increased resistance of *Legionella* biofilms to glutaraldehyde, chlorine, iodine, and other chemical formulations used to treat evaporative cooling towers compared with planktonic cells was reported by Cargill et al. (1992), Green (1993) and Green and Pirrie (1993). Using a bioluminescent strain of *Listeria monocytogenes*, Walker et al. (1993) demonstrated progressive increases in resistance to phenol associated with attachment to stainless steel surfaces over a period of 50 h.

A number of hypotheses have been suggested to account for the observed differences in susceptibility between planktonic and adherent bacteria (Chapter 1.7; Brown and Gilbert, 1993). Costerton (1984) proposed that the increased resistance of sessile compared with planktonic organisms arises from the protective effect of biofilm growth in glycocalyx enclosed microcolonies, i.e., reduced access of the biocide to the bacteria. It is also possible that physical or chemical interaction of reactive agents such as CRAs with glycocalyx material may reduce availability to the microbial cells within the biofilm. Brown and Gilbert (1993), on the other hand, proposed that the increased resistance of the sessile compared with planktonic bacteria is related to their slower growth rates. Using *E. coli* grown in suspension and as biofilms at controlled growth rates, Evans et al. (1990) showed that resistance to cetrimide increased as the growth rate decreased, and that at any given growth rate the resistance of the biofilm was generally greater than that of the planktonic cells. From their results it was suggested that, not only adhesion, but also reduced growth rates associated with nutrient limitation probably contribute to the biocide resistance of biofilm cells. A fourth hypothesis is that adhesion triggers the expression of specific phenotypes which confer resistance to biocides.

APPLICATION OF LABORATORY TESTS FOR APPROVAL OF DISINFECTANTS

Because of the known deficiencies in the precision and practical relevance of current disinfectant test methods, it is vitally important that the application of these tests for approval and licensing of disinfectants be adequately controlled. This is necessary to ensure on one hand that the tests are not "abused", but on the other that they are not so rigidly applied as to themselves inhibit use of these products to optimum benefit. As part of their work committee CEN TC 216 has agreed to write an additional standard or standards which indicate the way in which tests should be applied and interpreted in relation to product labeling and product claims. Since there is evidence to suggest that in some situations laboratory tests, most particularly surface tests as currently designed, may be inappropriate in some situations, CEN TC 216 has agreed that the guidelines must allow that products which might fail a particular laboratory test may be approved for use if product claims can be supported by properly designed and adequately controlled "field tests". These should indicate the reasons why the product does not meet the requirements of standard laboratory tests. It is agreed that CEN TC 216 should prepare a norm to define how field trials under practical conditions should be carried out.

APPENDIX

Disinfectant products cited in this text include: NaDCC, sodium dichloroisocyanurate (Fichlor, Chlorchem Ltd, Widnes); Hibisol, chlorhexidine gluconate 0.5% w/v in 70% alcohol; Savlon liquid antiseptic, solution of 3% w/v cetrimide and 0.3% w/v chlorhexidine gluconate (Zeneca, Macclesfield); Dettol, solubilized preparation containing 5% w/v chloroxylenol (Reckitt and Colman Products, Hull); Dettox, solution containing 1.75% w/v quaternary ammonium compounds (Reckitt and Colman Products, Hull); Dettol Fresh, solution containing 4% w/v BAK, 0.5% EDTA (Reckitt and Colman Products, Hull); Peratol, solution containing 4.5% m/m peracetic acid and 23% m/m hydrogen peroxide (Albright and Wilson, Warley); Clearsol, clear soluble phenolic preparation containing 40% xylenol fraction (Albright and Wilson, Warley); Iodophor, Povidone-iodine solution BP containing 10% w/v AvI_2 (Betadine, Napp Laboratories Ltd, Cambridge); Cidex, solution containing 2% w/v glutaraldehyde (Cidex, Surgikos Ltd, Livingston); Virkon, powder containing a stabilized blend of peroxygen compounds, surfactant, organic acid in an inorganic buffer system (Antec International, Sudbury); Stericol, clear soluble phenolic preparation containing 18% xylenol fraction (Stericol, Sterling Industrial, Chapeltown).

REFERENCES

Al-Hiti, M. M. A., and Gilbert, P. 1980. Changes in preservative sensitivity of the USP antimicrobial agents effectiveness test microorganisms. *Journal of Applied Bacteriology* **49:**119-126.

Al-Hiti, M. M. A., and Gilbert, P. 1983. A note on inoculum reproducibility. *Journal of Applied Bacteriology* **55:**173-175.

Anon. 1981. Antiseptiques et disinfectants utilisé à l'état liquide, miscibles à l'eau et neutralisables. Détermination de l'activité bactericide. *Association Française de Normalisation* NFT 72-150; NFT 72-171.

Anon. 1984. Methods for determination of the antimicrobial value of QAC disinfectant formulations. BS 6471. London: British Standards Institution.
Anon. 1985. Determination of the Rideal-Walker coefficient. BS 541. London: British Standards Institution.
Anon. 1987. Estimation of concentrations of disinfectants used in "dirty" conditions in hospitals by the modified Kelsey-Sykes test. BS 6905. London: British Standards Institution.
Anon. 1988. Method of test for the antimicrobial activity of disinfectants in food hygiene. DD 177. London: British Standards Institution.
Anon. 1993. Chemical antiseptics and disinfectants — basic bactericidal activity — test method and requirement. Provisional European Norm, PrEN 1040, London: British Standards Institution.
Anon. 1994. Chemical antiseptics and disinfectants — basic bactericidal activity — test method and requirement. Provisional European Norm, London: British Standards Institution.
Anon. 1994a. Chemical antiseptics and disinfectants — basic fungicidal activity — test method and requirement. Provisional European Norm, PrEN 1275, London: British Standards Institution.
Anon. 1994b. Chemical antiseptics and disinfectants — quantitative suspension tests for the evaluation of bactericidal activity of chemical disinfectants and antiseptics for use in food, industrial, domestic and institutional areas — test method and requirement. Provisional European Norm, PrEN 1276, London: British Standards Institution.
Anon. 1994c. Chemical antiseptics and disinfectants — quantitative suspension tests for the evaluation of fungicidal activity of chemical disinfectants and antiseptics for use in food, industrial, domestic and institutional areas — test method and requirement. Provisional European Norm, PrEN 1650, London: British Standards Institution.
Anon. 1994d. Chemical antiseptics and disinfectants-quantitative suspension tests for the evaluation of bactericidal activity of chemical disinfectants and antiseptics used in the veterinary field — test method and requirement. Provisional European Norm, PrEN 1656, London: British Standards Institution.
AOAC, Association of Official Analytical Chemists. 1984. *Official Methods of Analysis,* 14th Ed. Arlington: Association of Official Analytical Chemists.
Babb, J. R., Bradley, C. R., Deveril, C. E. A., Ayliffe, G. A. J., and Melikian, V. 1981. Recent advances in the cleaning and disinfection of fibrescopes. *Journal of Hospital Infection* **2:**329-346.
Beck, E. G., Borneff, J., Grun, L., Gundermann, K.-O., Kanz, E., Lammers, T., Mulhens, K., Primavesi, C. A., Schmidt, B., Schubert, R., Weinhold, E., and Werner, H.-P. 1977. Recommendations for the testing and the evaluation of the efficacy of chemical disinfectant procedures. *Zentralblatt für Bakteriologie, Parasitenkunde, Infektionskrankheiten und Hygiene, I. Abteilung Originale, Reihe B.* **165:**335-380.
Berkelman, R.L., Lewin, S., Allen, J.R., Anderson, R.L., Budnick, L.D., Shapiro, S., Friedman, S.M., Nicholas, P., Holzman, R.S. and Haley, R.W. 1981. Pseudobacteraemia attributed to contamination of povidone-iodine with *Pseudomonas cepacia. Annals of Internal Medicine* **95,** 32-36.
Bloomfield, S. F., Arthur, M., Looney, E., Begun, K., and Patel, H. 1991. Comparative testing of disinfectants and antiseptic products using proposed European suspension test methods. *Letters in Applied Microbiology* **13:**233-237.

Bloomfield, S. F., and Looney, E. 1992. Evaluation of the repeatability and reproducibility of European suspension test for antimicrobial activity of disinfectants and antiseptics. *Journal of Applied Bacteriology* **73:**87-93.

Bloomfield, S. F., Arthur, M., Begun, K., and Patel, H. 1993. Comparative testing of disinfectants using proposed European surface test methods. *Letters in Applied Microbiology* **17:**119-125.

Bloomfield, S. F., Arthur, M., Van Klingeren, B., Pullen, W., Holah, J. T., and Elton, R. 1994. An evaluation of the repeatability and reproducibility of a surface test for the activity of disinfectants. *Journal of Applied Bacteriology* **76:**86-94.

Brewer, J.H., Heer, A.H., and McLaughlin, C.B. 1956. The control of sterilisation procedures with thermophilic spore-formers. *Bacteriological Proceedings* **56:**61.

Brown, M. R. W. 1977. Nutrient depletion and antibiotic susceptibility. *Journal of Antimicrobial Chemotherapy* **3:**198-201.

Brown, M. R. W., Allison, D.G., and Gilbert, P. 1988. Growth rate of sessile bacteria and sensitivity to drugs. *Chemiotherapia* **7:**229-232.

Brown, M. R. W., Collier, P. J., and Gilbert, P. 1990. Influence of growth rate on susceptibility to antimicrobial agents: modifications of the cell envelope and batch and continuous culture studies. *Antimicrobial Agents and Chemotherapy* **34:**1623-1613.

Brown, M. R. W., and Gilbert, P. 1993. Sensitivity of biofilms to antimicrobial agents, pp. 87-975 in L. B. Quesnel, P. Gilbert and P. S. Handley (eds.), *Microbial Cell Envelopes: Interactions and Biofilms.* Society for Applied Bacteriology Symposium Series, Number 22. Oxford: Blackwell Scientific Publications.

Cargill, K. L., Pyle, B. H., Sauer, R. L. and McFeters, G. A. 1992. Effects of culture conditions and biofilm formation on the iodine susceptibility of *Legionella pneumophila. Canadian Journal of Microbiology* **38:**426-429.

Carson, L. A., Favero, M. S., Bond, W. W., and Peterson, N. J. 1972. Factors affecting compararative resistance of naturally occurring and subcultured *Pseudomonas aeruginosa* to disinfectants. *Applied Microbiology* **23:**863-869.

Coates, D. 1977. Kelsey-Sykes capacity test: Origin, evolution and current status. *Pharmaceutical Journal* **219:**402-403.

Cole, E. C., Rutala, W. A., and Samsa, G. P. 1988. Disinfectant testing using a modified use-dilution method: collaborative study. *Journal of the Association of Official Analytical Chemists* **71:**1187-1194.

Costerton, J. W. 1984. The formation of biocide-resistant biofilms in industrial, natural and medical systems. *Developments in Industrial Microbiology* **25:**363-372.

Costerton, J. W., and Lashen, E. S. 1984. Influence of biofilm of biocides on corrosion-causing bacteria. *Materials Performance* **23:**13-7.

Cowen, R. A. 1978. Kelsey-Sykes capacity test: A critical review. *Pharmaceutical Journal* **220:**202-204.

Cremieux, A., and Fleurette, J. 1991. Methods of testing disinfectants, pp. 1009-1026, in S. S. Block (ed.), *Disinfection, Sterilization and Preservation,* 3d Ed. Philadelphia: Lea and Febiger.

Croshaw, B. 1981. Disinfectant testing — with particular reference to the Rideal-Walker and Kelsey-Sykes tests, pp. 1-15 in C. H. Collins, M. C. Allwood, S. F. Bloomfield and A. Fox (eds.), *Disinfectants: Their Use and Evaluation of Effectiveness.* London: Academic Press.

Davies, A. J., Desai, H. N., Turton, S., and Dyas, A. 1987. Does installation of chlorhexidine into the bladder of catheterised geriatric patients help reduce bacteriuria? *Journal of Hospital Infection* **9:**72-5.

Evans, D. J., Allison, D. G., Brown, M. R. W., and Gilbert, P. 1990. Effect of growth rate on resistance of Gram-negative biofilms to cetrimide. *Journal of Antimicrobial Chemotherapy* **26:**473-478.

Gilbert, P., and Brown, M. R. W. 1978a. Effect of R-plasmid RP1 and nutrient-depletion upon the gross cellular composition of *Escherichia coli* and its resistance to some uncoupling phenols. *Journal of Applied Bacteriology* **133:**1062-1065.

Gilbert, P., and Brown, M. R. W. 1978b. Influence of growth rate and nutrient limitation on the gross cellular composition of *Pseudomonas aeruginosa* and its resistance to 3- and 4-chlorophenol. *Journal of Applied Bacteriology* **133:**1066-1072.

Gilbert, P., and Brown, M. R. W. 1980. Cell-wall mediated changes in sensitivity of *Bacillus megaterium* to chlorhexidine and 2-phenoxyethanol, associated with the growth rate and nutrient limitation. *Journal of Applied Bacteriology* **48:**223-230.

Gilbert, P. 1988. Microbial resistance to preservative systems, pp. 171-194 in S. F. Bloomfield, R. Baird, R. E. Leak, and R. Leech (eds.), *Microbial Quality Assurance in Pharmaceuticals, Cosmetics and Toiletries.* Chichester: Ellis Horwood.

Gilbert, P., Collier, P. J., and Brown, M. R. W. 1990. Influence of growth rate on susceptibility to antimicrobial agents: biofilms, cell cycle, dormancy, anstringent response. *Antimicrobial Agents and Chemotherapy* **34:**1865-1868.

Gilbert, P., Caplan, F., and Brown, M. R. W. 1991. Centrifugation injury of gram-negative bacteria. *Journal of Antimicrobial Chemotherapy* **27:**550-551.

Graham, G. S., and Boris, C. A. 1993. Chemical and biological indicators, pp 36-69, in R. F. Morrissey and G. Briggs Phillips (Eds.), *Sterilization Technology: A Practical Guide for Manufacturers and Users of Health Care Products.* Van Nostrand Reinhold, New York.

Green, P. N. 1993. Efficacy of bioades on laboratory-generated *Legionella* biofilms. *Letters in Applied Microbiology* **17:**158-161.

Green, P. N., and Pirrie, R. S. 1993. A laboratory apparatus for the generation and biocide efficacy testing of *Legionella* biofilms. *Journal of Applied Bacteriology* **74:**388-393.

Hayter, S. E. 1973. Bacterial evaluation of disinfectants. *Process Biochemistry* **8:**16.

Holah, J. T., Higgs, C., Robinson, S., Worthington, D., and Spenceley, H. 1990. A conductance-based surface disinfection test for food hygiene. *Letters in Applied Microbiology* **11:**255-259.

Klemperer, R. M. M., Al-Dujaili, D. A., Lawson, J. J., and Brown, M. R. W. 1977. The effect of nutrient depletion upon the resistance of *Proteus mirabilis* to antibacterial agents. *Journal of Applied Bacteriology* **48:**349-357.

Klemperer, R. M. M., Ismail, N. T. A. J., and Brown, M. R. W. 1980. Effect of R-plasmid RP1 and nutrient depletion on the resistance of *Escherichia coli* to cetrimide, chlorhexidine and phenol. *Journal of Applied Bacteriology* **48:**349-357.

Kolter, R. 1992. Life and death in the stationary phase. *American Society of Microbiology News* **58:**75-79.

Leak, R. E. 1983. Some factors affecting the preservative testing of aqueous systems. PhD Thesis, University of London.

Marrie, T. J., and Costerton, J. W. 1981. Prolonged survival of *Serratia marcescens* in chlorhexidine. *Applied and Environmental Microbiology* **42:**1093-1102.

Pallent, L. J., Hugo, W. B., Grant, D. J. W., and Davies, A. 1983. *Pseudomonas cepacia* as contaminant and infective agent. *Journal of Hospital Infection* **4**:9-13.

Panlilio, A. L., Beck-Sague, C. M., Siegel, J. D., Anderson, R. L., Yetts, S. Y., Clark, N. C., Duer, P. N., Thomassen, K. A., Vess, R. W., Hill, B. C., Tablan, O. C., and Jarvis, W. R. 1992. Infections and pseudoinfections due to povidone-iodine solution contaminated with *Pseudomonas cepacia. Clinical Infectious Diseases* **14**:1078-1083.

Reybrouck, G. 1990. The assessment of the bactericidal activity of surface disinfectants. *International Journal of Hygiene and Environmental Medicine* **190**:479-510.

Reybrouck, G. 1992a. Evaluation of the antimicrobial activity of disinfectants, pp. 114-133, in A. D. Russell, W. B., Hugo, and G. A. J. Ayliffe, (eds.), *Principles and Practice of Disinfection, Sterilisation and Preservation,* 2nd Ed. Oxford: Blackwell Scientific Publications.

Reybrouck, G. 1992b. The assessment of the bacterial activity of surface disinfectants. IV The AOAC use-dilution method and the Kelsey-Sykes test. *International Journal of Hygiene and Environmental Medicine* **192**:432-437.

Rhodes, M. E. 1983. Disinfectant testing: Are current tests adequate? in *Chemical times and Trends,* Oct.

Rubino, J. R., Bauer, J. H., Clarke, P. H., Woodward, B. B., Porter, F. C., and Gill Hilton, H. 1992. Hard surface carrier test for efficacy testing of Disinfectant Collaborative study. *Journal of the Association of Official Analytical Chemists, International Volume.* **75**:635-645.

Russell, A. D., and Chopra, I. 1990. Understanding antibacterial action and resistance. Chichester: Ellis Horwood.

Russell, A. D., and Gould, G. W. 1988. Resistance of Enterobacteriaceae to preservatives and disinfectants. *Journal of Applied Bacteriology, Symposium Supplement* 167S-195S.

Russell A. D. 1992. Factors affecting the activity of antimicrobial agents pp. 89-113, in A.D. Russell, W. B. Hugo, and G. A. J. Ayliffe, (eds.), *Principles and Practice of Disinfection, Sterilisation and Preservation,* 2nd Ed. Oxford: Blackwell Scientific Publications.

Rutala, W. A., and Cole, E. C. 1987. Ineffectiveness of hospital disinfectants against bacteria: A collaborative study. *Infection Control* **8**:501-506.

Scott, E., Bloomfield, S. F., and Barlow, C. G. 1984. Evaluation of disinfectants in the domestic environment under "in use" conditions. *Journal of Hygiene, Cambridge* **92**:193-203.

Stickler, D., Dolman, J., Rolfe, S., and Chawla, J. 1989. Activity of antiseptics against *Escherichia coli* growing as biofilms on silicone surfaces. *European Journal of Clinical Microbiology and Infectious Diseases* **8**:974-978.

Stickler, D., and Hewett, P. 1991. Activities of antiseptics against biofilms of mixed bacterial species growing on silicone surfaces. *European Journal of Clinical Microbiology and Infectious Diseases* **10**:157-162.

Sykes, G. 1965. *Disinfection and Sterilization,* 2nd Ed. London: E. and F. N. Spon Ltd.

Van Klingeren, B., Leussink, A. B., and Van Wijngaarden, L. J. 1977. A collaborative study on the repeatability and reproducibility of the Dutch Standard-suspension-test for the evaluation of disinfectants. *Zentralblatt für Bakteriologie, Parasitenkunde, Infektionskrankheiten und Hygiene, I. Abteilung Originale, Reihe B* **164**:521-548.

Van Klingeren, B., and Mossel, D. A. A. 1978. Official evaluation of disinfectants in the Netherlands. *Zentralblatt für Bakteriologie, Parasitenkunde, Infektionskrankheiten und Hygiene, I. Abteilung Originale, Reihe B* **166**:540-541.

Van Klingeren, B., Leussink, A. B., and Pullen, W. 1981. A European collaborative study on the repeatability and the reproducibility of the standard suspension test for the evaluation of disinfectants in food hygiene. National Institute of Public Health and Environmental Hygiene. Report No. 35901001. Bilthoven: The Netherlands.

Van Klingeren, B. 1983. A two-tier system for the evaluation of disinfectants. Proefschrift, University of Utrecht.

Van Klingeren, B. 1990. Studies of the activity of surface disinfectants in the Dutch Quantitative Carrier Test. National Institute of Public Health and Environmental Protection. Report No. 358704002. Bilthoven: The Netherlands.

Vess, R. W., Anderson, R. L., Carr, J. H., Bond, W. W., and Favero, M. S. 1993. The colonization of solid PVC surfaces and the acquisition of resistance to germicides by water micro-organisms. *Journal of Applied Bacteriology* **74**:215-221.

Walker, A. J., Holah, J. T., Denyer, S. P., and Stewart, G. S. A. B. 1993. The use of bioluminescence to study the behaviour of *Listeria monocytogenes* when attached to surfaces. *Colloids and Surfaces A: Physicochemical and Engineering Aspects* **77**:225-229.

Werner, H. P., Reybrouck, G., and Werner, G. 1975. A comparison of the results of four national methods for the evaluation of disinfectants in two laboratories. *Zentralblatt für Bakteriologie, Parasitenkunde, Infektionskrankheiten, I. Abteilung Orginale, Reihe B.* **160**:368-391.

Wright, N. E., and Gilbert, P. 1987a. Influence of specific growth rate and nutrient limitation upon the sensitivity of *Escherichia coli* towards chlorhexidine diacetate. *Journal of Applied Bacteriology* **62**:309-314.

Wright, N. E., and Gilbert, P. 1987b. Antimicrobial activity of *n*-alkyl trimethyl ammonium bromides: influence of specific growth rate and nutrient limitation. *Journal of Pharmacy and Pharmacology* **39**:685-690.

Wright, J. B., Ruseka, I., and Costerton, J. W. 1991. Decreased biocide susceptibility of adherent *Legionella pneumophila*. *Journal of Applied Bacteriology* **71**:531-538.

3.5 Reproducibility and Performance of Endospores as Biological Indicators

Norman Hodges

DEVELOPMENT AND CURRENT STATUS OF STERILIZATION INDICATORS

The principle of using resistant endospores to check the performance of a sterilization process was established by Koch. He employed anthrax spores, dried on a string, to confirm the effectiveness of his steam sterilization procedures. It was, however, a recognition of the limitations of conventional sterility testing procedures in terms of their reliability for detecting low levels of organisms surviving a sterilization process that provided a major impetus to the adoption of bacterial spore monitors (biological indicators) as adjuncts to sterility tests in quality assurance programs. The first commercial biological indicator for the control of steam sterilization was introduced in 1957, shortly after the publication of a detailed description of the use of spores of *Bacillus stearothermophilus* for that purpose (Brewer, Heer and McLaughlin, 1956).

Sterilization processes are employed for many products in both the food and pharmaceutical industries, but both the bioburden and the range of organisms which contaminate and represent spoilage problems in the food industry are wider than those encountered with the manufacture of medical products and pharmaceuticals. There has, therefore, been a greater tendency in the food industry to employ test organisms (commonly spores of *Clostridium* species) which are particularly relevant to the product in question, rather than commercially available indicators containing *Bacillus* spores that are more commonly used in the pharmaceuticals industry.

Over the last 35 years the use of spores in the pharmaceutical industry has increased substantially, with specifications for biological indicators now contained in most pharmacopoeias in North America and Europe. Despite this, there

0-8493-4752-1/95/$0.00+$.50

have been, and still are, marked variations in the degree of their acceptance by national and international medicines regulatory authorities, with those in the United Kingdom noticeably less enthusiastic both historically (Rosenheim, 1973) and currently (Medical Devices Directorate, 1993) than those in the United States, continental Europe, and Scandanavia for the employment of biological indicators as a form of control for anything other than gaseous (formaldehyde vapor and ethylene oxide) sterilization. This reluctance undoubtedly originates from the knowledge (1) that for most other sterilization processes physical methods of monitoring and control are highly reliable and predictive of sterilty assurance level, and (2) that spores of appropriate resistance and stability are difficult to produce. Not surprisingly, the volume of published work on biological indicators for use in different applications is directly related to the importance of the sterilization process in question. Consequently, a relatively large amount of data is available on biological indicators applied to steam sterilization. This will be reflected in the present chapter.

The range of commercially available endospore indicators is now wide, with products from several manufacturers available in most Western countries. Spores have been employed, however, both as laboratory-produced and commercial products, in the study and control of a variety of nonconventional sterilizing/decontamination applications in addition to large-volume manufacturing processes; these include, for example, the surface sterilization of implants by ultraviolet light (Singh and Schaaf, 1989) and the testing of microwave ovens (Jeng et al., 1987) and dental sterilizers, the latter now being mandatory in several states as a consequence of concerns about the spread of HIV (Miller, 1993).

SELECTION OF ORGANISMS FOR USE AS STERILIZATION INDICATORS

Regardless of the application, spores to be used as biological indicators should possess a number of desirable characteristics:

1. Biological indicators should possess an appropriate level of resistance to the lethal agent/sterilization process for which the indicator is intended, and should exhibit first-order survival kinetics with no activation, shouldering, or tailing being apparent.
2. Spores for use as biological indicators should exhibit minimal germination prior to use, but maximal germination on transfer to recovery medium after use, so that survivors exhibit reproducible recovery under standardized conditions.
3. The resistance properties of the spores, the viability of the suspensions, and the proportion of the spores which germinate should be stable during storage.
4. The vegetative organism should be nonpathogenic and nonpyrogenic.
5. The organism should grow on simple media, preferably minimal salts, in conventional cultures and sporulate readily. The spores should be easily harvested and cleaned.

6. The spores should not aggregate in suspension, and should be amenable to incorporation on a carrier possessing appropriate characteristics.

This review is primarily concerned with the properties of spores which are used for the manufacture of the biological indicator rather than the design, construction and use of the indicator itself, but for the sake of completeness it should be emphasized here that the carrier will have a significant effect on indicator performance. It must permit easy access of the lethal agent to the spores (Gurevich, Quadri and Cunha, 1993), restrict accidental contamination with organisms from the surroundings and possess low thermal mass to restrict heat-up time lags (McCormick, 1988). Because viable counting will yield more useful information than the mere recording of positive (survival) or negative (kill) results, the indicator should disintegrate or dissolve readily in aqueous media to facilitate such counting. The traditional way of achieving the last characteristic has been to use absorbent paper as a carrier onto which aqueous spore suspensions are dried. This entails the risk that the spores will irreversibly bind to, or become entrapped within the paper matrix and so give a low result for viable counts on the supernatant after centrifugation of the paper fibers. This source of error may be overcome by complete homogenization and dispersion of the paper carrier (Everall and Morris, 1978) or, more satisfactorily, by use of a calcium alginate carrier which will completely dissolve in aqueous salt solutions but not in steam (Hanlon and Hodges, 1993).

There is no organism which fulfils all of the criteria listed above. *Clostridium* species are often the most relevant for the food industry, and *Clostridium sporogenes* is frequently employed in the design of food sterilization processes. However, the requirements of the U.S. National Institute of Health, that organisms to be used for routine monitoring of hospital sterilizers should be nonpathogenic, eliminated many clostridia from contention as the basis of a commercial product even if they possessed appropriate resistance. *Cl. sporogenes* ATCC 7955 (P.A. 3679) is recommended in the 1993 British Pharmacopoeia for the validation of steam sterilization, but not by the United States or European Pharmacopoeias.

Obligate thermophils which do not grow at 37°C are obviously the best choice as nonpathogens. In this respect *Bacillus stearothermophilus* has been used almost exclusively for the monitoring of steam sterilization in the pharmaceutical industry. *Bacillus coagulans* has also been recommended as an alternative (Jones and Pflug, 1981) because it has been found to exhibit reproducible resistance and possess a D_{115} value of 6 to 8 min, which, assuming a Z-value of 10°C, would provide a D_{121} of an appropriate magnitude (i.e., 1.5 to 2.0 min).

One of the most comprehensive studies on the suitability of different organisms for use as biological indicators was made by Hoxey, Soper and Davies (1985), who quantified the heat inactivation kinetics and germination indices of 20 strains of Bacillus spores produced on each of three defined media. In addition to marked differences in the extent of sporulation on the three media, it was apparent that the spore crops could be divided into groups, according to their ease of germination. Two out of the four crops which germinated least readily exhibited survivor plots of progressively increasing slope. All of the four crops which germinated easily showed heat survivor plots of decreasing slope. All seven crops of intermediate germination index showed first-order survival kinetics.

If spores with a high degree of dormancy are used for biological indicators an increase in colony count or a shoulder on the survivor plot might arise as a consequence of sublethal heating. Both this phenomenon and that of nonlinear heat inactivation kinetics would render a spore crop or species unsuitable as a candidate for the manufacture of a biological indicator because it would not be possible to quote a meaningful D value. Unfortunately there are several examples in the literature when studies on spores as biological indicators have been conducted without regard to these properties.

Brown (1985) has reviewed the possession of extreme heat resistance by bacterial spores. Figure 1 shows the relative resistances displayed by the common marker organisms. Despite the fact that *B. stearothermophilus* has been extensively used and studied, it is not necessarily the most heat resistant organism: *Clostridium thermosaccharolyticum* has been reported to exhibit similar or higher D_{121} values (Brown, 1985). Occasionally spores of aerobic mesophils have also been reported to possess D_{121} values which exceed 1.0 min, and so might be considered as potentially suitable for sterilization control. The more usual D_{121} values observed are 0.5 min or less (Odlaug, Caputo and Graham, 1981), and even the well-characterized resistant strains of mesophilic species are found to be unsuitable when they are compared with thermophils (Hoxey, Soper and Davies, 1985).

A variety of *B. stearothermophilus* strains have been examined as potential biological indicators. Strain NCTC 8919 (ATCC 12976) has been used to some extent in the past, but the strains which are employed by most manufacturers of biological indicators are ATCC 7953 and ATCC 12980 (Graham and Boris, 1993). It would appear, however, that substrains of these are likely to be in use, since it has long been recognized that the colonial appearance and heat resistance of the organisms generally isolated from commercial biological indicator products differ from those of the culture collection strains (Alcock and Brown, 1985; Buhlmann, Gay and Schiller, 1973).

There is consensus regarding the organisms to be used for validation/monitoring of dry heat, ethylene oxide, and radiation sterilization processes. Spores of *Bacillus subtilis* var *niger* are invariably employed for both dry heat and ethylene oxide, with ATCC 9372 being the most commonly used strain. Radiation sterilization is so ameniable to monitoring by physical dosimeters that biological indicators are usually considered to be unnecessary. However, *Bacillus pumilus* ATCC 27142 is usually considered to be the most suitable strain if biological indicators are required for such processes.

Sterilization by low-temperature steam formaldehyde (LTSF) is less frequently employed as a gaseous sterilization method than ethylene oxide, but its use is well established, particularly in Europe. There are existing German standards for LTSF biological indicators (Mecke, Christiansen and Pirk, 1991),

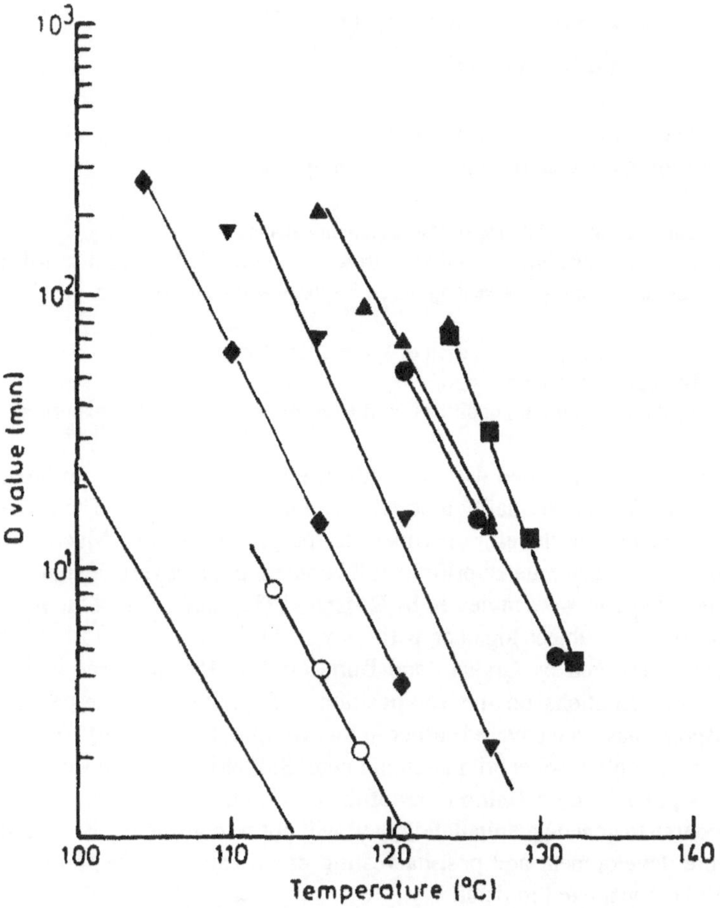

FIGURE 1 Heat resistance of *Clostridium thermosaccharolyticum* spores from forest bark compared with the resistance of other resistant spore-formers. No symbol, classical *Clostridium botulinum* resistance; (○), *Clostridium sporogenes*; (♦, ▲, and ■), three isolates of *Clostridium thermosaccharolyticum*; (▼) *Bacillus stearothermophilus* NCIB 8919; (●), *Desulphomaculum nigrificans* (From Brown, K. L., in *Fundamental and Applied Aspects of Bacterial Spores*, Academic Press, London, 1985, p. 276. With permission.)

and a European standard is in preparation for spores of *B. stearothermophilus* NCTC 10003 to be used for this purpose (Medical Devices Directorate, 1993). When the suitability of commercial biological indicators containing this species were recently compared for the control of LTSF processes, one out of six products exhibited a resistance which was too low and the remainder showed a resistance too high (Mecke, Christiansen and Pirk, 1991) It has been suggested that a combination of *B. stearothermophilus* with *B. subtilis* is likely to be advantageous because the former will reveal too low a concentration of formaldehyde gas and the latter will reveal insufficient humidification (Christiansen and Kristensen, 1992).

FACTORS AFFECTING BIOLOGICAL INDICATOR PERFORMANCE

The factors which influence the performance characteristics of endospores as biological indicators fall into five major categories

1. The intrinsic resistance of the organism selected.
2. Environmental influences during growth of the vegetative organism and sporulation, methods of harvesting the endospores, and manufacture of the biological indicator.
3. Post-harvesting modification of spore characteristics.
4. Biological indicator design.
5. Media and cultural conditions used to recover survivors after exposure.

This book is concerned with the relevance and reproducibility of inocula, and their application to quality assurance and testing programs. The first three of these factors are therefore pertinent to the present review. Much of the data which were accumulated prior to 1982 concerning recovery of spores exposed to lethal agents were reviewed by Russell (1982), and some of the more recent data on this subject together with current biological indicator designs have been considered by Graham and Boris (1993). The influence of vegetative growth conditions on the composition and performance properties of endospores has been covered earlier in this volume (Chapter 1.6). This topic will therefore only receive brief attention here. Similarly, the intrinsic resistance of endospores is, by definition, out of the control of the operator (other than by recourse to genetic manipulation) and will not be considered. Factors affecting spore development and post-harvesting modification of the spore properties will be considered in detail.

SPORE RESISTANCE RESULTING FROM ENVIRONMENTAL INFLUENCES DURING SPORULATION AND HARVESTING

Levels of Definition

In contrast to vegetative cells, bacterial spores unfortunately cannot conveniently be produced in large numbers from chemostats. Spore properties, therefore, will be influenced by the rate and extent to which the environment of a batch culture changes as logarithmic growth ceases and sporulation proceeds (Chapter 1.6). Thus, while it is possible to precisely define the medium composition in chemostat at steady state, in a batch culture the concentration of available nutrients changes simultaneously with parameters such as pH and redox potential. Partly because of this, the literature on bacterial spore production and properties, even that of the last ten years, is replete with ambiguity and anomaly. The developments, in terms of definition of cultural characteristics and properties of organisms to be used as inocula, which have been achieved in other areas of bacteriology are less in evidence with respect to endospores. There are conflicting reports in the literature about almost every feature of the environment which might be expected to have a bearing upon spore properties; thus culture age, sporulation temperature, medium composition, and pH have all been reported at various times to have either a marked effect or none.

It is obvious that the properties of a batch of spores are likely to be a reflection of the composition of the medium and the conditions under which the spores were formed. Attempts to achieve reproducibility are therefore likely to be facilitated by definition of the culture media (Chapter 1.6). There are several levels of such definition, of which the use of a complex medium with detailed specifications for the source of supply and the quality of the medium components is the first (El-Bisi and Ordal, 1956). An advance upon this is the use of a chemically defined (synthetic) medium which is developed empirically by adjustment of the concentration of medium components to achieve good sporulation and the desired spore properties (Anderson and Friesen, 1972). A third level is the use of media which are further defined in terms of the nature of the specific nutrient, the exhaustion of which limits vegetative growth and induces sporulation (Chapter 1.6). In such circumatances the degree of excess of the nonlimiting nutrients is also known (Hodges and Brown, 1975; Lee and Brown, 1975). The details of the production processes which are used in the commercial manufacture of biological indicators are unknown because they are considered to be proprietary (Graham and Boris, 1993), but the few indications available in published work suggest that the first level of definition is that which is most likely to be adopted in commercial manufacture.

B. stearothermophilus and *B. subtilis* are undoubtedly the most important and well-studied of the organisms used as biological indicators. Both of these are capable of growth and definition at all three levels. The benefits of enhanced medium definition in terms of improved reproducibility of heat and glutaraldehyde resistance and germination rate have been experimentally demonstrated (Hodges, Melling and Parker, 1980) and have been repeatedly stressed by a variety of other workers (Dadd, Stewart and Town, 1983; Hoxey, Soper and Davies, 1985; Friesen and Anderson, 1974).

Post-Harvesting Modification of Spore Resistance

Spores to be used in biological indicators must possess a level of resistance to the lethal agent in question which is within a relatively narrow specified range. In practice this is normally achieved by using a culture medium and growth conditions which have been developed empirically over a long period and have been shown to give a satisfactory product. Clearly the medium composition and environmental conditions are defined as precisely as possible and rigorously standardized in attempts to minimize batchwise variation (Chapters 1.2 and 1.6). If, however, the desired levels of resistance and dormancy are not achieved, then the potential exists, even after harvesting of the spores, for these characteristics to be modified to conform with the performance specifications. To be of practical value, however, any such treatment(s) must be capable of effecting both increases and decreases in D-value and z-value, and should be applicable to the properties of the spores as expressed within a bioogical indicator (i.e., dried onto an inert carrier or suspended in culture medium and sealed into an ampoule).

Gould (1985), in reviewing methods of modification for heat/chemical resistance and dormancy of endospores, listed a total of 16 treatments which are known to effect changes in these properties. The majority of these treatments, however, could only be used to reduce resistance. Two treatments only were capable of increasing resistance, viz., cation exchange and adjustment of equilibrium water content in the spores. Furthermore, only the former could realistically be applied to spores to be dried onto a carrier.

Adjustment of water content can be achieved by suspending spores in solutions of known water activity (Aw), but this is impractical because addition of the necessary solutes to the medium in sealed ampoules of spore suspension might influence the subsequent growth of any spores surviving the sterilizing agent.

The cation exchange properties of bacterial spores have been known for more than 30 years and have been reviewed (Gerhardt and Marquis, 1989). Exposure of cleaned spores to acid solutions (pH 1 to 4) at 25 to 60°C leads, depending upon the species, to the removal of minerals without any significant losses in viability. Subsequent exposure to basic solutions (pH 9), with or without the aid of elevated temperature, permits remineralization of the endospores. This allows spores to be produced which have an atypical balance of cations, and which may, in relative terms at least, be enriched with any one of several common mono- and divalent metal cations. Exceptionally, the content of any particular cation may, in real terms (Beaman and Gerhardt, 1986), be forced higher than that found in native spores. More commonly, full remineralization has been found to be difficult to achieve (Marquis et al., 1981).

Native spores are normally found to exhibit greater heat resistance than demineralized (acid-stripped or H-form) spores, which are usually also more sensitive than any other type (Figure 2). Those spores which have been enriched with manganese (Marquis et al., 1981) or calcium (Beaman and Gerhardt, 1986; Friesen and Anderson, 1974) may show greater thermal resistance than native spores. It should be noted, however, that the maximal resistance exhibited by native spores, through modification of vegetative growth environment, generally cannot be exceeded (Figure 2).

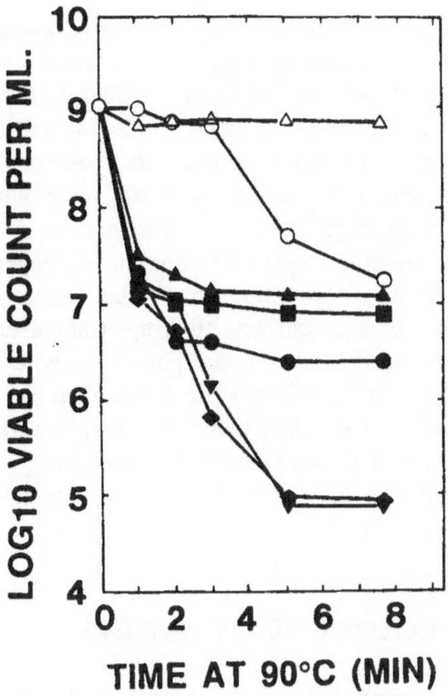

FIGURE 2 Heat sensitivities of various salt forms of *Bacillus megaterium*. (△), Manganese; (○), Native; (▲), Magnesium; (■), Calcium; (●), Potassium; (◆), Sodium, and (▼), protonated. (Adapted from Marquis, R. E., et al., in *Sporulation and Germination*, American Society for Microbiology, Washington, D.C., 1981, p. 266. With permission.)

Stability of Biological Indicators

The vast majority of published work on spores as sterilization indicators has been concerned primarily or exclusively with resistance. The factors which influence other important properties — such as spore viability, longevity, germination rates, and the stability of the resistance characteristics during long-term storage — have rarely been studied in any systematic fashion. There are several reports on the stability with respect to resistance during storage, but these often describe the effects of storage conditions, such as temperature and humidity, for a single spore crop rather than comparing the resistance stability of different spore crops cultured and allowed to sporulate in various media.

The reproducibility of biological indicator performance is clearly dependent upon the exercise of appropriate storage conditions between manufacture and use. Laboratory-produced spores generated from nutrient agar cultures (Smith, Pflug and Chapman, 1976) and commercially manufactured spores (Smith, Pflug and Chapman, 1976; Reich and Morien, 1982) have been shown to be most stable in terms of spore viability and heat resistance when stored, at 4°C and 0% relative humidity, although even under these conditions there is a significant fall in resistance within 1 year. When stored at ambient temperatures and 50% or more relative humidity, then the stability of the spore and its performance characteristics are markedly prejudiced. The heat resistance of *B. stearothermophilus* spores harvested from minimal salts, defined media has been shown to be stable during storage in aqueous suspension at 5°C for 2 years (Friesen and Anderson, 1974), and for at least 1 year when the spores were dried on paper over silica gel at room temperature.

Lyophilization might be expected to minimize changes in heat resistance during storage. This has been shown to be the case for *B. subtilis* spores obtained from nutrient agar cultures (Odlaug, Caputo and Graham, 1981) although the process itself resulted in a slight fall in spore viability. *B. stearothermophilus* spores, obtained from a glucose-limited culture grown in a defined minimal medium, were also stable after lyophilization, although, once again, the process resulted in a viability loss and, in this case, reduced heat resistance compared with that of the native spores immediately after harvesting (Alpin and Hodges, 1979).

QUICK-RESPONSE BIOLOGICAL INDICATORS

One of the most commonly cited criticisms of the use of biological indicators is the disadvantage of having to wait during the incubation period before the sterilized material can be released for sale or use. Consequently there has always been an incentive to produce indicators which give a reliable result in the shortest possible incubation period. This has resulted in an FDA guide to the procedure for validating incubation periods of less than 7 days (Graham and Boris, 1993). This incentive has been further heightened in recent years with the increasing use of flash sterilization processes in which exposure periods of 3.0 min to steam at 132°C are used for surgical instruments (Kotilainen and Gantz, 1987; McCormick, 1988).

Biological indicators normally reveal sterilization failure by the display of either turbidity or acid production in the culture medium used for incubation of the exposed indicator. There have been several recent approaches to the development of alternative means for detection of growth which accelerate the process. Measurement of APT, using firefly luciferase, permits detection within 5 h of biological indicators containing *B. stearothermophilus* spores which have survived a steam sterilization cycle (Webster et al., 1988). A more rapid detection time is possible through assay of the spore-bound enzyme, α-D-glucosidase in the same organism. This enzyme converts the nonfluorescent reagent 4-methylumbelliferyl-α-D-glucoside into a fluorescent product which can be detected within 1 h post-treatment (Vesley et al., 1992). Another method which has shown encouraging early results is the incorporation of a *lux* gene (coding for bioluminescence) into biological indicator organisms. This has been successfully achieved with *Bacillus megaterium* and *B. subtilis*. In

B. subtilis spores containing the *lux* gene, lethal levels of moist heat rapidly decreases both the extent of bioluminescence and the viable count (Stewart, Jassim and Denyer, 1992).

REFERENCES

Alcock, S. J., and Brown, K. L. 1985. Heat resistance of P A 3679 (NCIB 8053) and other isolates of *Clostridium sporogenes*, p. 261 in G. J. Dring, D. J. Ellar, and G. W. Gould (eds.), *Fundamental and Applied Aspects of Bacterial Spores*. Academic Press, London.

Alpin, S. J., and Hodges, N. A. 1979. Changes in heat resistance during storage of *Bacillus stearothermophilus* spores from complex and chemically defined media. *Journal of Applied Bacteriology* **46**:623-626.

Anderson, R. A., and Friesen, W. T. 1972. Growth and sporulation of *Bacillus stearothermophilus* in chemically defined media. *Australian Journal of Pharmaceutical Science* **NS1**:1-6.

Beaman, T. C., and Gerhardt, P. 1986. Heat resistance of bacterial spores correlated with protoplast dehydration mineralisation and thermal adaptation. *Applied and Environmental Microbiology* **52**:1242-1246.

Brewer, J. H., Heer, A. H., and McLaughlin, C. B. 1956. The control of sterilisation procedures with thermophilic spore-formers. *Bacteriological Proceedings* **56**:61.

Brown, K. L. 1985. Heat resistant anaerobe isolated from composted forest bark, p. 276 in G. J. Dring, D. J. Ellar, and G. W. Gould (eds.), *Fundamental and Applied Aspects of Bacterial Spores*. Academic Press, London.

Buhlmann, X., Gay, M., and Schiller, I. 1973. Test objects containing *Bacillus stearothermophilus* spores for monitoring of antimicrobial treatment in steam autoclaves. *Pharmaceutica acta Helvetica* **48**:223-244.

Christensen, E. A., and Kristensen, H. 1992. Gaseous sterilization, p. 557 in A. D. Russell, W. B. Hugo, and G. A. J. Ayliffe (eds.), *Principles and Practice of Disinfection, Preservation and Sterilization*. Blackwell Scientific Publications, Oxford.

Dadd, A. H., Stewart, C. M., and Town, M. M. 1983. A standardised monitor for the control of ethylene oxide sterilization cycles. *Journal of Hygiene, Cambridge* **91**:93-100.

El-Bisi, H. M., and Ordal, Z. J. 1956. The effect of certain sporulation conditions on the thermal death rate of *Bacillus coagulans* var. *thermoacidurans*. *Journal of Bacteriology* **71**:1-16.

Everall, P. H., and Morris, C. A. 1978. Quantitative recovery of spores from thermophilic spore papers. *Journal of Clinical Pathology* **31**:423-425.

Friesen, W. T., and Anderson, R. A. 1974. Effects of sporulation conditions and cation-exchange treatment on the thermal resistance of *Bacillus stearothermophilus* spores. *Canadian Journal of Pharmaceutical Science* **9**:50-53.

Gerhardt, P., and Marquis, E. 1989. Spore thermoresistance mechanisms, p. 43-59 in G. M. Smith, R. A. Slepecky, and P. Setlow, (eds.), *Regulation of Procaryotic Development*. American Society for Microbiology, Washington, D.C.

Gould, G. W. 1985. Modification of resistance and dormancy, p. 371 in G. J. Dring, D. J. Ellar, and G. W. Gould (eds.), *Fundamental and Applied Aspects of Bacterial Spores*. Academic Press, London.

Graham, G. S., and Boris, C. A. 1993. Chemical and biological indicators, p. 36-69 in R. F. Morrissey and G. B. Phillips (eds.), *Sterilization Technology: A Practical Guide for Manufacturers and Users of Health Care Products*. Van Nostrand Reinhold, New York.

Gurevich, I., Quadri, S. M., and Cunha, B. A. 1993. False-positive results of spore tests from improper clip use with the STERIS chemical sterilant system. *American Journal of Infection Control* **21**:42-43.

Hanlon, G. W., and Hodges, N. A. 1993. Quantitative assessment of sterilization efficacy using lyophilized calcium alginate biological indicators. *Letters in Applied Microbiology* **17**:171-173.

Hodges, N. A., and Brown, M. R. W. 1975. Properties of *Bacillus megaterium* spores formed under conditions of nutrient limitation, p. 550-555, in P. Gerhardt, H. L. Sadoff and R. N. Costilow (eds.), *Spores VI*. American Society for Microbiology, Washington, D.C.

Hodges, N. A., Melling, J., and Parker, S. J. 1980. A comparison of chemically defined and complex media for the production of *Bacillus subtilis* spores having reproducible resistance and germination characteristics. *Journal of Pharmacy and Pharmacology* **32**:126 -130.

Hoxey, E. V., Soper, C. J., and Davies, D. J. G. 1985. Biological indicators for low temperature steam formaldehyde sterilization: effect of defined media on sporulation, germination index and moist heat resistance at 110°C of *Bacillus* strains. *Journal of Applied Bacteriology* **58**:207-214.

Jeng, D. K., Kaczmarek, K. A., Woodworth, A. G., and Balasky, G. 1987. Mechanism of microwave sterilisation in the dry state. *Applied and Environmental Microbiology* **53**:2133-9.

Jones, A. T., and Pflug, I. J. 1981. *Bacillus coagulans*, FRR B8666, as a potential biological indicator organism. *Journal of Parenteral Science and Technology* **35**:82-87.

Kotilainen, H. R., and Gantz, N. M. 1987. An evaluation of three biological indicator systems in sterilization. *Infection Control* **8**:311-316.

Lee, Y. H., and Brown, M. R. W. 1975. Effect of nutrient limitation on sporulation of *Bacillus stearothermophilus*. *Journal of Pharmacy and Pharmacology* **27**:22P.

Marquis, R. E., Carstensen, E. L., Child, S. Z., and Bender, G. R. 1981. Preparation and characterisation of various salt forms of *Bacillus megaterium* spores, p. 266 in H. S. Levinson, A. L. Sonenshein, and D. J. Tipper (eds.), *Sporulation and Germination*. American Society for Microbiology, Washington, D.C.

McCormick, P. J. 1988. Biological indicators. *Infection Control and Hospital Epidemiology* **9**:504-507.

Mecke, P., Christiansen, B., and Pirk, A. 1991. The suitability of commercial bioindicators with spores of *B. stearothermophilus* for testing of efficacy of formaldehyde sterilizers. *Zentralblatt für Hygiene* **192**:25-32.

Medical Devices Directorate, Department of Health. 1993. Sterilization, disinfection and cleaning of medical equipment. H.M.S.O., London.

Miller, C. H. 1993. Update on heat sterilisation and sterilisation monitoring. *Compendium of Continuing Education in Dentistry* **14**:304-316.

Odlaug, T. E., Caputo, R. A., and Graham, G. S. 1981. Heat resistance and population stability of lyophilized *Bacillus subtilis* spores. *Applied and Environmental Microbiology* **41:**1374-1377.

Reich, R. R., and Morien, L. L. 1982. Influence of environmental storage relative humidity on biological indicator resistance, viability and moisture content. *Applied and Environmental Microbiology* **43:**609-614.

Rosenheim, The Lord. 1973. Report on the prevention of microbial contamination of medicinal products. H.M.S.O. London.

Russell, A. D. 1982. Inactivation of bacterial spores by thermal processes (moist heat), p. 30 in *The Destruction of Bacterial Spores.* A. D. Russell, (ed.), Academic Press, London.

Singh, S., and Schaaf, N. G. 1989. Dynamic sterilisation of titanium implants with ultraviolet light. *International Journal of Oral and Maxillofaciary Implants* **4:**139-146.

Smith, G. M., Pflug, I. J., and Chapman, P. A. 1976. Effect of storage time and temperature and the variation among replicate tests (on different days) on the performance of spore disks and strips. *Applied and Environmental Microbiology* **32:**257-263.

Stewart, G. S. A. B., Jassim, S. A. A., and Denyer, S. P. 1992. Engineering microbial bioluminescence and biosensor applications, p. 403 in R. R. Rapley and M. R. Walker (eds.), *Molecular Diagnostics.* Blackwell Scientific Publications, Oxford.

Vesley, D., Langholz, A. C., Rohlfing, S. R., and Foltz, W. E. 1992. Fluorimetric detection of a *Bacillus stearothermophilus* spore-bound enzyme, α-D-glucosidase, for rapid indication of flash sterilization failure. *Applied and Environmental Microbiology* **58:**717-719.

Webster, J. J., Walker, B. G., Ford, S. R., and Leach, F. R. 1988. Determination of sterilization effectiveness by measuring bacterial growth in a biological indicator through firefly luciferase determination of ATP. *Journal of Bioluminescence and Chemiluminescence* **2:**129-133.

3.6 Development and Production of Vaccines

Andrew Robinson, Howard S. Tranter, Christopher N. Wiblin, and Peter Hambleton

INTRODUCTION

The production of an effective bacterial vaccine relies heavily upon process reproducibility. This is, in turn, dependent upon standardized, high-quality inocula that are well characterized and regularly monitored. Such cultures, maintained as primary seed sources, must remain viable and pure, and must retain the characteristics for which they have been selected (e.g., toxin production). Although a wide variety of methods exist for culture preservation (Chapter 2.1; Snell, 1991), there is no universally applicable method that will successfully preserve all microorganisms. It may therefore be difficult to select a method suitable for a particular need. While freeze-drying may be used successfully to preserve those characteristics of *Bordetella pertussis* required for vaccine production, the same process can result in nontoxigenic cultures, or cultures of reduced toxicity, of *Clostridium botulinum*. Cultures of spore-forming bacteria used in the production of vaccines (e.g., *Bacillus anthracis*) may be adequately maintained at 4°C but may lose antigenicity/toxicity if frozen or freeze-dried. Some of the factors to be considered when producing reproducible high quality stock cultures for production of vaccines are shown in Table 1. Many of the problems indicated in the table can be overcome by the operation of a master and submaster seed stock system.

Stability of the stock culture is of paramount importance in the manufacture of vaccines, since studies of host-pathogen interactions over several decades have shown that the pathogen is far from being a constant entity and, in many cases, exhibits profound changes in its macromolecular composition. This marked variability confers two major advantages on the pathogen for the establishment and development of the disease process.

First, variability enables the pathogen to adapt to different environmental niches encountered during pathogenesis (Finlay and Falkow, 1989; Williams, 1988). For example, initial colonization of the mucosal surface may depend on the presence of fimbriae; invasion and intracellular survival may depend on

0-8493-4752-1/95/$0.00+$.50

TABLE 1
Factors to be Considered When Preserving Vaccine Stock Cultures

Culture survival and viability	Minimize loss, maximize viability
Purity	Minimize chances of contamination
Changes in genotype	Prevent or minimize mutations, selection of subpopulations or loss of plasmids, phages, etc.
Expense	Storage in liquid N_2 or freeze-drying may result in high capital costs
Number of seed stocks	Large seed stocks usually required for vaccine production
Maintenance of seed stocks	Liquid N_2 may need to be frequently replaced
Biological hazards	Preservation of bacterial pathogens must be carried out safely

various invasins and toxins; survival in the blood stream may depend on the presence of an external capsule; and growth and survival at certain sites may depend on the production of uptake mechanisms for specific nutrients. Thus, whereas certain virulence factors may be required for specific stages or sites of infection, at other stages or sites they may prove to be an unnecessary burden for the pathogen.

Second, variability plays a prime role in enabling the pathogen to evade the host's developing immune response. A pathogen may therefore possess mechanisms to alter its antigenic profile with respect to a vital virulence component, such that organisms which possess an immunologically distinct but functionally similar component, will have a selective advantage as the immune response to the original component develops.

This inherent variability of the bacterial pathogen poses major problems for the design and production of effective vaccines. The composition of bacteria grown *in vitro* for vaccine production may be quite different from that expressed by the organism *in vivo*, so that components present in vaccines may not be expressed (or may be masked) during *in vivo* growth. Furthermore, certain components expressed *in vivo* may be highly immunogenic but require careful and deliberate manipulation of the growth environment to ensure their expression *in vitro*.

This review presents examples of how variation in the characteristics of bacterial pathogens, including toxicity loss and genetic instability, may affect the design and production of vaccines, and how close assessment and quality assurance are required to ensure that the final vaccine produced is effective and of the intended composition.

VARIATION OF BACTERIAL COMPOSITION

There are essentially two mechanisms whereby bacterial composition can be varied to confer advantages in the pathogenic process. Phase variation endows bacteria with the ability to express, or not, a particular component (i.e., an on-off mechanism), whereas antigenic variation allows bacteria to vary the antigenic composition of a particular component.

Phase Variation

The ability to switch production of a particular component on or off is widespread among bacterial pathogens. In the following section some examples which have implications for vaccine production are described.

Fimbriae (Pili)

Adhesion to a mucosal surface is usually the initial stage of pathogenesis for most bacterial diseases and in many cases this adhesion is mediated, at least as a primary point of contact, by long filamentous surface antigens called fimbriae. The adhesive component of fimbriae is often found at the tip of the filament, the main shaft of the fimbriae being merely a means of delivering this adhesin to the site of attachment (Jones et al., 1992). Adhesion is a major, vaccine-preventable, stage of pathogenesis, since it enables the pathogen to resist normal physical clearance mechanisms, (e.g., coughing, mucocilliary flow, urinary flow, etc.) and establish infection. After the initial colonization, however, the expression of fimbriae at secondary sites of infection may be an energy-consuming burden for the bacteria, or even act to the detriment of the pathogen by mediating attachment to host macrophages that bear the appropriate receptor molecules. Thus, synthesis of fimbriae is subject to phase variation allowing expression to occur at the appropriate stages in the disease process. The genetic control of fimbrial phase variation has been mainly studied in *Escherichia coli* and *Neisseria meningitidis* (Kroll, 1990; Meyer, 1990; Saunders et al., 1993). If fimbriae are the intended vaccine components, the phase variation has to be considered to ensure that seed stock selection, and growth conditions during vaccine manufacture, are compatible with a high level of fimbrial expression.

An additional complexity is that bacteria may produce several types of fimbriae that while having similar functions, are expressed independently in an attempt by the bacteria to thwart the host's developing immune response. For example, *Bordetella pertussis* produces two major types of fimbriae, Fim2 and Fim3, which have subunit molecular weights of 22.5 kDa and 22 kDa, respectively (Robinson, Ashworth and Irons, 1989). Fim2 and Fim3 are, as far as is known, functionally identical but immunologically distinct. Strains of *B. pertussis* can express either, both, or neither type of fimbriae. Epidemiological studies in the 1960s led to the recommendation that whole-cell pertussis vaccines should possess both types of fimbriae in order to provide protection against all serotypes of *B. pertussis* (Medical Research Council, 1959). In pertussis vaccine manufacture, therefore, appropriate selection of seed stocks that express and maintain both fimbrial types is essential. Fimbrial expression can be assessed by agglutination of bacteria with monoclonal antibodies specific for each fimbrial type.

Coordinated Phase Variation

Phase variation may also affect the coordinated regulation of several different virulence components. A prime example of this is again with the pathogen *B. pertussis*, which has the ability to express or repress several protein components to yield either virulent or avirulent organisms respectively (Coote, 1991). This process of antigenic modulation is fully reversible and dependent on the growth environment. The variable proteins, which include fimbriae, heat labile toxin, adenylate cyclase, filamentous hemagglutinin, and pertussis toxin, are considered to be virulence components which may contribute to the protection induced by whole-cell pertussis vaccine. Growth of the bacteria in the presence of either nicotinic acid or high concentrations of magnesium sulphate, or alternatively at 25°C, yields bacteria which express none of the above components, and which are effectively avirulent and nonprotective. The environmental signaling and genetic mechanisms for such major phase variation have now, to some extent, been elucidated (Coote, 1991). For production of whole cell or acellular pertussis vaccines, it is essential to select growth conditions which permit full expression of the above components (i.e., 35 to 37°C, low concentrations of nicotinic acid and magnesium sulphate). In addition, care should be taken to ensure that mutations do not arise whereby the expression of the virulence components is irreversibly repressed.

Specific Nutrient Uptake Systems

Growth of the pathogen *in vivo* may depend on the ability to utilize specific nutrients supplied by the host, leading to the induction of nutrient uptake mechanisms. An excellent example of such variability is the utilization of various iron sources by pathogens. When grown *in vitro*, inorganic iron is usually in plentiful supply, as ferrous or ferric salts, which do not require complex uptake mechanisms. *In vivo*, however, free inorganic iron is not available and the pathogen has to compete with the host's ability to bind iron to high-affinity iron-binding proteins (Griffiths, 1987). Thus, iron is usually only available to the bacterial pathogen *in vivo* when it is complexed to transferrin in blood, lactoferrin in secretions, or ferritin inside host cells. Pathogenic bacteria have evolved two main strategies to compete with the host's iron-binding proteins (Griffiths, 1987; Williams and Griffiths, 1992). Firstly, pathogens can produce small iron-sequestering molecules, called siderophores, which effectively take iron from the host's proteins and, via specific bacterial receptor proteins, transport it into the bacteria. Alternatively, specific receptors on the bacterial surface can directly interact with the host's iron-binding proteins and transfer the bound iron into the bacteria. In both of these uptake mechanisms specific receptors are required on the bacterial surface and these receptors are only expressed *in vitro* when growth is restricted by the supply of iron (Chapter 1.2). Such proteins are also produced during growth *in vivo* as shown by the presence of antibodies to the proteins following infection.

The proteins responsible for iron uptake *in vivo* are potential vaccine components since they perform an essential growth function without which infection would cease. In particular, the transferrin-binding proteins of *Neisseria meningitidis* have been considered to be potential vaccine antigens which would be effective against all serogroups of the meningococcus (Williams and Griffiths, 1992). For the production of these and any other iron-uptake proteins, it is essential that the bacteria be grown in the appropriate iron-restricted environment (Chapter 1.2).

Antigenic Variation

The ability of bacteria to vary the antigenic composition of specific components is a powerful means of avoiding the immune response. Excellent examples of this elegant avoidance mechanism are provided by the pathogenic *Neisseria* species, *N. meningitidis* and *N. gonorrhoeae*.

Pili (or Fimbriae)

For both of the pathogenic *Neisseria* species, pili are required for colonization of mucosal surfaces. In each case, however, the pili exhibit pronounced intrastrain variability, even during the course of infection of the same individual (Tinsley and Heckles, 1986). The mechanism of such pilus variation has to some extent been elucidated and depends on the exchange of nucleotide sequences within the expressed pilin gene with sequences from several silent pilin genes (Meyer, 1990; Saunders et al., 1993). The variable regions of the pili which are coded for by these interchangeable sequences appear to be surface exposed and immunodominant (Heckles, 1989).

Major Outer Membrane Proteins (OMP)

Both *N. meningitidis* and *N. gonorrhoeae* express major proteins, which have the potential to be effective vaccine components (Heckles, 1993). Class 1 proteins of *N. meningitidis* are highly immunogenic proteins which function as porins. The structure and immunological properties of Class 1 proteins have been extensively studied. As with the gonococcal P1A and P1B porins, these proteins are highly homologous except for three variable regions which dictate the subtype specificity of the particular meningococcal strain. The variable sequences of these proteins have been shown to be expressed at the apices of loops which are exposed on the outer surface of the bacterial membrane (Van der Ley et al., 1991). Recent studies have demonstrated that point mutations can occur in a particular subtype epitope (P1, 16, McGuiness et al., 1991) of several meningococcal isolates, and that a single amino acid substitution can render the bacteria resistant to subtype-specific monoclonal antibodies (McGuiness et al., 1991).

Similarly, Class 2 and Class 3 meningococcal proteins are essential porins with variable exposed epitopes that define the serotype of meningococcal strains (Van der Ley et al., 1991). Although both Class 1 and Class 2/3 proteins are variable, they make attractive vaccine candidates, since strains of meningococcus of one particular serotype and subtype (i.e., possessing one type of Class 1 and one type of Class 2 or Class 3 protein) are usually responsible for a particular epidemic of meningococcal disease. A vaccine consisting of Class 1 and/or Class 2/3 proteins may, however, protect only against the homologous serotype or subtype strains of meningococcus. Thus, several Class 1 or Class 2/3 proteins will need to be included in a vaccine if it is to protect against several prevalent strains. It is essential that meningococcal strains for vaccine production are screened to determine the stability of expression of the required OMPs.

Loss of Toxicity

The production of effective bacterial toxoid vaccines (e.g., tetanus, botulinum, or diphtheria toxoids) relies upon the secretion of high levels of toxins by vaccine strains. Vaccine potency may be improved by the use of highly purified toxin components which are subsequently inactivated by chemical methods. Purification of these components is made easier by the use of highly toxigenic cultures, and maintenance of these stocks is clearly important. Despite this, the toxicity of many bacterial pathogens may often be difficult to maintain without modifications to the bacterial genome. This is often the case in bacteria where the genes controlling toxin expression are contained on a mobile genetic element, such as a plasmid, transposon or bacteriophage. For example, it is well appreciated that high-toxin-producing strains of *C. botulinum* types C and D are difficult to maintain, since toxigenicity of both of these types is dependent upon the presence of a lysogenic bacteriophage and strains cured of these phages become nontoxigenic (Eklund, Poysky and Reed, 1972). Reinfection of nontoxigenic strains of these types with the relevant phages can render them toxigenic once more (Eklund and Poysky, 1974). In our experience, strains of *C. botulinum* types C and D must be carefully maintained in liquid nitrogen or frozen at –70°C to preserve toxicity and be monitored regularly in order to assess any decrease. Similarly, diphtheria toxin is encoded by a family of closely related bacteriophages which integrate into the chromosome of *Corynebacterium diphtheriae* and convert nontoxinogenic, nonvirulent bacteria into toxinogenic, highly virulent species (Pappenheimer, 1982). The secretion of the toxin appears to be iron-regulated (Murphy and Bacha, 1979) and the presence of excess iron in the growth medium will inhibit toxin production.

Genes for some other toxins (e.g., staphylococcal enterotoxin D, lethal factor of *B.anthracis*, and *C. botulinum* type G toxin) are plasmid-borne, and loss of such plasmids during preparation of inocula would result in loss of toxigenicity. Indeed repeated subculturing of *Clostridium tetani* in liquid media can lead to the production of nontoxic cultures due to loss of the plasmid carrying the toxin genes (Laird et al., 1980).

Care must also be exercised during the resuscitation and growth of inocula from toxigenic strains of bacteria. The requirement for microbial vaccines to be manufactured from strains grown in non-animal protein media is increasingly requested by regulatory authorities. The growth of bacteria in simple or protein-reduced media, however, can have a dramatic effect on the ability to produce toxin. Indeed some strains of *C. botulinum* type F produce 100- to 1000-fold less toxin in minimal media compared to brain heart infusion broth (H. King, personal communication).

Recombinant Strains

Recombinant DNA technology or non-recombinant DNA techniques, such as gene transfer or mutagenesis, may be usefully applied to the development of microbial strains suitable for vaccines. The major objectives of these techniques are either to increase the yield of a microbial product or antigen, or to produce antigens that are more immunogenic and/or have much reduced toxicity. In the first case, this may be achieved by increasing the expression of desired genes by (1) cloning genes into high-expression vectors, (2) inserting promoter genes upstream of the target gene, (3) introducing appropriate mutations to relieve genetic repression, or (4) increasing the gene dosage by cloning into multicopy plasmids or phage vectors. In the second case, it is useful to have some knowledge of the biologically active portions of the antigen concerned, in order to introduce mutations that result in a loss of toxicity without affecting the immunogenicity (e.g., CRM-197, a mutated form of diphtheria toxin, codes for a product which is not toxic but is immunologically indistinguishable from diphtheria toxin) (Rappuoli, 1983). Such direct and deliberate manipulation of genes could enable the restructuring of protein molecules that possess precisely defined mixtures of desirable chemical, physical, and biological properties.

Despite the enormous growth in the area of strain improvement over the years, particularly for the manufacture of biological products, the number of recombinant strains used for production of vaccines is low. The introduction of new genes or alteration of the host genome may not always be achieved without adversely affecting other strain characteristics, and recombinant strains must be carefully monitored to assess the effects of gene deletion or rearrangements, which may result in reduced or complete loss of product or product modification.

Genetic instability constitutes a major drawback to recombinant strains. Such instability, which may be caused by spontaneous mutation within the genome, recombination or deletion of plasmids or cloned genes, can give rise to variants that are at a selective advantage during production and thus supplant the original strain leading to loss of or reduced yields. Variation in recombinant strains may be detected using specific gene probes and may be prevented by adjusting environmental conditions, such that they reduce any selective advantage of the variants.

CONTROL OF VACCINE SEED STOCKS

It is evident from the preceding paragraphs that many bacterial vaccine components are variable with regard to both the level of their expression and their antigenic composition. From a practical point of view, this is of major importance for vaccine manufacture, since seed storage methods and inoculum growth characteristics required for optimal stable expression of the required components should be understood fully, and suitable monitoring and control systems put in place. If seed stocks and starter cultures do not maintain their required biophysical, biochemical, and immunological profiles, then unwanted characteristics are likely to carry through into process intermediates and the final vaccine product.

In many instances, however, storage and growth criteria are incompletely elucidated, if at all, and, particularly if the product has no chemically defined composition, undetected alterations can and do occur. The increasing use of recombinant seed sources may only add to the problems since (a) the product of genes expressed in foreign hosts may differ structurally, biologically and immunologically from their "natural" counterparts, and (b) the expression of alternative native or host-vector genes may be favored during production scale-up.

THE SEED LOT SYSTEM

Vaccine manufacture should, wherever possible, utilize a seed lot system. A fully accredited bacterial cell source is used to derive a homogenous master seed dispensed in known quantities and stored under appropriate conditions. This stock may be subjected to a full range of safety and quality control tests (see below). One vial or ampoule of master seed is then used to derive a homogenous submaster (working) seed stock, dispensed, stored and tested as before. One vial or ampoule of submaster seed is the starter material for each vaccine production run; when this is ultimately depleted a fresh submaster seed is produced from a second master seed vial and so on. Provided the master seed is stable, this system offers the best means of ensuring that sequential production runs commence with inocula having identical properties.

MONITORING OF THE SEED SOURCE

Many of the tests employed to monitor characterization of seed stocks have been developed and applied over many years to vaccines and other biological materials, and they remain as valid for the more advanced biotechnological products as for those made by more traditional techniques.

A brief description of the type of information required is given below; full details of requirements surrounding manufacture of biological products for human use are to be found in the directives issued by regulatory bodies (Report, 1989, 1991).

The bacterial strain itself requires the full range of traditional documented information — origin, source, history, purity, phenotypic and genotypic details, growth characteristics under defined conditions, and immunological properties. This information must also be ascertained for master and submaster seed stocks that may be derived from the original host cell stock. If the organism is used as host for a recombinant vector, then a far wider range of additional information will be required. For the vector itself, full details should be provided for the preparation of the nucleic acid segments; the genes involved; the origin of replication; the fusion product (if any); the characterization of flanking, promoter, enhancer, and terminator regions; and the presence of any antibiotic resistant genes. The vector nucleic acid should be analyzed by sequence analysis and restriction endonuclease mapping, and compared with the original gene sequence.

The same range of comparative tests should be performed on the host-vector seed stocks but, in addition, it is essential to provide evidence for and to monitor a number of other parameters during storage, recovery and use of seed stocks under production conditions — preferably beyond the usual production time scale. Plasmid stability, the physical state of the vector, the copy number, and the fidelity of the nucleotide sequence encoding the product should all be monitored. While structural instability of the vector is rare in continuous culture, vector segregational instability is a major potential problem in large-scale systems, caused by either defective partitioning or a drop in copy number. The former may be overcome by careful genetic construction including a partitioning sequence if necessary, but copy number depletion can be markedly affected by dilution rate, nutrient limitation (Chapter 1.2), changes in host cell growth rate, and replication system alterations. Genetic stability is often monitored by screening for associated antibiotic resistance, but the association may be broken. Furthermore, antibiotic resistance gives no indication of the copy number of the vector. This can be assessed by agarose gel electrophoresis (Warnes and Stephenson, 1991).

For integrated vectors, identification and characterization of sequences is more difficult. Information may be obtained by Southern Blot analysis of total cellular DNA, by verification the sequence of isolated mRNA that codes for the product, or most rapidly and efficiently by polymerase chain reaction (PCR) amplification techniques (Mullis et al., 1986).

Fermentation

It is recognized that ever-closer control of growth and fermentation is an essential requisite for manufacture of an organism or product with the desired characteristics, particularly where there is the possibility of variable expression of vaccine components or use of recombinant organisms, as described

TABLE 2
Techniques Used to Characterize Biological Products

Sodium dodecyl sulphate polyacrylamide gel electrophoresis	Peptide mapping
Isoelectric focusing	Monoclonal antibody binding
Capillary electrophoresis	Western blot analysis
Chemical analysis of polysaccharide vaccines	Circular dichroism
Amino acid composition	Mass spectrometry
Full or partial amino acid sequence	Nuclear magnetic resonance

above. The analysis for product and the fermenter control systems used should be rapid, allowing a quick response to changes in conditions. Western Blot and ELISA techniques are able to detect low levels of a specific antigen product in the presence of an excess of other compounds and, since they are quantifiable, allow direct measurement of specific production rates or immunogen expression throughout the growth period. This information is necessary to ensure that the nature and quality of the product remains within previously defined limits with respect to specified production parameters (Smith and Perry, 1990).

Off-gas analysis, performed by mass spectrometry linked directly back to computerized fermenter control systems, is now widely used to monitor and control fermentation, giving exact indications of culture conditions and taking appropriate actions should parameters be exceeded. Biosensors, too, though still in a developmental stage, should also be able to monitor a number of parameters and would be expected to find application in fermentation control systems in due course (Clarke et al., 1985; Wiblin et al., 1993).

Product Characterization

The above discussion is directed primarily at the control of seed stocks and their subsequent growth/fermentation stages in a vaccine production system. It must be remembered, however, that for all biological products, quality is only assured by a continuing and rigorous monitoring of the process through subsequent stages of harvesting, purification, inactivation, toxoiding, and formulation (where appropriate).

Some of the techniques employed for assessing purity and identity of bacterial products are listed in Table 2, but it should be borne in mind that the validity of many of these tests needs careful evaluation, since any heterogeneity noted may have no clinical relevance. Although efforts are being made to develop *in vitro* assays (Benford, 1992), an ultimate assessment of efficacy of biological materials almost always relies on the use of an *in vivo* system(s) (Jeffcoate, 1992).

REFERENCES

Benford, D.J. 1992. The use of animal cells as replacements for whole animals in the toxicity testing of chemicals and pharmaceuticals. In Spier, R.E. and Griffiths, J.B. (eds.) *Animal Cell Biotechnology*, Vol 5, Academic Press, London, pp. 97-121.

Clarke, D.J., Calder, M.R., Carr, R.J.G., Blake-Coleman, B.C., Moody, S.C. and T.A. Collinge. 1985. The development and application of biosensing devices for bioreactor monitoring and control. *Biosensors* **1**: 213-320.

Coote, J.G. 1991. Antigenic switching and pathogenicity: environmental effects on virulence gene expression in *Bordetella pertussis*. *Journal of General Microbiology* **137**: 2493-2503.

Eklund, M.W., Poysky, F.T. and S.M. Reed. 1972. Bacteriophages and toxigenicity of *Clostridium botulinum* type D. *Nature, New Biology* **235**: 16-18.

Eklund, M.W. and F.T. Poysky. 1974. Interconversion of type C and D strains of *Clostridium botulinum* by specific bacteriophages. *Applied Microbiology* **27**: 251-258.

Finlay, B.B. and S. Falkow. 1989. Common themes in microbial pathogenicity. *Microbiology Reviews* **53**: 210-230.

Griffiths, E. 1987. The iron-uptake systems of pathogenic bacteria. In Bullen, J.J. and Griffiths, E. (eds.), *Iron and Infection: Molecular, Physiological and Clinical Aspects*. John Wiley & Sons, Chichester, England, pp. 69-138.

Heckles, J.E. 1989. Structure and function of pili from pathogenic *Neisseria* species. *Clinical Microbiology Reviews* **2** Supplement: S66-S73.

Heckles, J.E. 1993. Meningococcal vaccines. *Journal of Medical Microbiology* **39**: 17-20.

Jeffcoate, S.L. 1992. New biotechnologies: challenges for the regulatory authorities. *Journal of Pharmacy and Pharmacology* **44**: 191-194.

Jones, C.H., Jacob-Dubuisson, F., Dodson, K., Kuehn, M., Sponim, L., Striker, R. and S.J. Hultgren. 1992. Adhesion presentation in bacteria requires molecular chaperones and ushers. *Infection and Immunity* **60**: 4445-4451.

Kroll, J.S. 1990. Bacterial virulence: an environmental response. *Archives of Disease of Childhood* **65**: 361-363.

Laird, W.J., Aaronson, W., Silver, R.P., Haleig, W.H. and M.C. Hardegree. 1980. Plasmid-associated toxigenicity in *Clostridium tetani*. *Journal of Infectious Disease* **142**: 623.

McGuinness, B., Clarke, I.N., Lambden, P.R., Barlow, A.K., Poolman, J.T., Jones, D.M. and J.E. Heckles. 1991. Point mutation in meningococcal *porA* gene associated with increased endemic disease. *Lancet* **337**: 514-517.

Medical Research Council. 1959. Vaccination against whooping cough — final report. *British Medical Journal* **1**: 994-1000.

Meyer, T.F. 1990. Pathogenic *Neisseria* — a model for bacterial virulence and genetic flexibility. *International Journal of Medical Microbiology* **274**: 135-154.

Mullis, K., Faloona, F., Scharf, S., Saiki, R., Horn, G. and H. Erlich. 1986. The polymerase chain reaction. *Cold Spring Harbour Symposium. Quantitative Biology* **51**: 263-273.

Murphy, J.R. and P. Bacha. 1979. Regulation of diphtheria toxin production. In Schlessinger, D. (ed.), *Microbiology — 1979*, American Society for Microbiology, Washington, D.C. pp. 181-186.

Pappenheimer, A.M. 1982. Diphtheria: studies on the biology of an infectious disease. *The Harvey Lecture Series* **76:** 45-73. Academic Press, New York.

Rappuoli, R. 1983. Isolation and characterisation of *Corynebacterium diphtheriae* non-tandem double lysogens hyperproducing CRM 197. *Applied and Environmental Microbiology* **45**: 560-564.

Report. 1989. Commission of the European Communities. 1989. Guidelines on the quality, safety and efficacy of medicinal products for human use. In *the Rules Governing Medicinal Products in the European Community*, Vol. III, Luxembourg, pp 39-88.

Report. 1991. Commission of the European Communities. 1991. The rules governing medical products for human use in the European Community. In *The Rules Governing Medicinal Products in the European Community*, Vol. I, Luxembourg, pp 17-86.

Robinson, A., Ashworth, L.A.E. and L.I. Irons. 1989. Serotyping *Bordetella pertussis* strains. *Vaccine* **7**: 491-494.

Saunders, J.R., Wakeman, J., Sims, G., O'Sullivan, H.O., Hart, C.A. and M. Virji. 1993. Piliation in *Neisseria meningitidis* and its consequences. *Journal of Medical Microbiology* **39**: 7-9.

Smith, B.J. and M. Perry. 1990. Analytical techniques for biotechnology. *Chemistry and Industry* **18**: 563-567.

Snell, J.J.S. 1991. General introduction to maintenance methods. In Kirsop, B.E. and Doyle, A. (eds.), *Maintenance of Microorganisms and Cultured Cells — A Manual of Laboratory Methods.* 2nd ed. Academic Press, London.

Tinsley, C.R. and J.E. Heckles. 1986. Variation in the expression of pili and outer membrane protein by *Neisseria meningitidis* during the course of meningococcal infection. *Journal of General Microbiology* **132**: 2483-2490.

Van der Ley. P. Heckles, J.E., Birji, M., Hoogerhout, P. and J.T. Poolman. 1991. Topology of outer membrane porins in pathogenic *Neisseria* spp. *Infection and Immunity* **59:** 2963-2971.

Warnes, A. and J.R. Stephenson. 1991. The effect of nutritional limitation on the stability and expression of recombinant plasmids. In Grange, J.M., Fox, A. and Morgan, N.L. (eds.), *Genetic Manipulation: Techniques and Applications.* Society for Applied Bacteriology Technical Series **28**, pp 295-313.

Wiblin, C.N., Hambleton, P., Melling, J. and M. Scawen. 1993. The impact of new medicines development on pharmaceutical analyses. *Journal of Pharmacy and Pharmacology* **45**, (Suppl. 1): 374-380.

Williams, P. 1988. Role of cell envelope in bacterial adaption to growth *in vivo* in infections. *Biochimie* **70:** 987-1011.

Williams, P. and E. Griffiths. 1992. Bacterial transferrin receptors — structure, function and contribution to virulence. *Medical Microbiology and Immunology* **181**: 301-322.

3.7 Screening for Novel Antimicrobial Activity/ Compounds in the Pharmaceutical Industry

Peter Gilbert and Michael R.W. Brown

INTRODUCTION

During the so-called "Golden era" of antibiotic research, from the mid 1940s to the mid-1960s, the rate of discovery of new microbial metabolites did not change significantly (Berdy, 1974). There was a dramatic increase between 1972 and 1978, during which time the discovery rate jumped from 180 to 340 new antibiotic agents per year (Perlman, 1977). The fact that it has become increasingly difficult to find antibiotics which are both novel and useful is evidenced by the drop in clinical success rate from over 5% before 1960 to less than 1% today.

The objective of any screening program, be it for new pharmacological agents, biocides, or antibiotics, is to examine a large number of candidate substances and rapidly identify those with desired actions which are manifested at sufficiently low concentrations. Thus, Fleming, Nisbet and Brewer (1982), in reviewing the drug discovery process, describe a screening program of over 10,000 secondary metabolites of microbial origin, involving over 400,000 cultures over a 10-year period (up to 1961), which led to the discovery of only three utilizable drugs. More recently, the techniques and results of a 25-year screening program were appraised by Woodruff, Hernandez, and Stapley (1979). In this, 21,830 isolates were examined in 1 year with only two being selected as having therapeutic potential. In such instances specificity of the assay, or its direct relevance and ability to predict outcome in the clinic, is often sacrificed in the interests of speed and throughput. Thus, antibacterial screens are often diffusion assays, MIC or MBC determinations performed with serial dilutions of the compounds against a range of organisms often growing in a common, nutrient-rich medium. If particular activity is sought, and where this

0-8493-4752-1/95/$0.00+$.50

can be related to single or related sets of enzymes (i.e., β-lactamase, transpeptidase, superoxide dismutase), then *in vitro* assays of enzyme inhibiton might be performed.

MIC determinations are generally performed in nutrient rich media such that results may be obtained rapidly. The same medium is often used for a variety of organisms in order to simplify the methodology. Most of the cultures will, therefore, be growing very rapidly during the test, and only those compounds which have the greatest activity against these fast-growing cells will be selected as the compounds of interest. This is unfortunate since the sensitive cells in this instance will often be unrepresentative of *in vivo* and *in situ* where slow-growing or even dormant phenotypes will constitute the main problem (Chapters 1.1, 1.2, 1.4, and 1.6).

While the literature abounds with reports of antimicrobial susceptibility increasing with increasing cellular growth rate (Brown and Williams, 1985; Brown, Collier and Gilbert, 1990; Gilbert, Collier and Brown, 1990) where the level of such dependence differs between groups of agents, there are few, if any, reports of the converse relationship. It has been argued that such observations reflect the organisms and methods used in the primary screening processes rather than any fundamental property of cells or biocidal agents. Thus, it is not necessarily that agents possessing good activity against slow-growing cells, but poorer activity towards fast-grown ones do not exist; rather, they may have been rejected by traditional screening processes. As greater and greater activities are sought in the primary screens, then ever more compounds with activity favoring the slow-growing cells will be eliminated at an early stage (Brown, Costerton and Gilbert, 1991). Where studies have been conducted which compare the susceptibility of slow- with fast-growing cells for various antimicrobial agents then marked differences in relative activity emerge (Cozens et al., 1986; Ashby et al., 1994). β-lactam antibiotics acting preferentially against penicillin binding protein (PBP) 7, rather than PBP1, PBP2, and PBP3, were highlighted by a slow-growing screen using the chemostat and cells expressing a dormant phenotype (Tuomanen et al., 1986; Tuomanen and Schwartz, 1987). Similarly, of the carbapenems, imipenem, but not meropenem appears to have preferential activity against slow-growing/dormant cells (Ashby et al., 1994).

There is a certain inertia against the use of susceptibility screens other than the traditional MIC and MBC. This inertia reflects the manner in which many clinical laboratories perform susceptibility testing. Often they will conduct disk-diffusion assays, or similar, using media which facilitate rapid growth of the inoculum. This is well justified in that it speeds diagnosis and the administration of treatment and will highlight resistant strains. Unfortunately, any antibiotic agent that does not perform well in such an assay will generally fail to be prescribed, irrespective of its probable performance *in vivo*. Potential antibiotics must therefore not only act effectively *in vivo*, but they must also be seen to act in clinical microbiological laboratories (Chapter 3.9).

SCREENING STRATEGIES

Methodological approaches to the screening of compounds for utilizable antimicrobial activity are, as has been described above, often a compromise between throughput and predictivity. As the screening methods and preparation of the challenge inoculum becomes more complex, then the throughput efficiency of the process becomes reduced. Sometimes it is expedient to separate primary screens, aimed at achieving a high throughput of substances, from more selective secondary screens aimed at particular antimicrobial targets. Primary screens for antimicrobial activity have, in the past, relied heavily upon the traditional diffusion assays and MIC/MBC determinations conducted against a wide range of test organisms. There is little evidence in the literature of approaches which adopt more selective screening strategies (Hamill, 1977). In contrast, screens for pharmacological activity have tended to examine for activity against selected cell-free enzymes tissues. This might reflect the costs and moral issues surrounding the performance of large-scale animal testing, but the approach has, nontheless, been highly successful and could therefore be applied to antibacterials. This would be particularly useful if the chosen cell-free target were representative of a desired phenotype (e.g., biofilms) which is difficult reproducibly to create in whole cell inocula.

PRIMARY SCREENS

Primary screens of antimicrobial activity may either be conducted in cell-free systems, and investigate activity against known enzymes and processes, or they might utilize intact, whole cells. In both instances it is hoped that the tests will be predictive of activity *in vivo*.

TARGET-DIRECTED, CELL-FREE SCREENS

Cell-free systems have been designed and exploited within antimicrobial screens, which search for activity against particular targets. Such screens have sought activity directed against individual enzymic steps in cell wall synthesis (i.e., transpeptidase and carboxypeptidase activity; Frere, 1977), carrier systems (i.e., isoprenoid alcohols involved in the transport and assembly of wall precursors across the cell membrane; Somma, Merati and Paventi, 1977), and detoxification pathways (i.e., catalase, superoxide dismutase). Additionally, in the search for potentiators of antibiotic action, activity directed towards drug-inactivating enzymes such as β-lactamases and transacetylases etc. has been directly sought (Hood, 1982).

Provided that the enzymes used are critical to the growth and survival of the target organisms *in vivo*, such screens might lead to pharmacologically active chemotherapeutic agents. The choice of target, however, has often been made from knowledge of the mechanisms of action of established chemotherapeutic agents. Since these were generally first recognized by their activity in fast-growing screens, then choices of *in vitro* target based on such compounds will reflect the physiology of fast growing cells (Brown, Costerton and Gilbert, 1991). It is essential that targets which reflect the expressed *in vivo* physiologies are selected. In such respects fundamental studies of microbial physiology in various controlled environments, representative of *in vivo*, are prerequisite to the design of target-directed screens.

Activity detected within *in vitro* screens such as these does not necessarily translate to activity *in vivo*. This might be for a number of reasons (Fleming, Nisbet and Brewer, 1982). First, the chosen enzyme might not be active/relevant *in vivo*, or the affected processes might involve a number of alternative enzymes with differing susceptibility. Second, while the drug might be highly specific and active in a cell-free system, it might be unable to access these enzymes in the intact cell. Third, demonstration of activity against a specific target does guarantee specificity of action. Such screens will therefore fail to identify compounds of nonspecific toxicity.

Increasing the Predictivity of Cell-Free Screens

Often it is an inappropriate choice of target that leads to low utility of selected compounds *in vivo*. In order to make such approaches to screening more predictive of activity, physiologies must be identified that are not only unique to the target organims but also expressed *in vivo/in situ*. Many of the earlier chapters in this volume have considered the nature of the *in vivo/in situ* phenotype (Chapters 1.1, 1.2, 1.4, and 3.1) and its transcriptional control (Chapters 1.3 and 1.6). Factors such as slow growth rate, the adoption of nutrient-restricted phenotypes, dormancy, and growth in association with surfaces will alter antimicrobial susceptibility through modification/repression of target and changes in envelope structure and permeability (Brown, Collier and Gilbert, 1990). Thus, antibiotics developed for their activity against "laboratory" cultures are often ineffective in the clinic when directed against slow-growing, chronic infections (Gilbert, Brown and Costerton, 1987; Brown, Allison and Gilbert, 1988), particularly those associated with the surfaces of in-dwelling medical devices and prostheses (Bisno and Waldvogel, 1989). Fundamental studies of the physiology of slow-growing iron-limited attached cells are prerequisite to the selection of screening processses which will be predictive of such activity. It is only in recent years that the attention of microbial physiologists and geneticists has become focused towards this process. In this respect a number of emerging areas are likely to impinge on future *in vitro* screens.

Dormancy and Slow Growth Rates

Much is now known about the genetic switching of dormancy and expression of slow-growth/starvation phenotypes (Chapter 1.3). Dormancy is characterized by a marked decrease in metabolism and cessation of cell division, accompanied by the absence of cell wall septum formation, cell elongation and other processes essential for nucleod replication. Kaprelyants et al. (1993) describe 40 to 80 genes, including survival genes, thought to be involved in induction of dormancy. Any or each of these, or indeed their regulator genes, might prove to be suitable targets for antimicrobial development.

Cell-Density-Dependent Physiology

Cell-density-dependent physiologies have been recognized which are mediated through families of transcriptional activators such as *LuxI/R* (Chapter 1.6). Such transcriptional control is likely to determine the physiology of microcolonies and biofilms (Stuart and Williams, 1993; Meighan and Dunlop, 1993). Supercoiling of the DNA in cell dormancy, conjugal transfer (Piper et al., 1993), and the production of secondary metabolites (Williams et al., 1992), extracellular polymers and polysaccharides together with high-level production of virulence factors such as proteases and siderophores (Gambello and Iglewski, 1991; Gambello, Kaye and Iglewski, 1993; Evans, Brown and Gilbert, 1994) are now thought to be under this form of genetic control (Meighen and Dunlop, 1993; Claiborne, Winans and Greenberg, 1994). Targeting the autoinducers themselves or their binding sites might lead to the development of effective anti-biofilm agents.

As the various physiologies and phenotypes of slow-growing biofilms are dissected out and understood, then cell-free, *in vitro* screens will undoubtedly be developed which will be predictive of activity against such recalcitrant infections. However, until such a time there is no alternative other than to employ representative whole bacterial cells for primary screens.

Whole Cell Screens (Planktonic)

Whole cells employed as inocula in screening processes must reflect as accurately as possible the physiologies observed *in vivo*. Suspension test MICs and MBCs are wholly inadequate in this respect, unless some attempt is made to utilize media which facilitate slow rates of growth and impose nutrient restriction by iron insufficiency (Tuomanen, 1986). It is not sufficient simply to inoculate body fluids, such as urine, CAPD fluids, or plasma, and to expect these to function as predictive growth media. This is because *in vivo* such fluids are subject to homeostasis; thus as microorganisms grow and utilize nutrients then these will be replaced. *In vivo*, oxygen levels will remain relatively high and glucose levels will be maintained at ca. 10 m*M*, but *in vitro* both the oxygen and carbon substrate will be depleted. For a discussion on choice of culture media to represent various habitats and body sites see Chapters 1.1 and 3.1.

In many natural habitats organisms grow extremely slowly, if at all (above). Problems are encountered in the design of screening programs which look for activity against dormant phenotypes. First, as slower growth rates are achieved during the screen then longer times will be required to secure an MIC end point. If truly dormant cells are employed then MIC end points will be impossible to achieve. For these reasons screening programs which detect bactericidal end points must be employed for slow-growing/dormant inocula. A simple approach has been to conduct classical MBC determinations against logarithmic- and stationary-phase cells (Widmar et al., 1991). A number of studies have described methods for induction of the stringent response genes and cell dormancy (Tuomanen, 1986). Such cells must be broken out of dormancy before their viability can be demonstrated. Recovery processes must be validated to ensure the full recovery of sublethally injured, dormant cells. In a recent study, Ashby et al. (1994) reported differences in minimum effective concentration, determined spectrometrically, between organisms challenged in a dormant/nongrowing state and challenged during active growth. Dormancy was induced by the method of Tuomanen (1986) where an overnight culture of *Escherichia coli* C600 (thr, leu, thi), grown in minimal medium, was diluted 1000-fold into media containing reduced levels of leucine (<5 mg/l). This prematurely induces the stationary phase and activates the stringent response (Chapter 1.3). Ashby et al. (1994) tested a wide variety of β-lactam, carbapenem, and cephalosporin antibiotics together with gentamycin, erythromycin, ciprofloxacin, and indolmycin. Results were expressed as a Minimum Effective Concentration Ratio between nongrowing and actively growing cells (Table 1). Indices such as the Minimum Effective Concentration Ratio, developed by Ashby et al. (1994) reflect the degree of dependence of activity, for a given agent, upon growth rate. Some antibiotics were relatively unaffected in their activity by nongrowing phenotypes (i.e., gentamycin, cefminox, and cefmetazole), whereas others were significantly affected (cefpirome, cefotaxime, and ciprofloxacin). Imipenem showed preferential activity directed towards the nongrowing cells.

Whole Cell Screens (Attached Cells and Biofilms)

Many of the models which have been developed for the establishment of biofilm populations in the laboratory are elaborate in construction and intensive in operation. They are not therefore suitable for use within primary screening processes. Two of these techniques, the modified Robbins device (Nickel et al., 1985), and the Constant Depth Biofilm Fermenter (Wimpenny,

TABLE 1
Relative Activities of Antibiotics Against Actively Growing and NonGrowing Planktonic Cells and Against Biofilms Grown on Urinary Catheter Disks

Antibiotic	MIC[a] (μg/ml)	Planktonic (MECR)[b]	Sessile Planktonic Index[c]
		Cephamycins	
Cefminox	2	1.0	1.8
Cefoxitin	2	2.0	1.9
Cefotatan	0.5	32	2.5
Cefmetazole	1	1.0	1.5
		Cephalosporins	
Cefotaxime	<0.03	16	5.4
Ceftazidime	0.25	16	3.2
Cefoperazone	0.25	8	1.7
Cefpirome	<0.03	>32	3.2
		Penicillins	
Piperacillin	2	—	<2.2
		Carbapenems	
Imipenem	0.12	0.5	1.2
Meropenem	<0.03	8.3	1.6
		Various	
Gentamycin	0.5	1.0	1.1
Erythromycin	32	—	1.4
Indolmycin	32	—	2.2
Ciprofloxacin	<0.03	33.3	1.2

[a] 18 h MIC in Mueller Hinton Broth, 5×10^5 cfu/ml inoculum.
[b] Minimum effective concentration (MEC) non-growing cells/MEC actively growing cells.
[c] $I_{50\%}$ leucine incorporation sessile cells/$I_{50\%}$ leucine incorporation planktonic cells.

Data from Ashby et al., *Journal of Antimicrobial Chemotherapy*, 33, 443-452, 1994. With permission.

Peters and Scourfield, 1989) will generate multiple test pieces for use as inocula, but permit no more than one antimicrobial to be tested at any time against the actively growing biofilms. Test biofilms must therefore be generated and transfered to a secondary exposure medium. Such devices are more suited to kinetic studies of the effects of antimicrobials against biofouling and thick biofilms such as those associated with dental plaque (Chapter 3.8) than to primary screens.

A number of methods have been developed and used, however, which either allow organisms to attach onto a surface and fix them by drying, or allow natural biofilms to develop by submerging the test pieces in liquid growth medium (Prosser et al., 1987).

Drying organisms onto test surfaces now forms part of the European Surface Disinfection Test (Chapter 3.3). While such methods detect the ability of disinfectants to kill organisms desiccated within organic soil, as well as the ability of formulations to cleanse soiled surfaces, it is unlikely that in such a situation the test strains have had any opportunity to express attachment specific phenotypes. Primary screens for antibiotic agents are probably much better directed towards systems where the organisms are *grown* onto the test surface.

A number of of studies have employed biofilms that have been allowed to develop onto test pieces suspended within chemostats (Keevil et al., 1987; Anwar et al., 1989; Anwar, Dasgupta and Costerton, 1990). While such methods avoid the generation of a stationary phase within the adherent population through continuous supply of fresh medium, they require constant attention and exert strong selective pressures on the challenge organism (see Chapters 1.4 and 3.1). For use in primary screening programs, the sophistication of the chemostat is probably not required. A simplified approach, developed by Prosser et al. (1987), places small volumes of culture onto silicone disks and incubates in a humid environment for 18 to 24 h. The incubated disks are washed, to remove loosely attached cells, and transferred to the test. Large numbers of test pieces can be generated in this manner, facilitating their use in a large primary screen. Following transfer to the test, end points can be determined as an MBC, determined by removing each test piece and washing to remove residual antibiotic incubation in fresh culture medium, or inhibition of growth and metabolism (analogous to an MIC) might be determined by potentiometric monitoring of the cultures. An alternative is to monitor the incorporation of radiolabeled substrates into the test piece. In this manner Ashby et al. (1994) measured the concentrations of various antibiotics which were required to inhibit the incorporation of tritiated leucine ($I_{50\%}$) into planktonic populations of *E. coli* C600 and biofilms established on urinary catheter disks (Table 1). They expressed the results as a Sessile Planktonic Index of activity ($I_{50\%}$ sessile cells/$I_{50\%}$ planktonic cells).

Results showed a good correlation between the activity against nongrowing cells and activity against biofilms. This supports the hypothesis that slow growth rates contribute to the resistance properties of biofilms (Brown, Allison and Gilbert, 1988).

The test surface for biofilm development might form part of the container in which the exposures to the antimicrobial are conducted. A very simple screening approach for antibiofilm property might therefore be to culture various organisms within microtiter trays containing appropriate culture media. During growth a proportion of the cells will attach to the polystyrene (Styrofoam) walls and base of the wells and begin to form biofilm. After a period of growth the planktonic cultures may be washed away, and each well filled with a different concentration of the antimicrobial(s) under test. After exposure of the attached population to the biocides, for designated times, the plates are once again emptied and washed with sterile medium. These wells may be rechallenged with broth and incubated. All three incubation stages could be continuously monitored for growth using a microtiter plate reader. Activity against the initial attached population can be determined either as a total kill end point, where no recovery is seen, or it might be taken as the incubation time following removal of antimicrobial before a predetermined optical density is established in the well. Using such an approach, alterations in the starter-medium such as C:N ratio will influence the extent of exopolsaccharide deposition. Likewise, by varying the duration of the attachment/biofilm development phase, the susceptibilities of young and mature, dormant films might be compared. In this respect young and old biofilms, generated in chemostat culture, have been reported to have very different susceptibilities to antibiotics agents (Anwar, Strap and Costerton, 1992).

IN VITRO MODELS FOR SECONDARY SCREENING OF ANTIBIOTICS

Secondary screens should enable a larger number of compounds to move forward from the primary screen than would be possible if they were to go directly into animal models. Secondary screens may therefore be technically more intensive and time-consuming in their operation than primary ones. With respect to modeling recalcitrant infections one might look for growth systems which provide attached bacteria, growing slowly under iron limitation (Chapter 3.1). A number of different approaches to the controlled growth of bacterial populations have been considered elsewhere in this volume (Chapters 1.4, 1.6, and 3.7). These approaches include chemostats for the growth of planktonic populations (Chapter 1.4), and for modeling attached growth, the Perfused Biofilm Fermenter (Gilbert et al., 1989), the Constant Thickness Biofilm Fermenter (Wimpenny et al., 1989), the Robbins device (McCoy et al., 1981), the modified Robbins device (Nickel et al., 1985), RotoTorque fermenters (Characklis, 1990), and submerged tiles in chemostats (Keevil et al., 1987; Anwar et al., 1989, 1990).

The majority of these *in vitro* models allow biofilms to form, with time, on various test surfaces while subjecting them to controlled shear stresses, temperature, and nutrient supply. The models are well adapted to kinetic studies of the effects of biocides and antibiotics on the process of biofilm formation, invaluable for the design and testing of antifouling and antibiofilm agents for the enviromental protection industries etc. (Chapter 3.8). Nevertheless, prophylactic approaches, such as these, applied to the control of biofilm infections are unlikely to be adopted in the clinic. Application of such models for the testing of antibiotic substances is therefore restricted to the provision of test-pieces for inclusion in other tests. In this respect, since the models allow biofilms to develop over several days then appropriate control over the age, physiology and thickness of the biofilm communities should be applied. The Constant Thickness Biofilm Fermenter (CTBF) and the Perfused Biofilm Fermenter, on the other hand, are pseudo-steady-state models of biofilm development and facilitate study of the effects of biocides and antibiotics upon the steady-state growth kinetics of developed, and developing, biofilms.

While the CTBF achieves a pseudo–steady state (Chapter 1.6), this is achieved by the physical removal of cells at the surface of the biofilm. Biofilms are allowed to form within recesses of defined depth on a test plate. The top surface of the plate is "cleansed" by scraper blades which remove excess cells as they are produced and distribute fresh medium to the recesses. The biofilms so produced are relatively thick (100 to 400 μm), have gradients of nutrients and oxygen established through them, and do not allow the free and natural dispersal of cells. The existence of gradients through these films renders the cell populations very heterogenous with respect to their phenotype. While an appropriate model of dental plaque, the biofilms generated in this model are otherwise unrepresentative of medical biofilms. Application of the CTBF to the screening and testing of agents for the biofouling and environmental protection industries, and for antiplaque agents, are discussed further in Chapter 3.8.

The chemostat and the Perfused Biofilm Fermenter (PBF) are both well suited for the establishment of steady-state, homogeneous populations of bacteria at controlled growth rates. The PBF establishes biofilms on the underside of a bacteria-proof membrane. Sterile medium is perfused through the biofilm from the sterile side. Cells associated with the biofilm are able to disperse into the perfusing medium. Steady states are developed at which all of the cells are growing at the same rate and for which the rate of growth is regulated by the rate of flow of medium (Gilbert et al., 1989). Cells dispersed spontaneously from these model biofilms have been shown to have properties which are distinct from those of the attached cells (Allison et al., 1990a, b). These dispersed cells are thought closely to model bacteremic cells shed from biofilms *in vivo*. Such cells are characteristically susceptible to antibiotics and biocides, whereas the biofilms from which they derive are resistant (Evans et al., 1990a, b; 1991; Duguid et al., 1992a, b). In such studies chemostats have been used to provide planktonic cells grown at equivalent rates to the biofilms. This enables the effects of growth rate and attachment to be differentiated (Gilbert et al., 1989).

Chemostat populations may be harvested directly, and cells from steady-state biofilms may be resuspended to provide single-cell suspensions suitable for bactericidal assays or MIC methodologies. In this respect it is worth noting that the moment cells are removed from the steady-state growth conditions provided by these models, then growth will cease and the cells, while retaining some of the phenotype of the original culture, will enter their stationary phase. An alternative approach, which avoids such problems, is directly to expose intact, growing biofilms or steady-state chemostats to antibiotics.

Exposure of Chemostats to Antibiotics

Antibiotics may be added directly to culture vessels or to the media reservoirs. If the concentration of agent achieved in the fermenter is greater than the MIC, then the culture will be diluted from the vessel at a rate equivalent to the dilution rate. If, in addition to this, killing of the population is obtained then the viable count in the culture, measured in the eluate, will decrease faster than the dilution rate. The difference between the dilution rate and the rate of decrease in viable count can therefore be used to calculate the rate of killing.

On removal of the agent from the fermenter, or on its dilution to below MIC by the addition of fresh medium, then cell density will recover in the fermenter at a rate equivalent to $D-\mu_{max}$. Differences between the predicted recovery rate and the observed recovery can be used to calculate postantibiotic effects. In such a manner Cozens et al. (1986) evaluated the effectiveness of a number of β-lactam antibiotics against slow-growing populations of *Escherichia coli*. They reported that the agents ceftoxidine and ceftriaxone had little or no activity against slowly growing cultures (Cozens et al., 1986; Tuomanen et al., 1986). Expression of PBPs was noted to be highly growth-rate dependent, with PBP7 hardly being expressed in conventionally grown cultures (Turnowsky et al., 1983, Tuomanen et al., 1986). The β-lactam CGP 17520 was particularly effective against slow-growing cultures with activity directed against PBP 7 (Cozens et al., 1986; Tuomanen and Schwartz, 1987).

The effects of inhibitors, at concentration < MIC, may be quantified/qualified by their effect upon the steady-state kinetics of a chemostat. Steady state will be achieved provided that μ in the presence of inhibitor is less than the rate of dilution (D). In such systems competitive inhibition of the growth inhibitory substrate will lead to a reduction in biomass, whereas noncompetitive inhibition will cause a decrease in μ_{max} (Pirt 1972) Such techniques allow the estimation of the effects of sub-MIC upon selection of drug-resistant mutants, virulence factor production, and combination therapy.

Antibiotic Perfusion of the Biofilm Fermenter

The number of cells eluted from the PBF reflects the viability status and growth of the attached population. This has been used to monitor the performance of an antibiotic (Duguid et al., 1992a, b) included within the perfusing medium. In such instances inhibition of growth and division or bactericidal effects will cause the numbers of viable cells eluted from the biofilm to decrease. The degree and rate of recovery of viable cells to the eluted medium following removal of the agent can be used to indicate the extent of killing achieved on the biofilm and also the postantibiotic effects upon the survivors. The use of secondary dilution chambers prior to perfusion of antibiotic through the biofilm enables various pharmacokinetic parameters of antibiotic administration to be evaluated (Gilbert 1985; Ashby et al., 1994).

REFERENCES

Allison, D.G., Brown, M.R.W., Evans, D.J. and Gilbert, P. 1990a. Surface hydrophobicity and dispersal of *Pseudomonas aeruginosa* from biofilms. *FEMS Microbiology Letters* **71**, 101-104.

Allison, D.G. Evans, D.J. Brown, M.R.W. and Gilbert, P. 1990b. Possible involvement of the division cycle in dispersal of *Escherichia coli* from biofilms. *Journal of Bacteriology* **172**, 1667-1669.

Anwar, H., van Biesen, T., Dasgupta, M.K., Lam, K. and Costerton, J.W. 1989. Interaction of biofilm bacteria with antibiotics in a novel *in vitro* chemostat system. *Antimicrobial Agents and Chemotherapy* **33**, 1824-1826.

Anwar, H., Dasgupta, M.K. and Costerton, J.W. 1990. Testing the susceptibility of bacteria in biofilms to antibacterial agents. *Antimicrobial Agents and Chemotherapy* **34,** 2043-2046.

Anwar, H., Strap, J.L. and Costerton, J.W. 1992. Establishment of aging biofilms: possible mechanism of bacterial resistance to antimicrobial chemotherapy. *Antimicrobial Agents and Chemotherapy* **36**, 1347-1351.

Ashby, M.J., Neale, J.E., Knott, S.J. and Critchley, I.A. 1994. Effect of antibiotics on non-growing planktonic cells and biofilms of *Escherichia coli. Journal of Antimicrobial Chemotherapy* **33,** 443-452.

Berdy, J. 1974. Recent developments of antibiotic research and classification of antibiotics according to chemical structure. *Advances in Applied Microbiology* **18,** 309-402.

Bisno, A.L. and Waldvogel, F.A. (Eds.) 1989. *Infections associated with indwelling medical devices.* American Society for Microbiology Press, Washington, D.C.

Brown, M.R.W. and Williams, P. 1985. The influence of environment on envelope properties affecting survival properties of bacteria in infections. *Annual Reviews in Microbiology* 39, 527-556.

Brown, M.R.W., Allison, D.G. and Gilbert, P. 1988. Resistance of bacterial biofilms to antibiotics: A growth rate related effect? *Journal of Antimicrobial Chemotherapy* **22**, 777-789.

Brown, M.R.W., Collier, P.J. and Gilbert, P. 1990. Influence of growth rate on the susceptibility to antimicrobial agents: modification of the cell envelope and batch and continuous culture studies. *Antimicrobial Agents and Chemotherapapy* **34**: 1623-1628.

Brown, M.R.W., Costerton, J.W. and Gilbert, P. 1991. Extrapolating to life outside the test-tube. *Journal of Antimicrobial Chemotherapy* **27,** 565-567.

Characklis, W.G. 1990. Laboratory biofilm reactors, pp 55-89 in *Biofilms* (Charackalis, W.G. and Marshall, K.C. Eds.) John Wiley & Sons, New York.

Claiborne, W., Winans, S.C. and Greenberg, E.P. 1994. Quorum sensing in bacteria: the *LuxR-LuxI* family of cell density responsive transcriptional regulators. *Journal of Bacteriology* **176,** 269-275.

Cozens, R.M., Tuomanen, E., Tosch, W., Zak, O., Suter, J. and Tomasz, A. 1986. Evaluation of the bactericidal activity of β-lactam antibiotics upon slowly growing bacteria cultured in the chemostat. *Antimicrobial Agents and Chemotherapy* **29,** 797-802.

Duguid, I.G., Evans, E., Brown, M.R.W. and Gilbert, P. 1992a. Growth-rate-independent killing by ciprofloxacin of biofilm-derived *Staphylococcus epidermidis*; evidence for cell-cycle dependency. *Journal of Antimicrobial Chemotherapy* **30**, 791-802.

Duguid, I.G., Evans E., Brown, M.R.W. and Gilbert, P. 1992b. Effect of biofilm culture upon the susceptibility of *Staphylococcus epidermidis* to tobramycin. *Journal of Antimicrobial Chemotherapy* **30**, 803-810.

Evans, D.J. Allison, D.G., Brown, M.R.W. and Gilbert, P. 1990a. Effect of growth rate on resistance of Gram-negative biofilms to cetrimide. *Journal of Antimicrobial Chemotherapy* **26**, 473-478.

Evans, D.J., Brown, M.R.W., Allison, D.G. and Gilbert, P. 1990b. Susceptibility of bacterial biofilms to tobramycin: role of specific growth rate and phase in the division cycle. *Journal of Antimicrobial Chemotherapy* **25**, 585-591.

Evans, E., Duguid, I.G., Brown, M.R.W. and Gilbert, P. 1991a. Surface properties and adhesion of *Staphylococcus epidermidis* in batch and continuous culture. *Abstracts of the 91st. Annual Meeting of the American Society for Microbiology* **D-53.**

Evans, D.J., Allison, D.G., Brown, M.R.W. and Gilbert, P. 1991b. Susceptibility of *Pseudomonas aeruginosa* and *Escherichia coli* biofilms towards ciprofloxacin: effect of specific growth rate. *Journal of Antimicrobial Chemotherapy* **27**,177-184.

Evans, E., Brown, M.R.W. and Gilbert, P. 1994. Iron cheletor, exopolysaccharide and protease production in *Staphylococcus epidermidis*: a comparative study of the effects of specific growth rate in biofilm and planktonic culture. *Microbiology* **140,** 153-157.

Fleming, I.D., Nisbet, L.J. and Brewer, S.J. 1982. Target directed antimicrobial screens, pp 107-130 in *Bioactive Microbial Products: Search and Discovery* (Bu'Lock, J.D., Nisbet, L.J. and Winstanley, D.J. Eds.) Academic Press, London.

Frere, J.-M 1977. Mechanism of action of of β-lactam antibiotics at the molecular level. *Biochemical Pharmacology* **26,** 2203-2210.

Gambello, M.J. and Iglewski, B.H. 1991. Cloning and characterisation of the *Pseudomonas aeruginosa LasR* gene, a transcriptional activator of elastase expression. *Journal of Bacteriology* **173,** 3000-3009.

Gambello, M.J., Kaye, S. and Iglewski, B.H. 1993. *LasR* of *Pseudomonas aeruginosa* is a transcriptional activator of the line protease gene *(apr)* and an enhancer of exotoxin A expression. *Infection and Immunity* **61**, 1180-1184.

Gilbert, P. 1985. The theory and relevance of continuous culture to in-vitro models of antibiotic dosing. *Journal of Antimicrobial Chemotherapy* **15** (Suppl.), 1-6.

Gilbert, P., Brown, M.R.W. and Costerton, J.W. 1987. Inocula for antimicrobial sensitivity testing: A critical review. *Journal of Antimicrobial Chemotherapy* **20,** 147-154.

Gilbert, P., Allison, D.G., Evans, D.J., Handley, P.S. and Brown, M.R.W. 1989. Growth rate control of adherent bacterial populations. *Applied and Environmental Microbiology* **55,** 1308-1311.

Gilbert, P., Collier, P.J. and Brown, M.R.W. 1990. Influence of growth rate on susceptibility to antimicrobial agents: biofilms, cell cycle, dormancy and stringent response. *Antimicrobial Agents and Chemotherapy* **34,** 1865-1868.

Hamill, R.L. 1977. General approaches to fermentation screening. *Journal of Antibiotics* **30** (Suppl.), s164-s173.

Hood, 1982. Inhibitors of antibiotic inactivating enzymes, pp 131-145 in *Bioactive Microbial Products: Search and Discovery* (Bu'Lock, J.D., Nisbet, L.J. and Winstanley, D.J. Eds.) Academic Press, London.

Kaprelyants, A.S., Gottshal, J.C. and Kell, D.B. 1993. Dormancy in non-sporulating bacteria. *FEMS Microbiological Reviews* **104,** 271-286.

Keevil, C.W., Bradshaw, D.J., Dowsett, A.B. and Feary, T.W. 1987. Microbial film formation: dental plaque deposition on acrylictiles using continuous culture techniques. *Journal of Applied Bacteriology* **62,** 129-138.

McCoy, W.F., Bryers, J.D., Robbins, J.D. and Costerton, J.W. 1981. Observations on biofilm formation. *Canadian Journal of Microbiology* **27,** 910-917.

Meighen, E.A. and Dunlop, P.V. 1993. Physiological, biochemical and genetic control of bacterial luminescence. *Advances in Microbial Physiology* **112,** 1-67.

Nickel, J.C., Ruseska, I., Wright, J.B. and Costerton, J.W. 1985. Tobramycin resistance of *Pseudomonas aeruginosa* cells growing as a biofilm on urinary catheter material. *Antimicrobial Agents and Chemotherapy* **27**, 619-624.

Perlman, D. 1977. The roles of the *Journal of Antibiotics* in determining the future of antibiotic research. *Journal of Antibiotics* **30** Suppl., S133-S137.

Piper, K.R., Beck von Bodman, S. and Farrand, S.K. 1993. Conjugation factor of *Agrobacter tumifasciens* regulates Ti plasmid transfer by autoinduction. *Nature (London)* **362,** 448-450.

Pirt, S.J. 1972. Prospects and problems in continuous flow culture of microorganism. *Journal of Applied Chemical Biotechnology* **22,** 55-64.

Prosser, B., Taylor, D., Dix, B.A. and Cleeland, R. 1987. Method of evaluating effects of antibiotics upon bacterial biofilms. *Antimicrobial Agents and Chemotherapy* **31,** 1502-1506.

Somma, S., Merati, W. and Paventi, T. 1977. Gardimycin, a new antibiotic inhibiting peptidoglycan synthesis. *Antimicrobial Agents and Chemotherapy* **11,** 396-401.

Stuart, G.S.A.B. and Williams, P. 1993. Shedding new light on food microbiology. *American Society for Microbiology News* **59,** 241-246.

Tuomanen, E. 1986. Phenotypic tolerance: the search for β-lactam antibiotics that kill non-growing bacteria. *Reviews of Infectious Diseases* **8,** s279-291.

Tuomanen, E., Cozens, R., Tosch, W., Zak, O. and Tomasz, A. 1986. The rate of killing of *Escherichia coli* by β-lactam antibiotics is strictly proportional to the rate of bacterial growth. *Journal of General Microbiology* **132,** 1297-1304.

Tuomanen, E. and Schwartz, J. 1987. Penicillin binding protein 7 and its relationship to lysis of non-growing *Escherichia coli. Journal of Bacteriology* **169,** 4912-4915.

Widmar, A.F., Wienster, A., Frei, R. and Zimmerli, W. 1991. Killing of non-growing and adherent *Escherichia coli* determines drug efficacy in device related infections. *Antimicrobial Agents and Chemotherapy* **35,** 741-746.

Williams, P., Bainton, N.J., Swift, S., Chhabra, S.R., Winson, M.K., Stewart, G.S.A.B., Salmond, G.P.C. and Bycroft, B.W. 1992. Small molecule-mediated, density dependent control of gene expression in prokaryotes: bioluminescence and the biosynthesis of carbapenem antibiotics. *FEMS Microbiology Letters* **100,** 161-168.

Wimpenny, J.W.T., Peters, A. and Scourfield, M. 1989. Modeling spatial gradients, pp 111-127 in *Structure and Function of Biofilms* Charackalis, W.G.and Wilderer, P.A. Eds. John Wiley, Chichester.

Woodruff, H.B., Hernandez, S. and Stapley, E.D. 1979. Evolution of antibiotic screening programme. *Hindustan Antibiotics Bulletin* **21,** 71-84.

3.8 Screening for Novel Compounds/Activity in the Environmental Protection Industries

Hilary M. Lappin-Scott, Jana Jass, and J. William Costerton

INTRODUCTION

Bacteria have many useful properties that may be harnessed to industrial processes. These include the production of antibiotics, enzymes for effecting biotransformations, and a variety of pharmaceutical compounds, together with the use of microorganisms as vehicles for the synthesis of bioengineered products. In such processes the growth of these bacteria is well controlled and optimized for the process in hand. It is not necessary under such circumstances to duplicate *in vivo/in situ*. In many industrial processes, however, and sometimes within manufacturing plants, microorganisms can grow in an uncontrolled manner as contaminants or nuisances. In such circumstances they often grow attached to surfaces, fouling them with their growth products and eventually destroying the materials' surface, either through obstruction of the flow of materials over them or through biocorrosion. The control of biofouling/biocorrosion microorganisms forms the basis of this chapter. Methods for their control have, to date, focused on tests which deploy antimicrobial agents against pure cultures of vegetatively growing planktonic bacteria. This chapter highlights the disadvantages and inappropriateness of such methods and offers suggestions for the design of more relevant laboratory testing procedures.

THE INDUSTRIAL PROCESSES

In all of the natural environments studied to date, including rivers, lakes, soils, and estuaries, bacteria have been shown to attach to surfaces and grow within biofilms as sessile populations of cells. In the same manner bacteria

0-8493-4752-1/95/$0.00+$.50

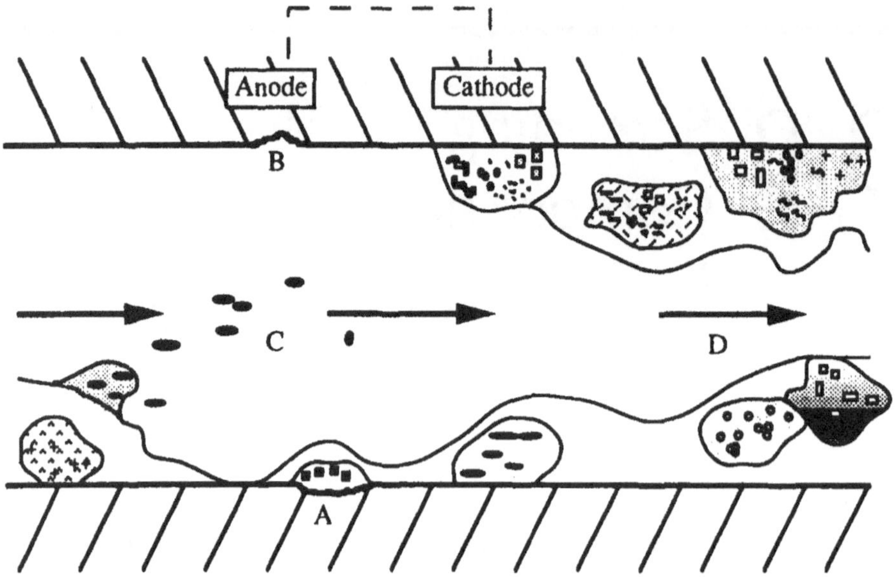

FIGURE 1 A schematic diagram of biofilms in an industrial water system. Biofilms can cause surface deterioration due to bacterial processes (A) and the formation of a corrosion cell caused by the production of metabolites and biopolymers with different charges produced by adjacent microcolonies (B). Biofilms may be reservoirs for potential pathogenic bacteria that could detach from the microcolonies to contaminate the water system (C). Reduced efficiency of the pipes may be caused by physical blockage by thick biofilm formations (D).

attach to a wide range of man-made surfaces and grow as biofilms (Chapter 1.6). Indeed, any surface that is wetted or submerged in water, such as heat exchangers, cooling towers, water tanks, and pipelines within industrial processes, offers the possibility of microbial attachment and may result in surface deterioration, contamination of the industrial processes by potentially pathogenic/spoiling microorganisms, physical blockage of the conduit, and/or impairment of the function of the surface (Figure 1; Lappin-Scott and Costerton, 1989). It is important to eliminate or to control this microbial growth, as the resulting damage to the industrial process is often very costly. Several factors need to be considered before devising the relevant testing procedures.

SCREENING OF ANTIMICROBIALS FOR INDUSTRY

Contamination of industrial waters often introduces microorganisms into environments that can provide growth nutrients to sustain their development. In considering methods to screen for novel antimicrobial activity and for the ability of developed strategies to control such growth, three specific factors are noteworthy:

1. Industrial waters may support the growth and/or survival of mixed communities of bacteria, and such growth may differ from that of similar organisms in monoculture.
2. If growth nutrients are not available then bacteria can exist in dormant forms, as spores or starved bacteria (Somnicells, Chapter 1.5).
3. Bacteria will generally attach to any available surface and grow as the sessile phenotype.

Each of these factors, relating to the survival and growth of bacteria within industrial waters, may affect their response to antimicrobial agents and treatment regimens. Each of these considerations will therefore be considered separately.

Mixed Cultures

The efficacy of biocides is usually tested using pure cultures. This emphasis in microbiology originates in the work of Koch (1881). It is, however, generally considered that such techniques overlook the importance of communities of microorganisms which are comprised of mixed species (Bull and Slater, 1982). In natural environments bacteria are rarely found in monocultures. Even within extreme environments (in terms of pH, temperature, pressure, and salinity), mixed microbial communities are typical. Within mixed communities, the individual species interact and frequently derive benefit from their association with one another. Indeed such associations may extend their overall metabolic capability (Slater, 1981). For example, mecoprop, a phenoxyalkanoic herbicide, has a complex chemical structure and contains chlorinated substituents. To date there have been no reports of the ability of pure cultures to degrade this herbicide. Interacting communities of bacteria, however, are able fully to degrade mecoprop (Figure 2).

In industrial waters, microorganisms exist within mixed, interacting communities. Many of the problems associated with growth of microorganisms on pipelines and other metal surfaces are caused by interacting communities of bacteria. In order, therefore, to prepare relevant inocula for the testing of novel biocides for industry, it is necessary to use mixed cultures of bacteria rather than pure cultures, and to allow these to establish growth niches that replicate those *in situ*. The best method is to collect the water itself, together with scrapings from the contiguous surfaces, and use them as an inoculum into an appropriate *in vitro* growth model and testing system.

Starved Bacteria

Most industrial processes use water either as a raw material or as a coolant. Generally a higher quality water is used as an ingredient than is required for cooling and other purposes (Geesey, 1987). Whatever the water quality, none of the processes involved is capable of maintaining a sterile environment. While most workers realize that even the highest quality waters may become

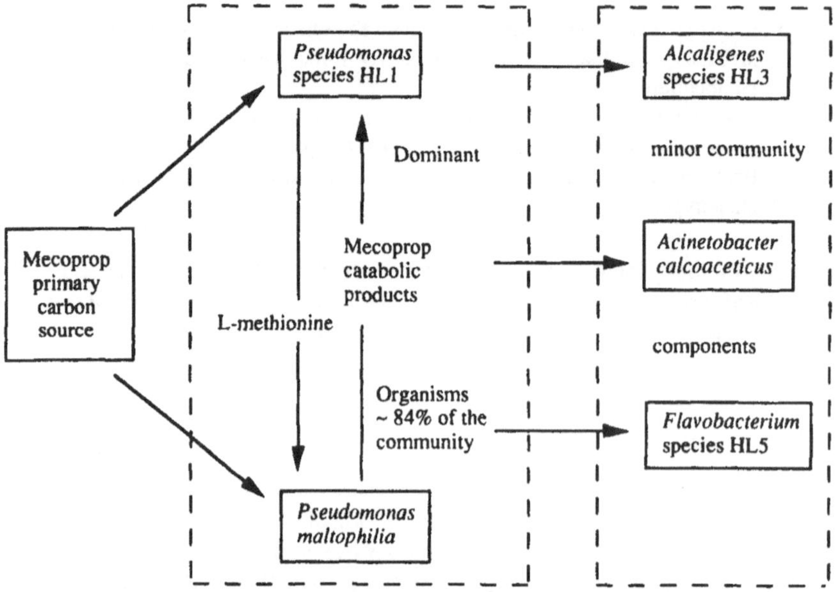

FIGURE 2 An example of a community of microoganisms interacting to degrade the herbicide, Mecoprop, in a natural soil environment. (Data from H. M. Lappin, Ph.D. Thesis, University of Warwick, U.K. 1984. With permission.)

contaminated, they often wrongly assume that such water is so pure that it will not support microbial growth or surface attachment involving active metabolism. In the absence of sufficient nutrients bacteria do not necessarily die. There is now a wealth of research describing survival of bacteria in low-nutrient environments. The reader is directed to a recent review on the subject of starvation survival (Kaprelyants, Gottschal and Kell, 1993) and also to Chapter 1.5 of the present volume. In essence, many bacteria do not die when all of the available growth nutrients have been utilized, rather they adopt a series of responses that assist their general survival (Novitsky and Morita, 1976; Chapter 1.3). These responses include the utilization of storage compounds, the formation of periplasmic spaces, breakdown of nonessential cellular components, reductions in endogenous metabolism, the production of specific proteins, reductions in cell size, and strategic reductions in population viability (Lappin-Scott and Costerton, 1990). In total these responses produce bacterial cells that are able to withstand starvation for indefinite periods. In many respects they are analogous to spores in that they are hardy structures which are able to resuscitate and grow when favorable growth conditions are reestablished.

An important consequence of the starvation-survival response in bacteria that contaminate industrial processes is that such organisms respond to antimicrobial agents in a manner completely different from that of their vegetative counterparts. Most of the chemical antimicrobial agents currently used to kill industrial contaminants have been developed against vegetative bacteria. If starved bacteria have a different response from that of the vegetative bacteria then they may survive such treatments. Studies of the efficacy of antibiotics and biocides have observed a reduced rate and extent of killing of slow-growing bacteria (Brown, Allison and Gilbert, 1988). To date, however, the ability of industrial biocides to eradicate starved bacteria has been little studied.

One recent investigation, recently undertaken in our laboratories, has involved the sulfate-reducing bacteria (SRB). Within the oil industry, SRB are nuisance organisms in that they grow within subterranean rock formations and the pipelines of the topside facilities to produce hydrogen sulfide, a toxic, corrosive gas (Hamilton, 1985). Thermophilic SRB, isolated from North Sea oil production waters, are able to exist as starved, dormant cells within seawater, but become fully resuscitated when given growth nutrients (Lappin-Scott, Bass and Sanders, 1992; Bass, Lappin-Scott and Sanders, 1993). Lappin-Scott and Sanders (1993) exposed vegetative and starved forms of SRB to different concentrations of glutaraldehyde. The effectiveness of the treatment was measured by monitoring metabolic activity over a 24 h post-treatment period. The metabolic activity of the vegetative cells decreased rapidly following exposure to glutaraldehyde, whereas that of their starved counterparts remained at over 50% of the basal level. This study demonstrated that even a chemically highly reactive biocide such as glutaraldehyde is not effective at eliminating starved bacteria and emphasised the need to consider the nutritional status of bacterial cells prior to preparing challenge inocula for biocide testing. If biocides are developed only against vegetative rather than starved bacteria, then their effectiveness in the field will be unpredictable.

Sessile versus Planktonic

The attachment of bacteria to surfaces has been dealt with in detail in Chapter 1.7 of this volume. This topic, therefore, will be only briefly dealt with in the context of industrial processes.

Bacteria grow within industrial tanks, pipes, and machinery, and often contaminate the industrial processes themselves. Growth of bacteria on heat exchangers reduces the efficiency of the cooling surfaces, with associated financial losses to industry, by acting as an insulating layer (McCoy, 1987). An additional problem, also responsible for huge financial losses, is that of microbially induced corrosion within industrial pipelines and tanks, and is responsible for huge financial losses. The bacteria growing on these surfaces, and in their physical proximity, promulgate chemical reactions between the bacteria and the substratum. Thus, during biocorrosion, SRB reside within the anaerobic regions of the biofilm and establish electrochemical corrosion cells between the biofilm and the surface metal. Again, suitable test methods must be deployed which duplicate such growth and provide inocula which are appropriate for effective biocide development against both the sessile and the planktonic phenotype.

TEST METHODS

Historically, the testing and screening of antimicrobial agents has been undertaken using pure cultures of vegetative, planktonic bacteria which have been grown in nutritionally rich media. Since this is atypical of most natural environments, then these testing methods screen antimicrobial agents only for their relative activity. The data generated cannot be extraploated to give direct information on *in situ* activity. Thus, compounds with activity against only sessile populations will be overlooked and compounds with especially good activity against fast-growing planktonic cells might be wrongly selected for further development.

In response to such problems, different methods have been developed to screen and test agents against microbial physiologies and communities which more closely resemble those found *in situ*. Each of the various testing methods uses different procedures to establish and maintain the test biofilms. Accordingly, each approach produces biofilms that are representative of different ecological niches and differ from one another in terms of physiology and metabolism. The established procedures include the use of reactors, such as the RotoTorque and the Constant Depth Film Fermenter, and biofilm sampling systems, such as the Robbins device, and the insertion of coupons within batch and continuous cultures. The application of methods, such as the perfused biofilm fermenter (Gilbert et al., 1989), which offer the possibility of greater control over the expressed phenotype, have been described in earlier sections (Chapters 1.7 and 3.7). All of these experimental approaches exercise different levels of control over biofilm formation and growth under conditions which model a number of environmental niches. These factors probably have a significant impact on the efficacy of the biocides tested.

In general, biocides to control contamination in industry should be tested in the context of their intended use; that is, if they are to be used against surface-attached bacteria, they should be tested against cells grown in this manner. The testing procedure should model the real situation as closely as possible. These considerations should influence the choice of test procedure, and we now briefly review the main approaches. We have described three factors that we consider important in developing suitable inocula for test methods, namely the use of mixed cultures representing *in situ* and growing both as sessile and planktonic populations under low nutrient or starvation conditions. Such organisms must therefore be used as the initial inocula for establishing the following *in vitro* test systems.

Biofilm Reactors

RotoTorque® Biofilm Reactor

The RotoTorque, a rotating annular reactor, produces biofilms on surfaces which are subject to different frictional resistance to attachment and maintains

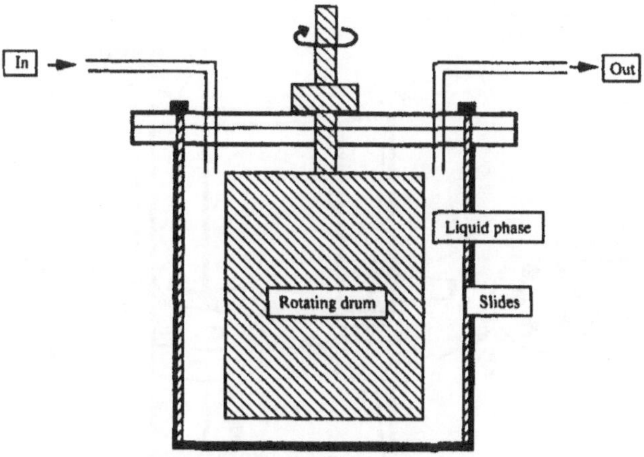

FIGURE 3 A diagram of the RotoTorque, a continuous flow reactor for forming biofilms under different shear stress. It is composed of an inner rotating drum and an outer glass cylinder into which glass slides are inserted, separated by a liquid phase containing microbial communities.

conditions which approximate to an open, continuous culture (Characklis et al., 1982). The RotoTorque consists of a rotating inner drum and a stationary outer drum operating as a continuous flow chemostat reactor (Figure 3). The outer drum contains 4 to 12 removable slides as an integral part of the inside wall. The slides permit sampling of the biofilm within the reactor. Fluid is recycled through the reactor to maintain a constant fluid shear stress which is independent of the nutrient flow into the system. These reactors maintain biofilm thickness by the controlled application of frictional force. The bulk fluid is kept to a minimal level such that planktonic growth is negligible and the bulk of the metabolic activity within the reactor can be attributed to the sessile population (Characklis, 1990).

Biofilm accumulation within the reactor can be monitored indirectly by measuring the drag force on the inner rotating cylinder. This is monitored by a torque transducer mounted onto the shaft between the motor and the cylinder. When the fluid shear stress in a clean reactor is calculated, the change in the fluid shear stress when a biofilm is present provides information about biofilm accumulation. Conversely, the effect of biocide treatment upon the extent of biofilm development can be easily monitored by the reduction in force required to turn the inner cylinder (Characklis, 1990). In a study by Characklis et al. (1980), periodic chlorination treatment of a biofilm showed that a temporary drop in frictional resistance occurred after treatment. This indicated that, during chlorination, there was a partial detachment of biofilm but that the biofilm recovered and built up before the next chlorination treatment.

This biofilm generator is well suited for systems where variable and high fluid shear stress is a major factor during biofilm formation. Industrial situations where this would be considered important are recirculating water systems such as pumps, cooling plants, and swimming pools.

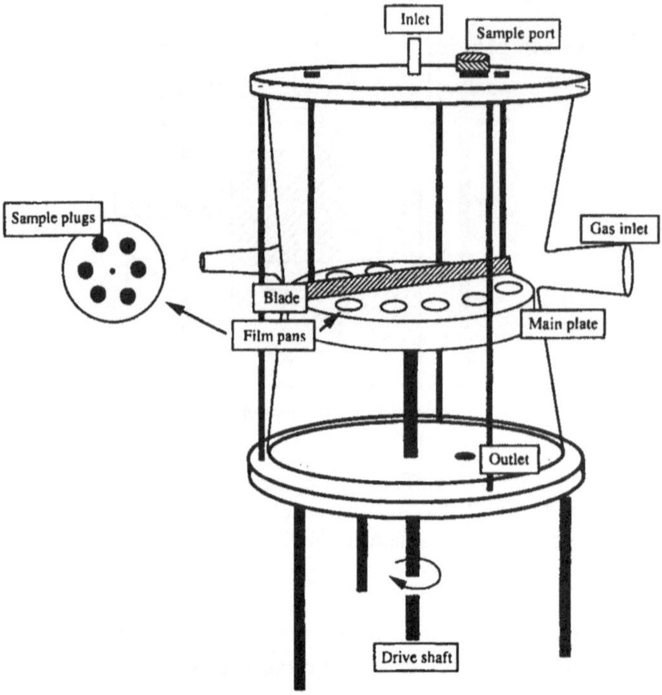

FIGURE 4 The constant-depth biofilm fermenter forms biofilms of a predetermined thickness by scraping excess biofilm with a blade. The fermenter is constructed with a main plate containing a number of film pans into which six surfaces have been recessed. A blade is held stationary over the surface while the main plate rotates beneath allowing the biofilms to be scraped off. Medium is fed to the biofilms by a port in the top of the fermentor and an outlet is at the bottom.

Constant Depth Film Fermenter

The Constant Depth Film Fermenter has been used by Coombe, Tatevossian, and Wimpenny (1982) and was further developed by Peters and Wimpenny (1988) as a method for studying the formation and treatment of dental plaque. The fermenter has also been applied to the study of biofilm formation by freshwater communities. The main features of the Constant Depth Film Fermenter are that it maintains a biofilm, contained within a recessed chamber, at a predetermined and constant depth by mechanically scraping the cells from the surface. The biofilm is developed onto one of six surfaces recessed into a film pan (Figure 4). There are 15 film pans on a main disk within the fermenter. The main disk, driven by a motor, rotates beneath a stationary, angled Teflon® blade which scrapes excess biofilm from the top of the film pans (Wimpenny, Peters and Scourfield, 1989). Nutrient is fed through an inlet at the top of the fermenter leading down to the surface of the rotating main disk where it is dispersed uniformly by the blade to the biofilms. Excess and spent medium and biomass is voided through the base of the fermenter. The top of the fermenter has a sampling port through which film-pans can be removed aseptically for sampling. Mixed community dental biofilms have been developed, using this system, to a thickness of 20 mm using mercury amalgum hydroxyapatite and polymethylmethacrylate resin as colonizing surfaces (Wimpenny, Peters and Scourfield, 1989).

Kinniment and Wimpenny (1990) used the Constant Depth Film Fermenter to determine the response to various biocides of model biofilms formed from microbial communities isolated from contaminated cutting fluids. They used viable counts and protein content to monitor the development of the biofilms. A concentration of 200 ppm of formaldehyde added to a biofilm formed over 50 h caused a decrease in both the protein content and viable cell density, within the biofilm, relative to untreated biofilm controls. The Constant Depth Film Fermenter provides a reproducible system for biofilm formation of a predetermined thickness, so that biocide efficacy can be tested. This system would be used for studying very thick biofilms such as dental plaque, sludge, and some industrial water systems.

Biofilm Sampling Devices

The Robbins Device

The Robbins device was first developed to provide multisampling facilities for the study of biofilm formation in tubular systems (McCoy et al., 1981; Ruseska et al., 1982). The tubular structure, made of admirality brass, contains stainless steel studs which screw into the device (Figure 5). Each stud contains a removable test surface of 0.5 cm^2 from which the developed biofilm may be sampled for viability determinations and electron microscopy. The Robbins device is well adapted for use as a self-contained laboratory experiment but may also be fitted into the plumbing of industrial water systems (Figure 5) to monitor the performances of biocides *in situ*. By forming a biofilm within the Robbins device and then treating with different regimens of biocide, the survival of sessile bacteria and the efficacy of the treatments could be determined. These investigative procedures involve both laboratory-based studies of single or groups of biocides against sessile cells and *in situ* observations of biocide efficacy. Continued on-line monitoring of the biofilm show the effectiveness of the biocide, even after treatment is discontinued. With a large number of sample ports, time studies of the effects of biocide treatment can be undertaken (Costerton and Lashen, 1984).

The Robbins device was modified in order to facilitate its routine use for the laboratory testing of biocides and antibiotics (Nickel et al., 1985). Constructed from a perspex (Plexiglas) block, the modified Robbins device has a lumen through which a culture can be passed and contains 25 equally spaced sample ports into the lumen. The ports can receive studs, the faces of which form a

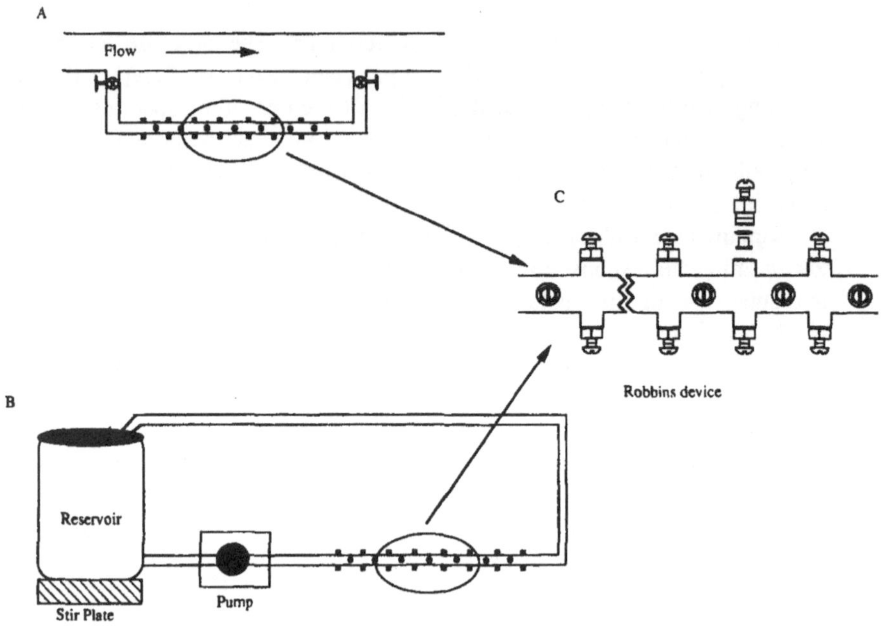

FIGURE 5 A schematic diagram of a Robbins device (C) showing a pipe with screws holding surfaces within the pipe. The system is strong enough to be used both within industrial plumbing systems where biofilm samples can be observed by redirecting the flow through the Robbins device rather than the main pipes (A). The Robbins device can also be used within the laboratory attached to a reservoir the culture pumped through via a peristaltic pump (B).

surface continuous with that of the lumen. Without removing the surfaces and potentially disrupting the biofilms, the sessile cells on the stud surfaces can be treated with antibiotics or biocides by pumping the solution through the lumen together with nutrient medium. Samples of the biofilm can be removed aseptically and, again without disrupting adjacent biofilms, assessed for bacterial viability. By substituting different materials onto the tips of the studs the effects of biocides against biofilms formed onto different surfaces can be assessed. Modified Robbins devices and other forms of tubular sampling devices have been attached to either batch cultures (Nickel et al., 1985) or chemostats (Camper, 1993; Green, 1993; Hoyle, Williams and Costerton, 1993; Jass, Sharp and Lappin-Scott, 1992) as a source of their inocula.

Tubular biofilm sampling devices of a similar design have been used by other researchers to monitor biofilm formation and biocide activity. For example, Green and Pirrie (1993) used different tubes as surfaces, connected to either a batch or a chemostat culture containing *Legionella* spp. By removing sections of the tubing and placing them into different biocide solutions, the efficacy of the biocides against the *Legionella* could be analyzed at different concentrations. This system also proved effective at comparing different biocides against sessile cells.

Submerged Test Pieces

Many researchers have used chemostats to control the growth of planktonic bacteria (Anwar et al., 1989; Keevil et al., 1987; Walker et al., 1991). Computer monitored chemostats provide full, automated control over the planktonic culture and thus provide reproducible planktonic inocula for biocide testing. By inserting a selected surface into a chemostat, biofilms will form on the test piece in a controlled and reproducible manner. Although the biofilm still maintains an heterogeneous environment, the growth rate of the planktonic cells can be maintained. Furthermore, for mixed cultures, species composition may be kept constant for the biofilm formation. There have been different chemostat set-ups to accomplish this. For example, Keevil et al. (1987) and Keevil, Mackerness, and Colbourne (1990), formed mixed consortia dental plaque biofilms on glass tiles by using a two-stage chemostat. The first stage provided the seed culture for the second stage. The test surfaces were inserted into the second stage for up to 28 days before sampling. Use of a two-stage fermenter ensures that the primary fermenter, which generates the inoculum for biofilm formation, is never exposed to the test conditions/substances. Using this approach they tested the biocide monochloramine, against both biofilm and planktonic cells by adding 0.2 mg/ml monochloramine to the fresh medium fed into the second stage of the chemostat. They reported that the biocide reduced both the biofilm and planktonic consortia by one order of magnitude (Keevil, Mackerness and Colbourne, 1990). Furthermore, not only was the biofilm community different after the treatment with the biocide, but the planktonic community was also different from the biofilm. Such systems allow simultaneous testing against both biofilm and planktonic cultures and facilitates comparative studies of the efficacy of biocides against each phenotype. Such studies reinforce the importance of testing biocides against mixed community biofilms rather than isolated monocultures.

CONCLUDING REMARKS

It is important that biocides and antimicrobial agents be properly assessed for their effectiveness before application in industrial or medical situations. The use of procedures based on pure cultures of vegetative, planktonic bacteria are unreliable in that they are not typical of the environments in which the biocide will be applied. The appropriate procedures involve the use of random samples of inocula obtained from the environment in which the biocides are to be used. Such a testing procedure may be more expensive than those using pure cultures, as biocides may have to be tested against a variety of inocula reflecting the variety of situations in which the biocide is expected to be effective. This might be a reasonable price to pay for improved effectiveness in use.

REFERENCES

Anwar, H., T. van Biesen, M. Dasgupta, K. Lam, and J. W. Costerton. 1989. Interaction of biofilm bacteria with antibiotics in a novel *in vitro* chemostat system. *Antimicrobial Agents and Chemotherapy* **33:**1824-1826.

Bass, C. J., H. M. Lappin-Scott, and P. F. Sanders. 1993. Bacteria that sour reservoirs: New concepts for the mechanism of reservoir souring by sulphide generating bacteria. *Journal of Offshore Technology* **1:**31-36.

Brown, M. R. W., D. G. Allison, and P. Gilbert. 1988. Resistance of bacterial biofilms to antibiotics: A growth-rate related effect? *Journal of Antimicrobial Chemotherapy* **22:**777-783.

Bull, A. T., and J. H. Slater. 1982. Microbial interactions and community structures, pp. 13-44 in A. T. Bull and J. H. Slater (eds.), *Microbial Interactions and Communities,* Vol. 1. Academic Press, London.

Camper, A. 1993. Biofilms in water distribution systems. *Fifth Annual Meeting Workshop on Biofilms.* University of Calgary, Canada.

Characklis, W. G., M. G. Trulear, N. A. Stathopoulos, and L. C. Chang. 1980. Oxidation and destruction of biofilms, pp. 349-368 in R. L. Jolley (ed.), *Water Chlorination.* Ann Arbor Science Publications, Ann Arbor.

Characklis, W. G., M. G. Trulear, J. D. Bryers, and N. Zelver. 1982. Dynamics of biofilm processes: Methods. *Water Research* **16:**1207-1216.

Characklis, W. G. 1990. Laboratory biofilm reactors, pp. 55-89 in W. G. Characklis and K. C. Marshall (eds.), *Biofilms.* John Wiley & Sons, New York.

Coombe, R. A., A. Tatevossian, and J. W. T. Wimpenny. 1982. Bacterial thin films as *in vivo* models for dental plaque, pp. 239-249 in R. M. Frank and S. A. Leach (eds.), *Surface and Colloidal Phenomena in the Oral Cavity: Methodological Aspects.* IRL, London.

Costerton, J. W., and E. S. Lashen. 1984. Influence of biofilm on efficacy of biocides on corrosion-causing bacteria. *Materials Performance* **23:**34-37.

Geesey, G. G. 1987. Survival of microorganisms in low nutrient waters, pp. 1-23 in M. W. Mittleman and G. G. Geesey (eds.), *Biological Fouling of Industrial Water Systems: A Problem Solving Approach.* Water Micro Associates, San Diego.

Gilbert, P., D. G. Allison, D. J. Evans, P. S. Handley, and M. R. W. Brown. 1989. Growth rate control of adherent bacterial populations. *Applied and Environmental Microbiology* **55:**1308-1311.

Green, P. N. 1993. Efficacy of biocides on laboratory-generated *Legionella* biofilms. *Letters in Applied Microbiology* **17:**158-161.

Green, P. N., and R. S. Pirrie. 1993. A laboratory apparatus for the generation and biocide efficacy testing of *Legionella* biofilms. *Journal of Applied Bacteriology* **74:**388-393.

Hamilton, W. A. 1985. Sulphate-reducing bacteria and anaerobic corrosion. *Annual Reviews in Microbiology* **39:**195-217.

Hoyle, B. D., L. J. Williams, and J. W. Costerton. 1993. Production of mucoid exopolysaccharide during development of *Pseudomonas aeruginosa* biofilms. *Infection and Immunity* **61:**777-780.

Jass, J., E. V. Sharp, and H. M. Lappin-Scott. 1992. Colonization of silastic rubber by *Pseudomonas fluorescens* and *Pseudomonas putida* using a chemostat and a modified Robbins device. *Abstracts of the 92nd General Meeting of the American Society for Microbiology.* Abstract **D-266.**

Kaprelyants, A. S., J. C. Gottschal, and D. B. Kell. 1993. Dormancy in non-sporulating bacteria. *FEMS Microbiology Reviews* **104:**271-286.

Keevil, C. W., D. J. Bradshaw, A. B. Dowsett, and T. W. Feary. 1987. Microbial film formation: Dental plaque deposition on acrylic tiles using continuous culture techniques. *Journal of Applied Bacteriology* **62:**129-138.

Keevil, C. W., C. W. Mackerness, and J. S. Colbourne. 1990. Biocide treatment of biofilms. *International Biodeterioration Bulletin* **26:**169-179.

Kinniment, S., and J. W. T. Wimpenny. 1990. Biofilms and biocides. *International Biodeterioration Bulletin* **26:**181-194.

Koch, R. 1881. Zur Untersuchung von pathogenen Organismen. *Mittheilungen aus dem Kaiserlichen Gesundheitsamte* **1:**1-48.

Lappin-Scott, H. M., and J. W. Costerton. 1989. Bacterial biofilms and surface fouling. *Biofouling* **1:**323-342.

Lappin-Scott, H. M., and J. W. Costerton. 1990. Starvation and penetration of bacteria in soils and rocks. *Experientia* **46:**807-812.

Lappin-Scott, C. J. Bass, and P. F. Sanders. 1992. Monitoring and detection of different growth states of thermophilic sulphate-reducing bacteria. *Proceedings of the UK Corrosion 92*, Vol. 3, Manchester, Europ. Fed. Corr., UK.

Lappin-Scott, H. M., and P. F. Sanders. 1993. Reservoir souring: Involvement of dormant thermophilic sulphate-reducing bacteria. *Proceedings of New Geoscience Technology in the Marketplace*, Department of Trade and Industry, London.

McCoy, W. F., J. D. Bryers, J. D. Robbins, and J. W. Costerton. 1981. Observations on biofilm formation. *Canadian Journal of Microbiology* **27:**910-917.

McCoy, W. F. 1987. Fouling biofilm formation, pp. 24-55 in M. W. Mittleman and G. G. Geesey (eds.) *Biological Fouling of Industrial Water Systems: A Problem Solving Approach.* Water Microbiology Associates, San Diego.

Nickel, J. C., I. Ruseska, J. B. Wright, and J. W. Costerton. 1985. Tobramycin resistance of *Pseudomonas aeruginosa* cells growing as a biofilm on urinary catheter material. *Antimicrobial Agents and Chemotherapy* **27:**619-624.

Novitsky, J. A., and R. Y. Morita. 1976. Morphological characterisation of small cells resulting from nutrient starvation of a psychrophilic marine vibrio. *Applied and Environmental Microbiology* **32:**617-622.

Peters, A. C., and J. W. Wimpenny. 1988. A constant-depth laboratory model film fermentor. *Biotechnology and Bioengineering* **32:**263-270.

Ruseska, I., J. Robbins, J. W. Costerton, and E. S. Lashen. 1982. Biocide testing against corrosion causing oil-field bacteria helps control plugging. *Oil & Gas Journal* March 8, 253.

Slater, J. H. 1981. Mixed cultures and microbial communities, pp. 1-24 in M. E. Bushell and J. H. Slater (eds.), *Mixed Culture Fermentations.* Academic Press, London.

Standards 1992a. National Committee for Clinical Laboratory Standards. *Methods for dilution antimicrobial susceptibility tests for bacteria that grow aerobically. Approved standard M7-A2.* National Committee for Clinical Laboratory Standards, Villanova, PA.

Standards 1992b. National Committee for Clinical Laboratory Standards. *Performance standards for antimicrobial disk susceptibility tests. Approved standard M2-A4.* National Committee for Clinical Laboratory Standards, Villanova, PA.

Walker, J. T., A. B. Dowsett, P. J. L. Dennis, and C. W. Keevil. 1991. Continuous culture studies of biofilm associated with copper corrosion. *International Biodeterioration Bulletin* **27:**121-134.

Wimpenny, J. W. T., A. Peters, and M. Scourfield. 1989. Modeling spacial gradients, pp. 111-127 in W. G. Characklis and P. A. Wilderer (eds.), *Structure and Function of Biofilms*. John Wiley & Sons, New York.

Working Party BSAC. 1991. Working Party on Antibiotic Sensitivity Testing of the British Society for Antimicrobial Chemotherapy. A Guide to Sensitivity Testing. *Journal of Antimicrobial Chemotherapy* **27**:(Suppl D).

3.9 Antimicrobial Susceptibility Testing Within the Clinic

Kenneth S. Thomson, Johan S. Bakken, and Christine C. Sanders

INTRODUCTION

The goal of *in vitro* antimicrobial susceptibility testing is to evaluate growth of a microorganism in the presence of various antimicrobial agents, and to translate the results into an accurate prediction of the likely treatment outcome with each agent. This should then serve as a reliable guide for the clinician in selection of appropriate therapy. Numerous studies have shown that this goal is only partially met by currently available susceptibility testing procedures. Test results indicating sensitivity of a pathogen to an antibacterial agent do not guarantee successful therapy with any particular agent because *in vitro* tests cannot take into account significant host factors that impact therapeutic outcome. These include the site of the infection, presence and nature of any underlying disease, and immune status of the patient. However, results of *in vitro* tests which indicate resistance to an agent are strongly predictive of therapeutic failure. The major strength of susceptibility tests, therefore, is in the detection of resistance and the identification of agents that are inappropriate for therapy (Sanders, 1991). It follows from this that while the goal of the susceptibility test is to discriminate between sensitive and resistant isolates, clinical laboratories should be most interested in performing those specific tests that optimize the detection of resistance. Failure to do so could mislead clinicians into prescribing inappropriate, possibly fatally inappropriate, drugs.

At present there is no perfect susceptibility test system. All current procedures are limited to some degree in their ability to detect both resistance and the potential for resistance to emerge. Microbiologists who understand the underlying principles and shortcomings of the testing systems they use should, however, be able to ensure that reports to clinicians contain both accurate information and meaningful interpretive comments. The purpose of this article is therefore to discuss the general principles, strengths and weaknesses, and

0-8493-4752-1/95/$0.00+$.50

TABLE 1
Clinically Important Bacteria and Predictability of Sensitivity to Commonly Utilized Antimicrobials

Organisms with Predictable Sensitivity	Organisms with Unpredictable Sensitivity
β-haemolytic streptococci	Staphylococci
Streptococcus pneumoniae[a]	Enterococci
Haemophilus influenzae[a]	Enterobacteriaceae
Neisseria meningitidis[a]	*Pseudomonas* species
Neisseria gonorrhoeae[a]	*Acinetobacter* species
Moraxella catarrhalis[a]	Other nonfermentative gram-negative bacilli
	Aeromonas species
	Vibrio species

[a] Special tests for specific antimicrobial resistance mechanisms and/or antimicrobial sensitivity tests should be performed in cases of serious, potentially life-threatening infections.

applications of the major susceptibility test systems (see Chapter 1.6 for biofilm test systems related to chronic infections).

The Need for Susceptibility Tests

Once an isolate, recovered on culture from an infected site, has been deemed to be a potential pathogen, the microbiologist must decide whether or not it should be subjected to susceptibility testing to antimicrobial agents. Routine susceptibility testing is, for most laboratories, limited to relatively nonfastidious bacteria which grow aerobically. Susceptibility testing of obligate anaerobes, highly fastidious bacteria, yeasts, fungi, and viruses is typically limited to major referral centers or academic institutions, and will not be discussed further.

Pertinent clinical information about the patient should accompany the clinical specimen sent for culture to allow the microbiologist to determine if antibiotic susceptibility testing is indicated. Some infections may be treated successfully without the assistance of susceptibility test results. These include mild to moderately severe infections that are going to be handled on an outpatient basis for which the etiologic agent is known or accurately predictable on clinical bases alone. Thus, antibiotics with predictable activity against the pathogen in question are typically prescribed under these circumstances (Table 1). For example, streptococcal tonsillitis is usually treated with penicillin, and susceptibility tests are not indicated since Group A streptococci are uniformly sensitive to penicillin. Susceptibility testing is indicated for infections due to organisms that are not predictably sensitive to the antimicrobials normally considered to be a reasonable choice for therapy (Table 1) (Jorgensen, 1993). This includes organisms recovered from most serious, potentially life-threatening, infections, those acquired during hospitalization or stay in an extended-care facility, and organisms for which empiric therapy has failed.

Selection of a Susceptibility Test Method

The actual methods used by laboratories for antimicrobial susceptibility testing should be determined by an analysis of the fiscal and personnel resources available, the workload and patient population to be served, the type of results desired by clinicians, and the expertise of laboratory personnel. There are three major groups of tests available based upon type of result generated (Table 2). Some methods are qualitative, reporting results as sensitive, intermediate (indeterminate) or resistant, and include most agar diffusion assays performed with disks or tablets impregnated with antibiotics. These cannot be used to ascertain *how* sensitive or resistant an organism is to a particular antibiotic, but they are usually reliable and cost-effective for routine use. Some methods are quantitative, reporting results as a minimal inhibitory concentration (MIC) and include agar and broth dilution procedures that test a broad range of drug concentrations. These tend to be more expensive to perform than qualitative tests, but in special clinical settings, quantitative results can be very important in selection of the most appropriate agent. The third major group of tests generate semiquantitative results. These include dilution procedures that test only a narrow range of drug concentrations and report results as an MIC, an MIC above the concentrations tested, or an MIC below the concentrations tested if the organism is inhibited within the range, not inhibited within the range, or inhibited by even the lowest concentration tested, respectively. These tests are often referred to as "breakpoint" methods since the range of concentrations tested is targeted near the clinically achievable concentrations. Breakpoint methods offer a compromise between qualitative and quantitative results.

Qualitative, quantitative, and semiquantitative methods may be manual, semiautomated, or fully automated. The larger the workload, the greater the advantage of automation. The higher the degree of automation, however, the lower the degree of freedom in individualizing the test for the particular needs of the laboratory and the patient population it serves.

THE INOCULUM

Whichever susceptibility test system a laboratory uses, a critical determinant of the clinical relevance of the results is the size of the inoculum employed. The inoculum for a susceptibility test should be such that it yields results which are therapeutically relevant and, ideally, reproducible. If both relevance and reproducibility cannot be attained, relevance (i.e., accuracy) is the more important goal.

TABLE 2
Advantages and Disadvantages of the Three Major Types of Susceptibility Test

Characteristic	Type of Test and Examples: Qualitative e.g., Disk diffusion, Stokes test	Quantitative e.g., Macrobroth (ma), microbroth (mi), agar (ag) dilution, E-test (et)	Semiquantitative e.g., Automated, semiautomated, breakpoint
Can be used for almost any bacterium	No	Yes	No
Highly flexible in range of drugs tested	Yes	$Yes^{ma, ag}$; $No^{mi, et}$	No
More than usual problems in resistance detection	No	No	Yes
Ability to detect contamination	Yes	Yes^{et}; $Sometimes^{ag}$; $No^{ma, mi}$	No
Ability to measure bactericidal activity	No	Yes^{ma}; $No^{mi, ag, et}$	No
Labor intensive	No	$Yes^{ma, ag}$; $No^{mi, et}$	No
Cost-effective in small lab	Yes	Yes^{mi}; $No^{ma, ag, et^{a}}$	Yes (some breakpoint tests); No (if any level of automation)
Cost-effective in large lab	Yes	$Yes^{mi, ag}$; $No^{ma, et^{a}}$	Yes
Ability to provide same-day results	No	No	Yes

[a] E-test may be cost-effective for specialized applications, but not for routine testing.

In current tests it is implied that the cells in the test inoculum are representative of all the cells at the infected site, each of which has the same antibiotic susceptibility. This is misleading, for cells of a single bacterial strain may be able to express multiple susceptibility phenotypes depending on gene expression and environmental conditions. For example, cells growing on artificial media may have higher growth rates and therefore increased antibiotic susceptibility compared to slower growing cells in nutrient- and/or iron-depleted tissues or biofilms that may be present at the site of infection. Resistant subpopulations (mutants), alterations in gene regulation, and plasmid loss or acquisition are also important determinants of antibiotic susceptibility. These can give rise to discrepancies between the susceptibility phenotypes being expressed by cells at the site of infection and the phenotypes in the small inocula that are used in susceptibility tests (Gilbert, Brown and Costerton, 1987). Susceptibility tests usually use much lower inocula than the populations of $\geq 10^9$ colony-forming units (CFU), typical of many infections, and the 10^6 to 10^9 CFU that should be tested if laboratories are to detect certain resistant subpopulations that may portend therapeutic failure (Greenwood, 1981; Jorgensen, 1993; Sanders, Thomson and Bradford, 1993).

To date no one has been able to define what is the therapeutically "correct" inoculum (Greenwood, 1981) or to produce a test in which conditions closely approximate those in human infections. Some regulatory bodies, recognizing that current *in vitro* test systems are very artificial, have chosen to ignore the importance of establishing the most relevant inoculum size for testing, and instead have concentrated on maximizing test reproducibility (Greenwood, 1981; Sanders, 1991; Working Party BSAC, 1991). This goal is best achieved by reducing the inoculum size, for this increases the homogeneity of the cell population and improves the clarity with which test end points can be determined. Unfortunately, this also increases the apparent activity of some antibacterial agents and reduces the likelihood of detecting resistance.

Regardless of the size of the inoculum chosen for testing, standardization of the inoculum is very important. Ten-fold or greater variations in inoculum size may significantly alter results (Jorgensen, 1991; Standards 1992a, b; Thornsberry, 1991). The inoculum may be standardized by growth of the test organism in an appropriate broth (see Chapters 1.1 and 1.2) or by direct suspension of cells in an appropriate liquid (Barry et al., 1979; D'Amato and Hochstein, 1982; Standards, 1992a, b; Working Party BSAC, 1991). Whichever approach is used, it is important to prepare the inoculum from at least four or five colonies to help ensure that all susceptibility phenotypes present in the cell population are represented (Thornsberry, 1991). However, there is some evidence that five colonies is less than adequate representation of the clinical situation (Thomson, File and Burgoon, 1989).

DISK DIFFUSION TESTS

The underlying principle of this test is the organism-antibiotic encounter that occurs when a lawn culture of bacteria on the surface of an agar-containing

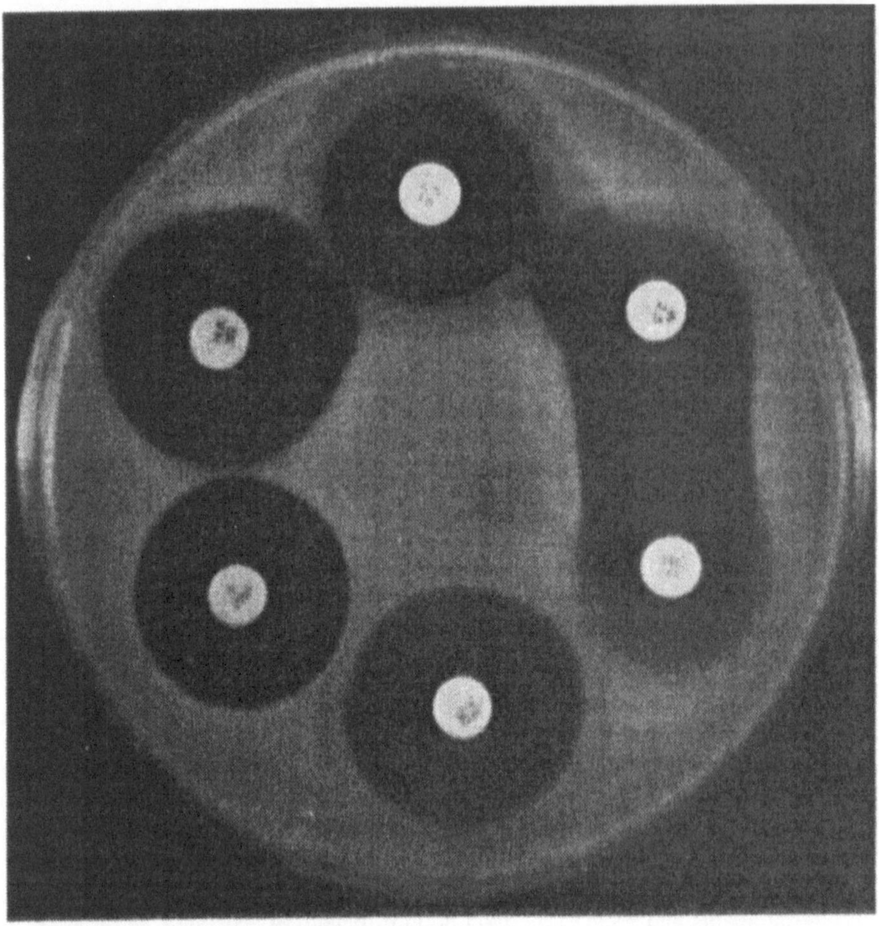

FIGURE 1 Disk diffusion susceptibility test showing large, clear inhibition zones around antibiotic-impregnated filter paper disks.

medium is exposed to an antibiotic concentration gradient in agar (Figure 1). The concentration gradient is produced by diffusion of drug from an antibiotic-impregnated filter paper disk (or other suitable source, e.g., tablet) located on the surface. After an appropriate incubation period, usually 16 to 18 h, an inhibition zone around the disk indicates if the drug is active against the organism. The size of the zone is determined by the organism's susceptibility, the inoculum density and growth rate, the diffusibility of the drug, the culture medium composition and depth, and the incubation conditions (Jorgensen, 1993; Standards 1992b; Working Party BSAC, 1991). As far as possible, the test conditions are controlled or standardized so that the organism's qualitative susceptibility can be interpreted on the basis of zone size.

Disk potency is important and should vary no more than 60 to 140% from the stated drug mass to avoid markedly influencing zone size and test interpretation (Isenberg, 1988). Regular testing with reference organisms (at least weekly) must be performed to ensure that all technical aspects of the test are under control (Standards, 1992b; Working Party BSAC, 1991). Should tests with reference organisms yield zone sizes outside acceptable ranges, appropriate corrective action is required.

For many laboratories the disk diffusion test is the susceptibility testing system of choice. This procedure is well standardized and controlled, reproducible, convenient, relatively inexpensive, flexible in the range and number of drugs that can be tested, and does not require special equipment. This test does have shortcomings. Slow growing organisms and obligate anaerobes cannot be tested reliably. Some drugs diffuse poorly in agar, producing small inhibition zones which may make it difficult to discriminate between sensitivity and resistance, (e.g., vancomycin, polymyxin B), and may necessitate use of an alternative method. Only bacteriostatic activity can be inferred from disk tests. Bactericidal activity cannot be evaluated. Although the qualitative results generated with this test are sufficient for most clinical situations, more quantitative information may be needed on occasion. For example, quantitative results may be desirable if the infection is at a site of limited drug perfusion, e.g., bone, central nervous system. If a precise MIC is desired, then a quantitative test must be performed because a sensitive result obtained in a diffusion test merely indicates that the MIC lies somewhere within the range of achievable serum concentration of the drug.

DILUTION TESTS

Quantitative, or dilution, tests are generally regarded as the standard by which other susceptibility testing methods are evaluated. This is because MICs can be related to achievable drug levels at an infected site and thus, in theory at least, be translated into predictions of therapeutic outcome. The MIC, usually measured in μg/ml or mg/L, after 18 to 20 h incubation, is the lowest drug concentration which inhibits growth of the test organism. Classically this is determined by exposing the test organism to a range of doubling concentrations of drug in an agar- or broth-based medium. The end point of the test is the lowest concentration at which there is an absence of visible growth. Inevitably some MICs lie close to one of the test concentrations, leading to MICs varying twofold on a day-to-day basis. This inherent twofold error of the test is tolerable in tests of relatively safe drugs, such as the β-lactams, but is a disadvantage in tests of toxic drugs, such as the aminoglycosides.

In contrast to disk diffusion tests, dilution tests can be used for extended incubation periods for slow-growing organisms. This is because the procedure is uninfluenced by growth rates and the diffusion properties of antibiotics. Contamination, however, may be much more difficult to detect, especially in broth tests. In-house preparation of media requires a special supply of laboratory drug powders. Drugs for human use are generally unsuitable because they contain additives which interfere with the test, or because they are biologically inactive until metabolized.

Agar Dilution Tests

MICs determined by the agar dilution procedure are expensive and labor-intensive, and cannot be used to detect bactericidal end points. They may, however, be cost-effective if many strains are tested. Usually a multipronged replicating device is used to deliver inocula of 10^4 colony-forming units (CFUs) of each organism to drug-containing plates, with up to 32 organisms being tested per plate (Jorgensen, 1993; Standards, 1992a; Working Party BSAC, 1991). Some drugs, such as the fluoroquinolones, adhere tenaciously to the replicator prongs and should be tested last to minimize problems of drug carryover.

A recent innovation which overcomes much of the inconvenience of the agar dilution test is the E-Test (AB Biodisk, Solna, Sweden). In this test, an exponential antibiotic concentration gradient is produced by placing commercially available drug-impregnated strips on agar plates. Up to six drugs may be tested on a single large plate which is inoculated with a lawn culture of the test organism. After an appropriate incubation period, usually overnight, MICs are determined from the point of intersection of the elliptical inhibition zone with calibration marks on the upper side of the strip (Figure 2). Comparative evaluations indicate that the E-Test is equivalent to the standard agar dilution procedure (Baker et al., 1991; Brown and Brown, 1991). This approach may appeal to laboratories which have only an occasional need to determine MICs. At present, E-Test strips are, however, relatively expensive.

One concern with agar-based quantitative tests is the apparently lower MICs determined with some organism/β-lactam drug interactions compared to those determined with broth-dilution methods (Bradford and Sanders, 1992; Kayser, Morenzoni and Homberger, 1982; Rylander et al., 1979). Broth dilution tests therefore appear to be more reliable indicators of β-lactamase-mediated drug resistance, possibly because of the more intimate drug–β-lactamase encounter and also because in broth tests a single resistant cell can multiply overnight to yield turbidity, whereas on agar it will produce a single colony only, which will usually be ignored.

Broth Dilution Tests

Broth dilution methods offer one important advantage over all other susceptibility tests. They permit subculture to facilitate determination of bactericidal activity. The minimal bactericidal concentration (MBC) of a drug is arbitrarily defined as the concentration associated with a 99.9% killing of the original inoculum, as determined by subculture onto drug-free medium (Stratton, 1993). An MBC may be of value to clinicians as a guide to therapy of an infection where bactericidal activity is desirable (e.g., bacterial endocarditis), or when used in conjunction with assays of antimicrobial concentrations in body fluids. Results obtained in broth dilution tests have the same inherent twofold error as agar dilution assays. It is important to remember, however, that the test end point, i.e., absence of visible growth, may allow as many as

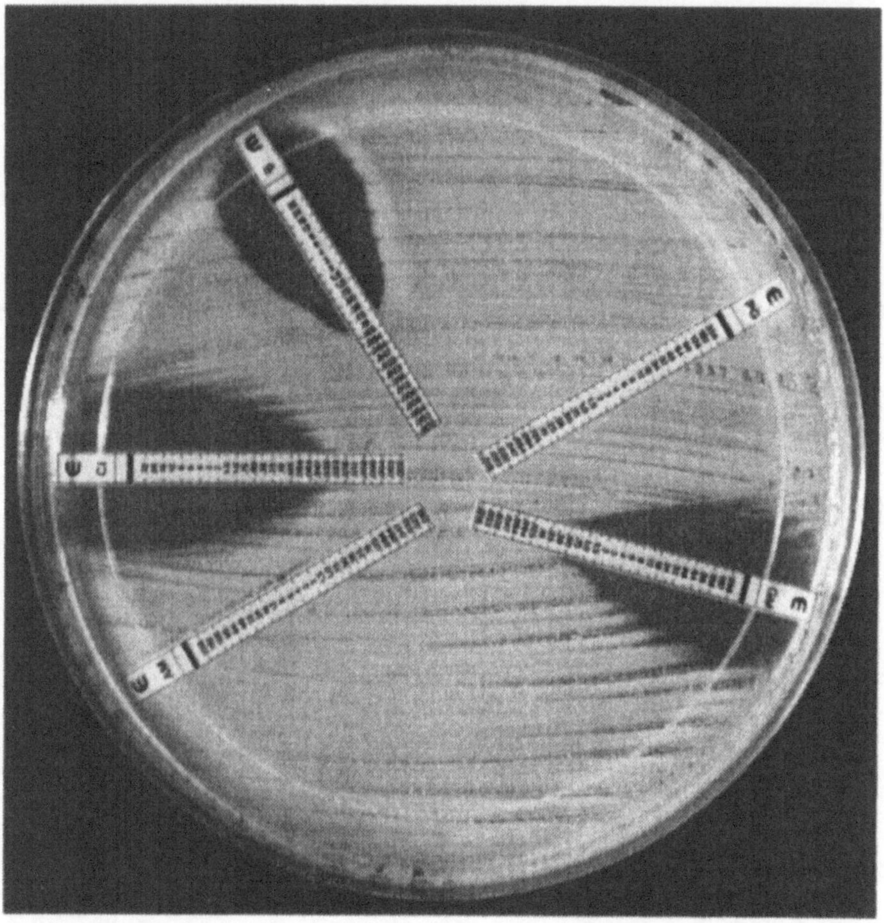

FIGURE 2 E-Test showing elliptical inhibition zones where organism was inhibited by drug diffusing from strip containing an exponential antibiotic concentration gradient.

10^6 to 10^7 CFU/ml to be present without being detected by the human eye. The MIC, therefore, does not necessarily reflect complete inhibition of growth of the inoculum.

Macrobroth dilution tests involve drug dilutions in tubes containing 1 to 3 ml of an appropriate broth (Figure 3). This is the most reliable susceptibility test for detecting resistance, but is too cumbersome and time consuming for routine use. The microdilution test is more economical in terms of cost, labor, space requirements, and reagents. It is performed in small plastic trays containing wells which contain volumes of 50 to 100 μl. This test is convenient, highly reproducible, and amenable to mechanization, but often limited in flexibility of the arrays of drugs that can be tested. Commercially available trays contain lyophilized or frozen panels of drugs. Manufacturers' standard panels may not, however, always provide the array of drugs that the laboratory desires to test. In this situation, the laboratory must either prepare its own trays or make

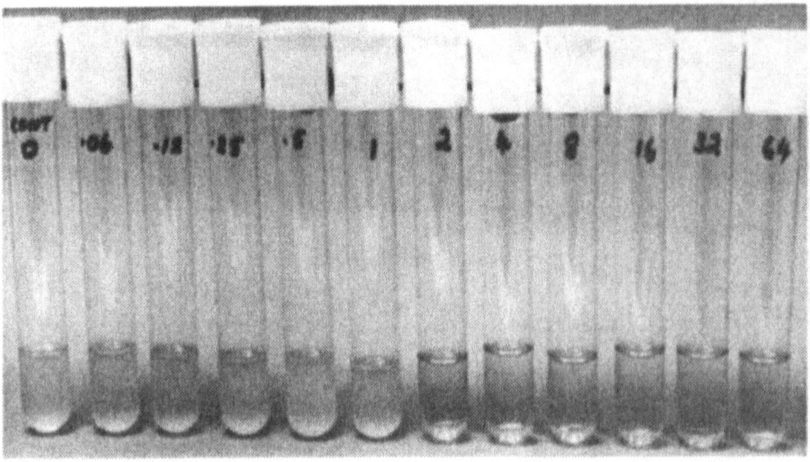

FIGURE 3 Macrobroth dilution MIC test in which tubes containing turbid broth indicate growth in the antibiotic-free control tube (0) and in the lower antibiotic concentrations (tubes labeled 0.06 to 1). The absence of visible growth (clear broth) in the tubes labeled 2 or greater indicates that the MIC of the test drug is 2 μg/ml (i.e., the lowest drug concentration with no visible growth).

special arrangements with a manufacturer to obtain customized, but more expensive, trays. When manufacturers experience delays in obtaining regulatory approval to include new drugs in their trays, alternative susceptibility testing methods may be required to provide clinicians with results for newly released or investigational compounds. Although both macrodilution and microdilution tests use an inoculum density of approximately 10^5 CFU/ml (Jorgensen, 1993; Standards, 1992a; Working Party BSAC, 1991), the differences in the test volume have a major impact on the numbers of CFUs actually tested. There can be up to a 3 $\log_{10}$ difference between the number of cells tested by each method, with the microdilution test inoculum of only 10^3 to 10^4 CFUs being less representative of cell populations involved in most infections. It is therefore not surprising that in some instances the microdilution test is less likely to detect clinically relevant resistance.

RECENT DEVELOPMENTS

Several automated tests have recently been developed to provide clinicians with earlier test results; i.e., within 6 to 8 h of inoculation. Results of these tests may be quantitative or semiquantitative depending upon the number and array of concentrations actually tested. Most of these tests involve automated monitoring of various indicators of microbial growth; e.g., turbidity, release of radioactive, or fluorescent metabolites. From comparisons of growth in drug-free and drug-containing wells, results are determined using algorithms developed with a complex database and software package. This approach to rapid generation of results, although highly desirable, has many shortcomings.

Side-by-side comparisons of results generated in these rapid, automated tests with those obtained in conventional dilution assays have often revealed significant discrepancies (Jorgensen, 1991; Louie et al., 1992; Schadow, Giger and Sanders, 1993; Thornsberry, 1991; Working Party BSAC, 1991; York, Brooks and Fiss, 1992). Usually, the rapid test's limitation is a failure to detect clinically relevant resistance. A thorough understanding of a specific automated testing system is essential for microbiologists to know when results are reliable, and when a more conventional, overnight susceptibility test procedure should be used.

ANTIBIOTIC SELECTION AND REPORTING OF RESULTS

ANTIBIOTICS TO BE TESTED

The site of infection, the pathogen isolated, and the availability of drugs on the hospital formulary will normally determine which antimicrobial agents should be evaluated. Thus, a knowledge of an antibiotic's spectrum as well as its pharmacokinetic properties is required for the selection of the most appropriate drugs to test. The most important pharmacokinetic determinant is drug access to the site of infection. *In vitro* antibiotic susceptibility alone provides insufficient information about *in vivo* usefulness of any given antibiotic. Antibiotics which achieve high concentrations at the site of infection should be included in the drug testing panels. Ideally a wide spectrum of choices should be included, with representation from multiple antibiotic classes. This will facilitate the choice of active drugs for therapy for patients who cannot receive first- or second-line therapy due to drug hypersensitivity, gastrointestinal intolerance, toxicity, or other reasons.

RESULTS TO BE REPORTED

Even though 20 or more antimicrobial agents may be included in susceptibility testing panels, appropriate choices for therapy are often restricted to only a few drugs. Test panels may contain a multiplicity of drugs for cost saving purposes and to accommodate testing of many different bacterial species. It is usually appropriate, therefore, for the microbiologist to screen results and report only those most relevant for the pathogen tested. In some instances, for example, *in vitro* results do not predict a successful clinical outcome. Species of *Salmonella* invariably appear sensitive to aminoglycosides *in vitro,* although these drugs are clearly ineffective therapeutically. Results should, therefore, always be screened and clinically irrelevant results deleted. In other instances, laboratories may choose to withhold reporting of results obtained with broad-spectrum, usually newer, more expensive agents unless the pathogen tested is sensitive only to those agents. Such selective reporting has cost savings advantages as well as the potential to delay the inevitable emergence of resistance that is seen with any new agent once it is used frequently in the hospital environment. Overall, the goal of the report of susceptibility tests is to provide the clinician with results for only those drugs which are reasonable choices for therapy. Reports that contain results for drugs that are inappropriate choices may mislead the unwary clinician, and treatment may result in unnecessarily increased morbidity, mortality, costs, or resistance.

Interpretation of Results

Although quantitative susceptibility results may be important in certain clinical situations, many clinicians have limited knowledge and interest about the numerical values that express the interaction between an antibiotic and bacterial pathogen. Many find it less confusing if results are expressed qualitatively as susceptibility categories (Jorgensen, 1993). When MICs are reported, it should be remembered that they are not absolute values. Different test conditions may greatly influence results. Furthermore, the actual clinical relevance of the MIC has not been unequivocally established, "except in the special case of gonorrhoea where it has been shown that failure of a particular single-dose regimen of procaine penicillin correlates with an increased MIC of penicillin" (Greenwood, 1981). There is, therefore, little justification to believe that the MIC is a more clinically relevant result. It is merely a more quantitative result. If, however, MICs are provided by a laboratory, then each report should also include the means for the clinician to translate the results into appropriate susceptibility categories.

The susceptibility categories sensitive, intermediate, or resistant are determined by how the actual MIC relates to predetermined susceptibility breakpoint values, which are themselves based upon drug levels that are clinically achievable. What the actual breakpoints are varies with locale. There is a striking lack of international consensus on breakpoint concentrations. This may reflect differences in philosophy, dosing regimens, and political influence (Bakken, 1992). The intermediate result is really an equivocal result reflecting the inherent error of the test utilized. It has no proven clinical relevance and should not be used as an indication for therapy if a pathogen is susceptible to alternate drugs of proven efficacy. It represents a buffer or safety zone between "true" resistance and "true" susceptibility. The breakpoint values for the United States have been determined by the National Committee for Clinical Laboratory Standards (NCCLS) and are revised and published at regular intervals in the Performance Standards for Antimicrobial Susceptibility Testing (Standards 1992a, b). The NCCLS guidelines build on a voluntary consensus process and rely on rigorous standardization of testing procedures. Recently the Working Party on Antibiotic Sensitivity Testing of the British Society for Antimicrobial Chemotherapy (BSAC) published testing guidelines for the United Kingdom (Working Party BSAC, 1991). In these it is stated that "the requirement in Britain is that the test should give the right answer, rather than that it should be performed in a rigorously standardized manner."

SUMMARY

Regardless of which susceptibility test is utilized by a laboratory, it is essential for the microbiologist to understand the limitations of the test and to ensure that results reported to the clinician do not exceed those limitations. In smaller laboratories, usually a single, qualitative test is sufficient to meet the need for antimicrobial susceptibility testing. In larger laboratories, or those serving academic or referral centers, a single qualitative method is also usually sufficient for routine susceptibility testing. A second backup system may, however, be needed for special circumstances. These would include instances where more detailed quantitative information is needed or where testing of more fastidious or highly resistant pathogens is needed. Although automated or semiautomated quantitative methods would be desirable for routine use in larger laboratories, they currently have major limitations in the numbers and types of organisms and drugs that can be tested reliably. Clearly, the selection and utilization of any antimicrobial susceptibility test system must be a dynamic process, changing constantly as the microbial pathogens and their drug susceptibility changes.

REFERENCES

Baker, C. N., S. A. Stocker, D. H. Culver, and C. Thornsberry. 1991. Comparison of the E test to agar dilution, broth microdilution and agar diffusion susceptibility testing techniques by using a special challenge set of bacteria. *Journal of Clinical Microbiology* **29**:533-538.

Bakken, J. S. 1992. The many faces of MIC breakpoints: do they tell the story like it is? *Clinical Microbiology Newsletter* **14**:94-96.

Barry, A. L., D. Amsterdam, M. B. Coyle, H. Gerlach, C. Thornsberry, and R. W. Hawkinson. 1979. Simple inoculum standardizing system for antimicrobial disk susceptibility tests. *Journal of Clinical Microbiology* **10**:910-918.

Bradford, P. A. and C. C. Sanders. 1992. Use of a predictor panel for development of a new disk for diffusion tests with cefoperazone-sulbactam. *Antimicrobial Agents and Chemotherapy* **36**:394-400.

Brown, D. F. J. and L. Brown. 1991. Evaluation of the E test, a novel method of quantifying antimicrobial activity. *Journal of Antimicrobial Chemotherapy* **27**:185-190.

D'Amato, R. F. and L. Hochstein. 1982. Evaluation of a rapid inoculum preparation method for agar disk diffusion susceptibility testing. *Journal of Clinical Microbiology* **15**:282-285.

Gilbert, P., M. R. W. Brown, and J. W. Costerton. 1987. Inoculum for antimicrobial sensitivity testing: a critical review. *Journal of Antimicrobial Chemotherapy* **20**:147-154.

Greenwood, D. 1981. *In vitro veritas*? Antimicrobial susceptibility tests and their clinical relevance. *Journal of Infectious Disease* **144**:380-385.

Isenberg, H. D. 1988. Antimicrobial susceptibility testing: a critical evaluation. *Journal of Antimicrobial Chemotherapy* **22**:(Suppl A)73-86.

Jorgensen, J. H. 1991. Antibacterial susceptibility tests: automated or instrument-based methods, pp. 1166-1172 in A. Balows, W. J. Hausler, Jr., K. L. Herrmann, H. D. Isenberg, and H. J. Shadomy (eds.), *Manual of Clinical Microbiology*, The American Society for Microbiology, Washington, D.C.

Jorgensen, J. H. 1993. Antimicrobial susceptibility testing of bacteria that grow aerobically, pp. 393-409 in J. A. Washington (ed.), *Infectious Disease Clinics of North America: Laboratory Diagnosis of Infectious Diseases* Vol. 7. W.B. Saunders Co., Philadelphia.

Kayser, F. H., G. Morenzoni, and F. Homberger. 1982. Activity of cefoperazone against ampicillin-resistant bacteria in agar and broth dilution tests. *Antimicrobial Agents and Chemotherapy* **22**:15-22.

Louie, M., A. E. Simon, S. Szeto, M. Patel, B. Kreiswirth, and D. E. Low. 1992. Susceptibility testing of clinical isolates of *Enterococcus faecium* and *Enterococcus faecalis*. *Journal of Clinical Microbiology* **30**:41-45.

Rylander, M., J.-E. Brorson, J. Johnsson, and R. Norrby. 1979. Comparison between agar and broth minimum inhibitory concentrations of cefamandole, cefoxitin, and cefuroxime. *Antimicrobial Agents and Chemotherapy* **15**:572-579.

Sanders, C. C. 1991. ARTs versus ASTs: where are we going? *Journal of Antimicrobial Chemotherapy* **28**:621-622.

Sanders, C. C., K. S. Thomson, and P. A. Bradford. 1993. Problems with detection of β-lactam resistance among nonfastidious gram-negative bacilli, pp. 411-424 in J. A. Washington (ed.), *Infectious Disease Clinics of North America: Laboratory Diagnosis of Infectious Diseases* Vol. 7. W.B. Saunders Co., Philadelphia.

Schadow, K. H., D. K. Giger, and C. C. Sanders. 1993. Failure of the Vitek AutoMicrobic system to detect beta-lactam resistance in *Aeromonas* species. *American Journal of Clinical Pathology* **100**:308-310.

Standards. 1992a. National Committee for Clinical Laboratory Standards. Methods for dilution antimicrobial susceptibility tests for bacteria that grow aerobically. Approved standard M7-A2. National Committee for Clinical Laboratory Standards, Villanova, Pa.

Standards. 1992b. National Committee for Clinical Laboratory Standards. Performance standards for antimicrobial disk susceptibility tests. Approved standard M2-A4. National Committee for Clinical Laboratory Standards, Villanova, Pa.

Stratton, C. W. 1993. Bactericidal testing, pp. 445-459 in J. A. Washington (ed.), *Infectious Disease Clinics of North America: Laboratory Diagnosis of Infectious Diseases* Vol. 7. W.B. Saunders Co., Philadelphia.

Thomson, R. B., T. M. File, and R. A. Burgoon. 1989. Repeat antimicrobial susceptibility testing of identical isolates. *Journal of Clinical Microbiology* **27**:1108-1111.

Thornsberry, C. 1991. Antimicrobial susceptibility testing: general considerations, pp. 1059-1064 in A. Balows, W. J. Hausler, Jr., K. L. Herrmann, H. D. Isenberg, and H. J. Shadomy (eds.), *Manual of Clinical Microbiology*, 5th ed. American Society for Microbiology, Washington, D.C.

Working Party BSAC. 1991. Working Party on Antibiotic Sensitivity Testing of the British Society for Antimicrobial Chemotherapy. A guide to sensitivity testing. *Journal of Antimicrobial Chemotherapy* **27**:(Suppl. D).

York, M. K., G. F. Brooks, and E. H. Fiss. 1992. Evaluation of the autoSCAN-W/A Rapid System for identification and susceptibility testing of gram-negative fermentative bacilli. *Journal of Clinical Microbiology* **30**:2903-2910.

Index

Ablation, 104
Achromobacter, 169
Acid phosphatase production, 23
Acinetobacter, 276
Acinetobacter baumanis, 169
Adhesin, 237
Aeromonas, 276
Agrobacterium tumefasciens, 70
n-Akyltrimethyl ammonium bromides, 41
Albumin, 210
Alcohol(s), 94
  antimicrobial activity of, 157
  Candida albicans sensitivity to, 198
  Enterococcus faecium sensitivity to, 196
  mean microbial effect of, 195, 210, 212
  Pseudomonas aeruginosa sensitivity to, 196, 198
  Staphylococcus aureus sensitivity to, 198
  in surface carrier tests, 196
Amardori compounds, 110
Amidinocillin, 73
Anaerobes, 102, 137, 140
Analysis, techniques for biological, 244
Annular reactor, 75
Antibiotic(s), 5, 42–43, 257, see also Antimicrobial agents
Antibiotic penetration, 66
Antibiotic production, 90
Antigenic modulation, 238
Antimicrobial agents, see also Biocide(s); Disinfectant(s)
  activity of, 40
    neutralization of, 157–158
    preservative, 157
    screening for, 6, 247–257
  screening for, 6, 247–257
    biofilm reactors in, 266–269
    cell-free, target-directed, 249–251
    in industry, 262–271
    methodological approaches to, 249
    methods for, 264–271
    mixed cultures in, 263
    primary, 249
    secondary, 255–257
    starved bacteria in, 263–265
    whole cell, 251–255
      attached, 252–255
      planktonic, 251–252
  susceptibility to, see Antimicrobial susceptibility
Antimicrobial susceptibility
  cell envelope changes and, 40
  in continuous culture, 40–43
  dormancy and, 248
  growth rate and, 6, 248
  testing for, see Susceptibility testing
AOAC methods, 190, 206
AOAC phenol coefficient tests, 191
appR gene, 23
Argon, 106
Arrhenius kinetics, 107
Ascorbic acid, 103
Aspergillus, 169
Aspergillus niger, 169, 170
  biocide sensitivity of, nutrient depletion and, 204
  USP test strains, 18–19
Atomic absorption spectroscopy, 16
Autoinducer, 68
Azobacter, 123

Bacillus, 21, 57, 102
Bacillus anthracis, 235, 240
Bacillus cereus, 123
Bacillus coagulans, 56, 223
Bacillus megaterium, 54
  centrifugation injury and, 123
  in disinfectant testing, 207
  sporulation media for, 51, 52
Bacillus stearothermophilus
  auxotrophic strains of, 53
  prototrophic strains of, 53
  spores, 221
    heat resistance of, 229, 230
    from mineralized salts, 229, 230
    stability of, 230
  sporulation, 51, 53–54
    heat resistance and, 56
    media for, 53
    nutrient depletion and, 55
    stationary phase cell density and, 55
  strains of, 224
Bacillus subtilis, 54, 57, 58, 224
Bacteremia, 63

Bacteria, see also specific organisms
  antibiotic resistance of, see Bacterial resistance
  antigenic variations in, 239–240
  compositional variations in, 236–241
    antigenic, 239–240
    phase, 237–239
  contaminant, detection of, 7
  cultured, see Culture(s)
  density-dependent inducers of, 4
  dormant phenotype, 248
  fimbriae of, 237–238
  growth rate of, see Growth rate
  industrial uses of, 261
  luminescent, 103
  nutrient uptake systems of, 238
  phase variation in, 237–239
    coordinated, 238
    fimbriae, 237–238
  recombinant strains of, 241, 243
  resistance of, see Bacterial resistance
  sulfate-reducing, 265
Bacterial resistance, 53, 286
  biofilms and, 62, 66
  detection of, 275
  mechanism of, 239
  sporulation temperature and, 54
Bacteroides melaninogenicus, 102
Batch culture, 13
  for disinfectant testing, 204, 206
  growth rate in, maximum, 32
  inocula grown in suspension, 135–142
  nutrient depletion in, 14–15
Bentonite, 111
Benzalkonium chloride, 195
  Candida albicans sensitivity to, 198
  Enterococcus faecium sensitivity to, 196
  Escherichia coli sensitivity to, 204
  mean microbial effect of, 196, 210, 212
  Pseudomonas aeruginosa sensitivity to, 196, 198
  Staphylococcus aureus sensitivity to, 198
  in surface carrier tests, 196
Beta-lactams, 252, 282
Biocide(s), 66
  industrial, 265
  resistance to, 213–214
  testing of, see Biocide testing
Biocide testing, 190, see also Disinfectant testing
  capacity, 190
  end-point, 190
  quantitative, 190
Biocorrosion, 261, 265
Biofilm(s), 4, 146
  antibiotic penetration and, 66
  bacterial resistance and, 62
  biocidal resistance of, 213
  biocide penetration and, 66
  cells as, 38
  definition of, 61
  exopolysaccharide in, 69
  genetic information in, 65
  growth as adherent, 7
  heterogeneity of, 65
  from industrial processes, 262
  long-term culture of, 39
  microniches within, 65
  models of, see Biofilm models
  oxygen tensions in, 65
  pH gradients in, 65
  protease in, extracellular, 69
  Pseudomonas aeruginosa as, 66
  resistance properties of, 64
  in screening for antimicrobials, 252–255
  siderophore in, 69
Biofilm models, 7, 71–76
  annular reactor in, 75, 266–267
  closed growth, 71–73
  constant thickness, 76, 146, 255–256, 268–269
  drug resistance of, 73
  open growth, 73–76, 266–267
    non-steady-state, 73–75
    steady-state, 75–76
  perfused, 76
  Robbins device in, 74–75, 255, 269–270
  RotoTorque fermenters, 255, 266–267
Biological indicators, 221
  B. stearothermophilus spores as, 221
  characteristics of, 222–223
  Clostridium spores as, 221
  manufacture of, 223, 225
  organisms for, 222–225, 230
  performance of, factors affecting, 226
  quick-response, 230–231
  stability of, 229–230
  of sterilization, 222
Biological oxygen demand, 37
Biomass, 34, 35–37
bolA gene, 24
Bordetella pertussis, 235, 237, 238
Botulinum toxoid, 240
Bronopol, 157, 206
  biofilm resistance to, 213
Brucella, 102
Brucella abortus, 106
Butane diols, 95
Butyrated hydroxyanisole, 106

Calcium, 140, 228
Campylobacter, 104
Campylobacter coli, 102
Campylobacter jejuni, 102
Candida, 169
Candida albicans, 152, 204
alcohol sensitivity of, 198
benzalkonium chloride sensitivity of, 198, 204
chlorhexidine sensitivity of, 198, 204
dettol sensitivity of, 198
in disinfectant testing, 195
in preservation efficacy tests, 169, 170
sodium hypochlorite sensitivity of, 198
thiomersal sensitivity of, 204
Capacity tests, 152, 190, 202
Carbapenems, 248
Carbon starvation, 25
continuous culture and, 36
Catalase, 160
Cefamandole, 73
Cefmetazole, 252, 253
Cefminox, 252, 253
Cefoperazone, 253
Cefotatan, 253
Cefotaxime, 253
Cefoxitin, 253
Cefpirome, 253
Cefsulodin
diffusion coefficient of, 66
Ceftazidime, 253
Ceftoxidine, 42
Ceftriaxone, 42
Cell death, linear regression analysis and, 153
Cell density, 4, 94, 121
Cell-density-dependent physiology, 251
Cell division cycle, 4, 14, 94
Cell envelope, 5, 40, 139
Cell growth, see also Growth rate
freeze-drying and, 103
lag phase of, 14, 90, 105
logarithmic phase of, 14, 136–141
cryosensitivity and, 94
harvesting cell in, 138
predictivity of, 138–141
reproducibility in, 136–138
surface-associated, 61
Cell permeability, 92
Cell surface, 14
Cell suspensions, 122–126, 129–130
Centrifugation, 121
alternative to, 126
cell injury from, 122–126
salt tolerance and, 123
Cephaloridine, 73
Cephalosporins, 252, 253
Cetrimide, 125
biofilm resistance to, 214
Escherichia coli resistance to, 206
Escherichia coli sensitivity to, 41, 42, 43
Proteus mirabilis resistance to, 206
Challenge test organisms, 152, 154
Chemostats, 33, 144, 257
computer-monitored, 33, 144, 257
Chemosterilization, 189
Chlorhexidine, 8, 73
antimicrobial activity of, 157
Candida albicans sensitivity to, 198
centrifugation and bactericidal activity of, 8
Enterococcus faecium sensitivity to, 196
Escherichia coli sensitivity to, 41, 204
mean microbial effect of, 195, 210, 212
Proteus mirabilis resistance to, 206
Pseudomonas aeruginosa sensitivity to, 196, 198, 206
Staphylococcus aureus sensitivity to, 198
in surface testing, 196, 212
Chlorine dioxide, 206
Cidex, 212
Ciprofloxacin, 66, 252, 253
Citrobacter diversus, 213
Clearsol, 212
Clostridium, 57, 102, 221
Clostridium botulinum, 51, 235, 240
heat resistance, 225
type C, 240
type D, 240
type F, 241
Clostridium sporogenes, 223, 225
Clostridium tetani, 56
Clostridium thermosaccharolyticum, 56, 224, 225
Cold-osmotic shock, 128–129
Cold-shock, 129
Constant Thickness Biofilm Fermenter, 76, 146, 255–256, 268–269
Contaminants, 7, 182
in creams and lotions, 170–171
detection of, 7
in dilution tests, 281
in eye cosmetics, 171
iron, 15–16
psychrophilic, 97
in shampoo, 170
in toothpaste, 170
Continuous culture, 32
basic assumptions regarding, 37–38
biofilms and, 38–39
carbon limitation in, 35–36
for disinfectant testing, 207

fermenter design of, 39
flocculation of, 38
foaming of, 38
growth rate and, 33
homogeneity of, 38–39
nutrient depletion and, 35–37
substrate concentration in, 33–34
Cooling, 92, 96
Corrosion, microbially induced, 261, 265
Corynebacterium, 102
Cosmetics, 163–184
definition of, 164
isolates from, 169
preservation efficacy and, see Preservation efficacy, in cosmetic and toiletry industries
shelf-life expiration dating of, 165
Creams and lotions, contaminants in, 170–171
Cryoprotection, 94–95
agents, 93
interaction between cell membrane and, 95
mechanism of action, 94–95
toxic effects of, 95
Cryosensitivity, 14
Cryotolerance, 94
csgA gene, 24
Culture medium, 14
carbon substrate in, 140
for cell suspensions, 122
definition of, 137
freeze-drying and, 103
liquid vs. solid, 146
low-iron, 15
nitrogen source in, 140
nutrient-depleted, 15–19
nutrient excess, 136
reducing iron contamination of, 15–16
solid vs. liquid, 146
spore formation and, 49–51
for sporulation, see Sporulation media
for susceptibility testing, 7
Culture(s), 83–85
biological variations in, 83
post-growth handling procedures and, 85, 121–130
cellular injury in, see Injury
centrifugation of, 122–126
dairy, lactic acid production in, 90
filtration of, 126–127
harvesting of, 122–128
washing and resuspension after, 128–130
master, 111–112, 175–178
medium, see Culture medium
origins of, 83–84
stock, see Stock cultures
wild populations in, 153
Curli, 22
Cyclic AMP, 25
Cytoplasm, 14
cyxAB gene, 24

Deep freezers, 96–97
Dental plaque, 62, 268, 271
Dental sterilizers, 222
Desferoxamine, 16
Desulphomaculum nigrificans, 225
Dettol, 195, 210
Candida albicans sensitivity to, 198
Enterococcus faecium sensitivity to, 196
mean microbial effect of, 195, 210, 212
Staphylococcus aureus sensitivity to, 198
in surface carrier tests, 196
Dextran, as freeze-drying excipient, 101
Dilution rate, 33, 35
Dilution tests, 281–284
agar, 282
antibiotic concetratin gradient in, 282
broth, 282–284
contamination and, 281
disadvantage of, 281
disk diffusion vs., 281
inherent error in, 281
inocula in, 284
macrobroth, 283
Dimethyl sulfoxide, 93, 94
Diphtheria toxoid, 240, 241
Disinfectant(s), 189, see also Biocide(s)
approval of, laboratory tests and, 214–215
mean microbial effect of, 195, 196, 210
testing of, see Disinfectant testing
Disinfectant testing, 191–202
batch culture for, 204, 205
conditions for, 204, 205
contact time in, 208
continuous culture for, 207
end-point, 191
hard surface carrier test, 192, 196
inter- and intralaboratory variability in results of, 198, 200, 202
intralaboratory variability in results of, 194, 197–198, 200, 202
nutrient depletion in, 204, 206, 207
operator skill in, 197
presence of interfering substances in, 208–209, 210
quantitative, 192–193
repeatability and reproducibility of, 191–202
temperature and, 208
test organisms for, 203–204, 206–207
test suspensions in, 194, 197, 207–208

variations between test organisms and products in, 197, 198
within-day and between-day variability in, 197–198
Disk diffusion tests, 279–281
Division cycle, see Cell division cycle
DNA ligase, 129
Dormancy, 84
antimicrobial susceptibility and, 248
characteristics of, 251
freeze-drying and, 109–110
induction of, 252
modification of, 228
nutrient-deprivation and, 13
resuscitation from, 91
dps gene, 24
Drying techniques, 111, 158, 175
D-value, 122

EDTA, see Ethylenediaminetetraacetic acid
Electromagnetic radiation, 109
Electrophoresis, gel, 16
ELISA technique, 244
Endonucleases, 129
End-point tests, 190
Enterobacter, 169
Enterobacter cloacae, 169
Enterobacter gergoviae, 169
Enterobacteriaceae, 276
Enterococcus, 276
Enterococcus faecium, 194, 196
Environmental protection industries, 261–271, see also Industry
Erwinia, 70
Erythromycin, 252, 253
Escherichia, 70
Escherichia coli, 17
benzalkonium sensitivity of, 204
biofilms of, 72, 213, 214
ceftoxidine and, 42
ceftriaxone and, 42
centrifugation injury and, 123, 124, 125
cetrimide resistance of, 206
cetrimide sensitivity of, 41, 42, 43
chlorhexidine diacetate sensitivity of, 8
chlorhexidine resistance of, 206
chlorhexidine sensitivity of, 41, 204
fimbriae, genetic control of, 237
freeze-drying and, 102
growth rate of, 25–26
methanol as cryoprotectant for, 95
n-alkyltrimethyl ammonium bromides and, 41
oxygen restriction and, 137
phenol resistance of, 206
in preservative efficacy tests, 152, 169, 170
rpoS and, 26
sigma factor in, 23
starvation and, 22–23, 26
stationary phase of, 22
thiomersal sensitivity of, 204
transformation efficiency of, centrifugation and, 126, 127
USP test strains of, 18
Ethylenediaminetetraacetic acid, 14, 41
Ethylene oxide, 224
European Community Cosmetics Directive, 163–165
Exopolysaccharide, 69
Extracellular protease, 69
Eye cosmetics, contaminants in, 171

Fatty acids, 5
Fermentation, 243–244
Fermenter, two-stage, 144
Fibronectin-binding filament, 22
Filtration, 126–127
Fimbriae, 237
Flavobacter, 169
Floccules, 38
Fluidized bed reactors, 38
Fluoroquinolones, 282
Food sterilization, 223
Foot and mouth disease, 103
Formaldehyde, low-temperature steam, 224
Freeze-drying, 90, 98–111, 124
ablation and, 104
advantages of, 111
alternatives to vacuum, 111
anaerobes and, 102
atmospheric, 111
containers for, 101
desorption drying, 100
disadvantages of, 111
electromagnetic radiation and, 109
equipment for, 98
excipients for, collapse temperature of, 100, 101, 107
free-radical damage and, 106
glow discharge testing and, 109
goal of, 105
heat and mass transfer in, 99
heat annealing processes in, 99
injury and, 124
Maillard reactions and, 108, 110
mutations caused by, 108
organism survival in, 101–105
primary drying, shelf temperatures during, 99
protectants, 103
rate of freezing prior to, 98–99

reconstitution in, 104–105
protracted lag phase after, 105
sealing gas in, 106
secondary drying, 100
spore formation and, 102
stability during and after, 105–108
stages of, 98
sublimation interface in, 99
suspending medium composition in, 103–104
vapor-phase rehydration in, 105–106
Freezing, 124, 175
cell permeability and, 92
in domestic deep freezers, 96–97
hypertonic solution during, development of, 91
of inoculum, 91–93
liquid/gas-phase nitrogen, 97
methods of cooling for, 95–96
P. aeruginosa sensitivity to, 93–94
solute behavior during, 91–92
storage temperatures for, 96–97
Fungi, 111

Gel electrophoresis, 16
Gene(s), 23–25
mutants, see Mutations
stability, see Genetic stability
stringent reponse, 252
transcription, 68
Genetic stability, 89, 90
Genome, 14
Gentamycin, 252, 253
glgS gene, 24
Glow discharge testing, 109
Glucose, as freeze-drying excipient, 101
Glutaraldehyde, 227, 265
Glycerol, 94
Glycocalyx, 63
as moderator of chemical resistance, 66–67
role of, 65–67
structure of, 65
Glycogen storage, 23
Glycolmonophenylethers, 125
Gram-negative bacteria, 40, see also specific organisms
cold-osmotic shock and, 129
in preservation efficacy tests, 169
Growth, see Cell growth; Growth rate
Growth factors, 140
Growth rate, 4
antibiotic activity and, 43
biofilms and, 7
colony density and, 8
culture medium and, 50
defined, 32
in excess nutrient media, 136
factors affecting, 16
infection and, outcome of, 13
inoculum and, 84
iron and, 4
maximized, 32
maximum specific, 13
nutrient deprivation and, 5–7, 13
osmolarity and, 4
pH and, 4
spore formation and, 50
susceptibility to chemical agents and, 6, 248

Haemophilus influenza, 72, 102, 276
Haemophilus suis, 103
Halogens, 157
Heat resistance, 23, 56, 224, 225
B. stearothermophilus, 227, 229, 230
C. sporogenes, 225
C. thermosaccharolyticum, 225
culture age and, 57–58
lyophilization and, 230
modification of, 228
Heat-sterilization, cell injury from, 123, 158
Helium, 106
Hibisol, 212
Host defenses, 5
Hydrogen peroxide, 57
Hydrogen sulfide, 265

Imipenem, 248, 252, 253
Immunogenicity, microbial, 43
Implants, infections of, 7
Implants, sterilization of, 222
Indolmycin, 252, 253
Industry, 261
microbial growth associated with, 261–262
screening of microbials for, 262–271, see also Antimicrobial agents, screening for
Influenza virus, 106
Injury, 121
centrifugation, 122–126
chemicals and, 124, 159
from cold-osmotic shock, 128–129
freeze-drying and, 124
freezing and, 124
heat-sterilization and, 123, 124
irradiation and, 124
metabolic, 123
physical, 123, 159
preservatives and, 124
structural, sublethal, 123
sublethal, 123, 158–160
secondary stress and, 159

structural, 123
temperature fluctuations and, 159
Inocula, 83–85
allenge, in preservation efficacy test, 168, 178, 180
in antimicrobial susceptibility testing, 277, 279
from batch culture, 135–142
biofilm, 146
challenge, in preservation efficacy test, 168, 178, 180
from continuous culture, 142–145
dilution tests in, 284
growth conditions of, 84
levels of, 181
logarithmic phase, 136–142
predicitivity of, 138–141
reproducibility of, 136–138
planktonic, 146
predictivity of
from continuous culture, 145
in logarithmic phase, 138–141
in stationary phase, 142
replicate plating, 89
reproducibility of, 89–90, 135–142
from continuous culture, 143–145
in logarithmic phase, 136–138
in stationary phase, 141–142
stationary phase, 141–142
Intercell signaling, 15
Intrauterine conception devices, 7
Invasins, 68
In vitro models, 255–257
of biofilm, 7
Iodine, 66
Iodine-polyvinylpyrollidone, 66
Iodophor, 212
Iron, 4
Iron-binding proteins, 238
Iron chelators, 16
Iron depletion, 16, 36–37
Irradiation, 124, 158

katE gene, 24
katF gene, 23
Kelsey-Sykes test, 191, 202
Klebsiella, 169
Klebsiella pneumoniae, 14, 169

Lactic acid production, 90
Lactobacillus, 102
Lactose, as freeze-drying excipient, 101
Lactose plus, as freeze-drying excipient, 101
Lecithin, 157
Legionella, 74
Legionella pneumophila, 214
Leptospira, 111
Linear regression analysis, 153
Lubrol W, 157
Lux gene, 230
LuxI gene, 70
LuxR gene, 70
Lyophilization, 230
Lysozyme, 160

Magnesium, 140, 229
Magnesium depletion, 14, 36–37
Maillard reactions, 108, 110
Manganese, 228
Mannitol, as freeze-drying excipient, 101
Master cultures, 111–112, 235
establishment and maintenance of, 175–178
mcc gene, 23
Mecoprop, 263, 264
Medium, culture, see Culture medium
Mercurials, 157
Meropenem, 248, 253
Metal cations, 5
Methanol, 95
Methylparabens, 206
Michaelis-Menten kinetics, 32
Microbial Quality Management, 163
Microcin B17 production, 25
Microorganisms, see Bacteria
Microwave ovens, sterilization of, 222
Milk, skimmed, 103
Mineral oil, 111
Minimum bactericidal concentration, 248
in dilution tests, 282
Minimum inhibitory concentrations, 122, 248, 282
in disk diffusion tests, 281
Mist desiccans, 103
Molar yield, 136
Monochloramine, 271
Moraxella catarrhalis, 276
Mutations, 23–24
diphtheria toxoid, 241
directed, 26
freeze-drying and, 90
selection of, 25
starvation and, 25, 26
subculturing and, 90
Mycobacterium, 102, 213
Myxococcus, 21

National Committee for Clinical Laboratory Standards, 286
Neisseria gonorrhea, 102
antigenic modulation of, 239

freeze-drying protectants, 102
outer membrane protein expression by, 239
predictable sensitivity of, 102
Neisseria meningitidis, 237
antigenic modulation of, 239
outer membrane protein expression by, 239
predictable sensitivity of, 276
transferrin-binding proteins of, 238
Nitrilotriacetate, 67
Nitrogen, liquid, 96
Nitrogen source, 140
nur gene, 23
Nutrient depletion, 4, see also Starvation
bacterial response to, 21–22
in disinfectant testing, 204, 206, 207
dormancy and, 13
resistance and, 53
spore formation and, 53–54

osmB gene, 24
Osmolarity, 4
Osmotic shock, 93
gram negative bacteria and, 129
protection against, 23
otsAB gene, 24
Oxidative stress, protection against, 23
Oxygen deprivation, 37

P. fluorescens, 102
P. maltophilia, 102
Parabens, 157
Pasteurella, 102
Pasteurella haemolytica, 16
Penicillin-binding protein, 248
Penicillium, 169
Pentoses, 103
Peptone-glucose plugs, 111
Peracetic acid, 57
Peratol, 210, 212
Perfused Biofilm Fermenter, 146, 255, 256, 257
Peroxide, 160
Pex proteins, 23
pH, 4, 56–57
Phenol(s), 41, 157, 206
Phenotypic flexibility, 5
Phospholipid(s), 5, 40, 139
Phosphorus, 140
Picornaviruses, 102
Piperacillin, 253
Planktonic cells, sessile vs., 62–63, 69, 73
in screening for antimicrobials, 251–255, 265
Plasmid, 90
Plasmid DNA, superhelical density of, 22
Plate diffusion assays, 8
Polymixin, 43
Polymixin B, 14
Polyoma viruses, 102
Polyvinylpyrollidone, as freeze-drying excipient, 101
Porin protein, 40
Porins, 239, 240
Potassium, 140, 229
Poxviruses, 102
Preservation efficacy, 89
challenge test, 165, 167, 168
inoculation in, 168, 172–174, 178, 180–181
number of, 182
selection of organisms for, 168–172, 179
consumer-use testing and, 182
in cosmetic and toiletry industries, 163–184
criteria for, 149
end products and, biochemical properties of, 90
genetic stability and, 90
pharmacopoeial requirements for, 149, 150–153
limitations of, 150–152
modifications in, 150–152
testing for, 149–160
antimicrobial activity and, 156–157
capacity tests and, 152
challenge tests in, 152, 167–172
in cosmetic and toiletry industries, 163–184
diagrammatic representation of, 166
first-generation, 149
limitations of, 150–152
linear regression analysis in, 153
nutritional status of challenge inocula in, 154
objective of, 167
organic matter and, 156–157
in pharmaceutical industry, 149–160
pharmaceutical product and, nature of, 155–156
pharmacopoeial requirements for, 149, 150–153
recovery systems and, 158–159
repeat-inoculations in, 152
test strains for, 154–155, 168–175, 179
variation in results of, 153–160
wild strain cultivation in, 153–154
Preservatives, 124, 157
Procaine penicillin, 286
Propane diols, 95
Prostheses, 62
infections associated with, 63
Prosthetic devices, 7
Protease, extracellular, 69

Protein(s), 22, 23, 24
envelope-associated, 5
iron-binding, 238
outer membrane, 239
penicillin-binding, 248
pex, 23
porin, 40
transferrin-binding, 239
Protein degradation, 22
Protein synthesis, 23
Proteus mirabilis, 17
cetrimide resistance of, 206
chlorhexidine resistance of, 206
intercellular communication of, 68
phenol resistance of, 206
Pseudomonas, 276
Pseudomonas aeruginosa, 158, 169, 203
alcohol sensitivity of, 196, 198
benzalkonium chloride sensitivity of, 196, 198, 204
biofilms of, 66, 72, 73
perfused, 76
resistance of, 76
bronopol sensitivity of, 206
centrifugation injury and, 123, 124
chlorhexidine sensitivity of, 8, 196, 198, 204, 206
chlorine dioxide resistance of, 206
in disinfectant testing, 194, 195, 196
EDTA sensitivity of, 14, 41
freeze-drying and, 102
freezing and, 93–94
glutaraldehyde resistance of, 206
iron-depletion and, 16
magnesium-depletion and, 14
methylparaben sensitivity of, 206
nutrient requirements for, 17
phenol sensitivity of, chlorinated, 41
in preservative efficacy tests, 152, 169, 170
sodium hypochlorite against, 196, 198
in surface carrier tests, 196
tobramycin resistance of, 73
transcriptional activation and, 70
USP test strains of, 18
Pseudomonas cepacia, 169, 213
Psychrophilic contaminants, 97
Puff-drying, 111
Pyruvate, 160

Quantitative tests, 190
Quaternary ammonium compounds, 157, 206
Quorum sensing, 68

Recombinant DNA technology, 241
Recovery systems, 158–159
Repeat-inoculation tests, 152
Residence time, 38
Resistance, see Bacterial resistance
Resuscitation, 90
Rhizobia, 70
Rickettsiaceae, 102
Rideal-Walker tests, 190, 191, 202
Robbins device, 74–75, 255, 269–270
modified, 74, 255
rpoS gene, 23
Escherichia coli and, 26
mechanism of action, 24
mutants of, 23

Saccharomyces cerevisiae, 102
Salmonella, 68, 102, 158, 285
Salt tolerance, 90
centrifugation and, 123
Sample reconstitution, 90
Savlon, 212
Schiff bases, 110
Screening for new antibiotics, see Antimicrobial agents, screening for
Sealing gas, 106
Serratia marcescens, 123, 213
Serum, 103
Sessile cells, planktonic vs., 62–63, 69, 73, 251–255, 265
Shampoo, contaminants in, 170
Shigella, 102
Siderophore, 69
Siderophores, 238
Sigma factor, 23
Silica gel, 96
Slime molds, 68
Sodium, 229
Sodium dichloroisocyanurate, 210, 212
Sodium glutamate, 108
Sodium hypochlorite, 195
Candida albicans sensitivity to, 198
Enterococcus faecium sensitivity to, 196
mean microbiocidal effect of, 196, 212
Pseudomonas aeruginosa sensitivity to, 196, 198
Staphylococcus aureus sensitivity to, 198
Sorbic acids, 157
Sorbitol, as freeze-drying excipient, 101
Spirillum atlanticum, 105
Spore(s), 13
B. stearothermophilus, 221
germination rate of, 227
glutaraldehyde resistance of, 227
heat resistance of, 224, 225, 227, 229, 230
from mineralized salts, 229, 230

stability of, 230
in steam sterization, 223
B. subtilis, 227
Clostridium, as biological indicators, 221
Clostridium sporogenes, in food sterilization, 223
demineralized vs. native, 228
dormancy of, 224
formation, see Sporulation
heat resistance, 224
production of, 222
Spore resistance, 53
mineralization and, 228, 229
post-harvesting modification of, 227–228
Sporulation, 21–22
B. stearothermophilus, 51
Bacillus stearothermophilus, 51
C. botulinum, 51
cell division cycle and, 50
culture medium and, 49–51, see also Sporulation media
freeze-drying and, 102
growth rate and, 50
nutrient depletion and, 53–54
pH and, 56–57
temperature and, 54, 56
Sporulation media, 51, 52, 57
Sporulation pH, 57
Spray-drying, 111
Staphylococcal enterotoxin D, 240
Staphylococcus, 276
Staphylococcus aureus, 16, 158, 203
alcohol sensitivity of, 198
benzalkonium chloride sensitivity of, 198, 204
biofilms of, 72
chlorhexidine sensitivity of, 198, 204
dettol sensitivity of, 198
in disinfectant testing, 195
freeze-drying and, 102, 103
in preservative efficacy tests, 152, 169, 170
sodium hypochlorite sensitivity of, 198
USP test strains, 18
Staphylococcus epidermis, 66
biofilms of, 68, 69, 73
biocidal resistance of, 213
perfused, 76
centrifugation injury and, 123
in preservation efficacy tests, 169
tobramycin resistance of, 73
vancomycin resistance of, 73
Starvation, 22–23, 129–130
Escherichia coli and, 26
mutations and, 25, 26
population changes during, 25–26
screening for antimicrobial activity and, 263–265
Stationary phase, 22
biomass at, 35
cell density in, 55
cell size in, 22
curli and, 22
description of, 141
entry into, indicators of, 23
fibronectin-binding filament in, 22
gene expression in, regulation of, 23–25
glycogen storage in, 23
inocula from, 141–142
microcin B17 production in, 25
oxygen restriction and, 136
Pex proteins in, 23
protein degradation in, 22
protein synthesis and, 23, 24
trehalose production in, 23
Steady-state biomass, 35–37
Stericol, 210
Sterilization, 221
of dental sterilizers, 222
ethylene oxide, 224
food, 223
implant, 222
indicators of, 222–225
low-temperature steam formaldehyde, 224
of microwave ovens, 222
Stock cultures, 7
asynchronous, 109
cold storage of, without ice formation, 97–98
master, see Master cultures
preparation of, 178
preservation of, 89
by cooling, 90–93, see also Freezing
cooling rates for, 92
by drying techniques for, 111
by freeze-drying, see Freeze-drying
thawing rates for, 92
stability of, vaccine production and, 235
thawing rates for, 93
vaccine, 235–236
working, 112
Streptococci, beta-haemolytic, 276
Streptococcus faecalis, 104
Streptococcus pneumoniae, 276
Streptococcus pyogenes, 102
Streptomycin, 43
Stressed cells, 158–160, see also Injury, sublethal
Subculturing, 90
Sucrose, as freeze-drying excipient, 101
Sulfur, 140
Superoxide scavengers, 160

Surface carrier test, 192, 196, see also Surface testing
repeatability and reproducibility of, 192, 196, 199
Surface colonization, 63, 64–71, see also Biofilm(s)
Surface testing, 209, 211–214
mean microbicidal effect in, 212
Surfactants, ionic, 158
Susceptibility, see also Antimicrobial susceptibility
towards chemical biocides, 41
Susceptibility testing, 275-287
culture for, 7–8
divison cycle and, 14
goal of, 275
inocula for, 277, 279
liquid vs. solid media for, 7–8
methods for, 275
advantage and disadvantges of specific, 278
algorithms for, 284
antibiotic selection for, 285
automated, 284
dilution assays vs., 285
dilution tests, 281–284, see also Dilution tests
automated vs., 285
disk diffusion tests as, 279–291
interpreting results of, 286
quantitative, 277, see also Dilution tests
reporting results of, 285–286
selection of, 277
need for, 276–277
plate diffusion assays for, 8
solid vs. liquid media for, 7–8
therapeutic outcome and results of, 275, 285

Temperature, 4, 54, 56
Tetanus toxoid, 240
Therapeutic enzymes, 90
Thermophils, obligate, 223
Thiomersal, 204
Tobramycin, 43, 66
Toiletry industry, preservation efficacy in, 163–184
Toothpaste, contaminants in, 170
Transcriptional activation, 68–71
Transferrin, 16
Transferrin-binding proteins, 239
treA gene, 24
Trehalose, 23
Trehalose, as freeze-drying excipient, 101
Treponemes, 111
Turbidostat, 143
Turbiostats, 37
Tween 80, 157

USP preservative challenge test, 6–7
USP test strains, 17–19

Vaccine(s), 235
bacterial toxoid, 240
pertussis, 235
potency of, 240
production of, 237
recombinant DNA in, 241
stock culture stability and, 235
quality control of, 244
seed stocks, 242–244
genetic stability of, 243
master, 242
monitoring of, 242–243
submaster, 242
system for, 242
Vibrio, 276
Vibrio fischerii, 68, 70
autoinducer, 70–71
Vibrio parahaemolyticus, 68
Virkon, 210, 212
Virulence factor production, 43

Washed cell suspensions, 121
Western blot technique, 244
Wild populations, 153–154
cultivation of, 153–154

xthA gene, 24

Yeast, 90, 210
Yersinia, 102

Zygosacch cerevisiae, 170